P9-AEY-897

Northeast Lakeview College
33784 0001 2197 0

DARWIN AND INTERNATIONAL RELATIONS

DARWIN AND INTERNATIONAL RELATIONS

On the Evolutionary Origins of War and Ethnic Conflict

Bradley A. Thayer

THE UNIVERSITY PRESS OF KENTUCKY

Publication of this volume was made possible in part by a grant from the National Endowment for the Humanities.

Scholarly publisher for the Commonwealth, serving Bellarmine University, Berea College, Centre College of Kentucky, Eastern Kentucky University, The Filson Historical Society, Georgetown College, Kentucky Historical Society, Kentucky State University, Morehead State University, Murray State University, Northern Kentucky University, Transylvania University, University of Kentucky, University of Louisville, and Western Kentucky University.

Editorial and Sales Offices: The University Press of Kentucky
663 South Limestone Street, Lexington, Kentucky 40508-4008
www.kentuckypress.com

08 07 06 05 04 5 4 3 2 1

Library of Congress Cataloging-in-Publication Data

Thayer, Bradley A.
Darwin and international relations : on the evolutionary origins of war and ethnic conflict / Bradley A. Thayer.
p. cm.
Includes bibliographical references and index.
ISBN 0-8131-2321-6 (hardcover : alk. paper)
1. International relations. 2. Human evolution. 3. War. 4. Ethnic relations. I. Title.
JZ1249.T48 2004
303.6'6'01—dc22 2003024586

This book is printed on acid-free recycled paper meeting the requirements of the American National Standard for Permanence of Paper for Printed Library Materials.

Manufactured in the United States of America

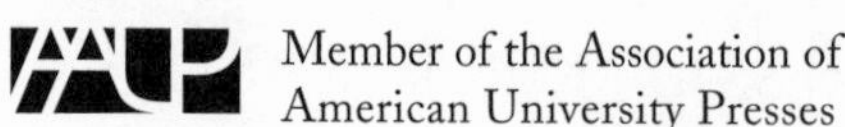

To my parents: Gerald L. and Erna B. Thayer

Contents

Tables

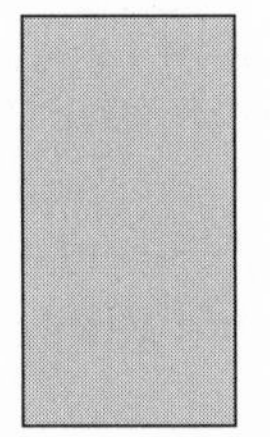

Preface

As a student of human behavior, I have always been puzzled by the lack of intellectual exchange between the life sciences and the social sciences. Both evaluate and discern the causes of human behavior in exceptional detail, and each has generated profound insights that would greatly aid the other, but they exist in largely separate worlds, unaware and untouched by the other, almost like the peoples of the pre-Columbian Old and New Worlds. It is ironic that this situation exists in the Information Age, where distance is dead and knowledge flows from a search engine like a virtual biblical flood. This book is an attempt to alter this situation, and end a division that needlessly hurts the life and social sciences by hindering their advance. In this book, I draw from the life sciences to generate insights for the social science discipline of international relations.

If we seek to explain the totality of human behavior, then the life and social sciences are required because humans are the product of the interaction of their evolution and their environment. Human behavior is the product of these equally important causes and cannot be reduced to an "essence," the idea that human behavior is solely the product of evolution or the environment. Such essentialism must be rejected if the scientific understanding of human behavior is to advance.

Each science has a piece of the human behavioral puzzle. Their unification gives scholars a richer and deeper explanation of human behavior and a more profound conception of what it is to be human. When brought together, they reveal a broader conception of humanity and human behavior than is possible when they are discrete. It is important to recognize that the life and social sciences are each the other's equal, and neither science is subordinate to the other—just as evolution and environment are not when we study human behavior. Moreover, each may benefit the other. While international relations is the focus of the present study, the life sciences may aid other disciplines—

anthropology, psychology, sociology—as well. As I will emphasize, the intellectual exchange between the life and social sciences is not a one-way street. Social science may assist the life sciences, too.

From my specific perspective as a scholar of international relations, the integration and use of insights provided by the life sciences is profound. It allows scholars to determine the origins of specific behaviors, such as egoism and ethnocentrism, and it permits us to understand why war would begin and why it may be found in other animals. A life science approach to the study of international relations is a new and important development. It allows old questions to be answered in new ways, and it improves existing theories and generates new knowledge. The life science approach also illuminates novel issues for study. It allows the discipline of international relations to be more sensitive to the impact of ecology, the consumption of natural resources, the continuing problem of famine, and the profound impact other forms of life, such as disease, have on humans and on the relations between states.

Some of these insights concern issues that may frighten us and that we wish would disappear, like Albrecht Dürer's terrifying image of "The Four Horsemen of the Apocalypse." Using the life sciences in the discipline of international relations does help us to understand each of the Horsemen in Dürer's work, Pestilence, War, Famine, and Death. This book in particular explains some of the phenomena embodied in the Horsemen: disease, war, and famine. Indeed, they must be studied in order to understand and mitigate the effects of each. But the use of the life sciences in international relations is not limited to these topics. These and many other issues often studied in international relations may be explored through the lens of the life sciences. Additional scholars in the social sciences will use the life sciences to illuminate other issues.

The study of war or ethnic conflict has no monopoly on the life sciences. Indeed, the life sciences also allow scholars of international relations to comprehend better the opposite of Dürer's image, "The Four Horsemen of Civilization" if you will, such as Health, Peace, Cooperation, and Altruism.

This book begins with the simplest of ideas from evolutionary theory. Many of these were developed by Charles Darwin, and have become known as evolution by natural selection. First, animal behavior is the result of the animal's genes and its environment. Second, neither genes nor environment is fixed over time. An animal's genes and the environment in which it exists evolve over time. Animals that are able to continue to survive and reproduce in changing circumstances, perhaps because of an adaptation that makes them slightly faster (or slower), taller (or shorter), or smarter (or even less intelligent), are fit; those that are not able to adapt become extinct. Third, what biologists term a species is the sum of these adaptations. Fourth, humans are animals. As animals, they have adapted to changing environmental circumstances over time. Darwin's ideas are the foundation of the life science approach.

Recognizing this, it is odd that the social sciences still study human behavior extensively but without any references to or acknowledgment of human evolution. Some scholars in social science disciplines may believe that human evolution is not particularly relevant for the study of human social behavior. They may hold that what came before major social upheavals like the French Revolution, the Industrial Revolution, or the rise of Modernity and Post-Modernity is not relevant due to the profound effects of these social revolutions and processes. After all, remember that Virginia Woolf told us in her essay "Mr. Bennett and Mrs. Brown" that "on or about December 1910 human nature changed," and this perhaps made human evolution much less important for explaining human behavior in modern times.

Of course, depending on the issue studied, such a belief may be spot on. For example, Darwin's ideas have very little to say about Plato's epistemology, why Robespierre was guillotined, Marx's conception of surplus value, or which model of economic development is best for Nigeria. Yet, this is certainly not true for other topics, such as those I address in this book. These and many other issues often studied in international relations may be explored through the lens of the life sciences. Moreover, although it may be hard to accept at first, even behaviors considered purely social are the result of the evolutionary process. Humans are the product of evolution, and we evolved behaviors that served us well in a dangerous environment where resources were scarce. So when human behavior is analyzed, there is no escape from the conjoined twins that are evolution and environment; neither may be maligned nor trumpeted without damaging the health of the other.

In 1932, Albert Einstein wrote that as a man he had just enough intelligence to be able to understand clearly how utterly inadequate that intelligence actually is when confronted with the world's complexity and wonder. In the course of writing this book, I often felt similar emotions, because it draws upon multiple disciplines. I was overwhelmed with what there is to know about the process of evolution, about how evolution affected and continues to influence humans, of ethology, of genetics, of archaeology, and of the history and practice of politics among nations. Despite the challenge of interdisciplinary scholarship, its rewards exceed its costs. We are at the beginning of a period of great intellectual progress. To my mind, it is the equal of the great Age of Exploration as the human genome is revealed, the human brain mapped, and as the life and social sciences learn from each other, increasingly no longer strangers to one another, no longer separate trees but only separate and interconnected branches in the tree of knowledge.

While international relations is not a discipline that is associated with Darwin, evolutionary theory, ethology, or the other disciplines upon which I draw to make my arguments, in time it will be, as all things human are affected by Darwin's arguments. For me, researching and writing this book was an intellectual feast because I have had the honor to meet and interact with scholars in many fields, including archaeology, anthropology, ethology, evolutionary theory, philosophy of science, primatology, and zoology. They have taught me much, "confronted me with what exists," as Einstein expressed it. In the course of our interaction, I hope I was able to give them insights into international relations and, more broadly, political science.

Such interdisciplinary work is a good thing I believe because it helps promote knowledge and to bridge what is increasingly becoming an untenable gap between the natural and social sciences. More selfishly, I found it simply fascinating to learn the work of and communicate with scholars in disciplines as far from my own as archaeology, evolutionary theory, and ethology. Finally, and this is both a hope and a reasoned expectation, as science advances, the use of evolutionary theory in international relations will become less curious and—I fully expect—will be used considerably. This will happen when more theorists of international relations and students of warfare and ethnic conflict understand how evolutionary theory can assist their arguments. Just as the Berlin Wall fell, so too will the barriers preventing the lack of intellectual exchange between the life and social sciences, to the benefit of all.

While writing this book I accumulated many debts, and it is truly a privilege at this time to thank those who have assisted me along the way. For their infinitely helpful comments and criticisms of this study, I am grateful to Mlada Bukovansky, Stephen Chilton, Craig Cobane, David Cole, Peter Corning, Owen R. Coté Jr., Leah Meredith Farrall, James Fetzer, Eleanor Hannah, Samuel M. Hines Jr., Tatiana Kostadinova, Christopher Layne, Sean M. Lynn-Jones, Michael Mastanduno, David "Doc" Mayo, Thomas Powers, Paul Sharp, David Sobek, Jennifer Sterling-Folker, Alexander Wendt, Janelle Wilson, Mike Winnerstig, and Howard Wriggins. Roger Masters, Steven Peterson, Jeffrey

Taliaferro, and Johan van der Dennen provided exceptional comments on the entire manuscript, and I thank them for their suggestions. It is also a pleasure to thank Henning Gutmann, who provided encouragement in the early stages of this project.

I presented elements of my argument in seminars at Dartmouth College, University of Queensland, Stockholm University, and University of Minnesota–Duluth, and I am grateful to all the participants in those seminars for their excellent comments that allowed me to sharpen and refine my arguments. Further presentations at the annual meetings of the Association for Politics and the Life Sciences, the American Political Science Association, and the International Studies Association assisted me as well, and for making these presentations possible I am grateful to Gary Johnson, Dan Lindley, Sean M. Lynn-Jones, Steven Peterson, and Al Somit.

In addition, I had a wonderful opportunity to make my argument concerning the causes of ethnic conflict to the Jerusalem Institute for Israel Studies and received helpful comments from some of their members, especially Ms. Ora Ahimeir. I thank Efraim Inbar and Avi Kober, of the Begin-Sadat Center for Strategic Studies at Bar-Ilan University; Monica Pataki and Linda Slutzky of the American Center, Jerusalem; and Michael Richards and Anne Walter of the U.S. embassy for hosting me in Israel. I am also indebted to them for their insights into that country's difficult security situation, and for their incomparable warmth, generosity, and assistance.

I am grateful to Nathaniel Fick, David Hawkins, Jeremy Joseph, Christopher Kwak, Craig Nerenberg, and Jordana Phillips for their able research assistance and comments along the way. Abigail Marsh gave me insight into the discipline of evolutionary psychology and its literature when I needed it most. I am indebted to officials at *Íslensk erfðagreining* (Decode Genetics) in Reykjavík for their assistance and instruction in genetics, particularly Eiríkur Sigurðsson. During my trips to Reykjavík, my adopted Icelandic family, Eyrún Rós Árnadóttir, Lucinda Árnadóttir, Valdís Ösp Ingvadóttir, and Kolbrún Lilja Antonsdóttir, provided exceptional hospitality, and I am grateful to them as well for allowing me to share their holidays, for educating me about the intriguing culture and history of Iceland, and for allowing me to witness the stunning beauty of their native land.

Dean Linda T. Krug of the College of Liberal Arts at the University of Minnesota–Duluth has been tireless in her support of this project and went out of her way to help me at critical junctures, and I am grateful for her help. Derik Shelor and Helen Snively did exceptional work copyediting the manuscript. Kathryn Anderson and Nuray Ibryamova also aided in this respect, despite the promise of only a karmic reward. Mark Braunstein, the visual resources librarian of the Art History Department of Connecticut College, who curates the college's Wetmore Print Collection, and Gene Kritsky were very helpful in the acquisition of the cover art. The interlibrary loan librarians at Baker Library, Dartmouth College, and at the University of Minnesota–Duluth, particularly Heather McLean, went beyond the call of duty to help me and never failed to come through for me, as did Bonita Drummond, Roger Petry, and Kathy Skelton, who greatly aided me in the course of completing this book. Robert Swanson quickly and efficiently compiled the index.

Stephen Wrinn, the director of the University Press of Kentucky, was all that I could want in an editor. His encouragement, help, and advice were valuable at many points in the long process of turning a manuscript into a book. David Cobb, Gena Henry, and Henrietta Roberts, also of the University Press of Kentucky, were a great aid as well, and answered my inquiries quickly and with good cheer.

I thank the MIT Press for allowing me to draw on my article "Bringing in Darwin: Evolutionary Theory, Realism, and International Politics," *International Security,* Vol. 25, No. 2 (Fall 2000), and from my correspondence in *International Security,* Vol. 26, No. 1 (Summer 2001). In addition, I am grateful to Palgrave Macmillan for the same privilege concerning my chapter "Ethnic Conflict and State Building," in Albert Somit and Steven A. Peterson, eds., *Human Nature and Public Policy: An Evolutionary Approach.*

Finally, I am more than pleased to acknowledge the generous support of the Earhart Foundation, and Ingrid Gregg and Antony Sullivan in particular. The support of the Earhart Foundation, quite simply, allowed me to complete this project in a timely fashion. I thank as well Chancellor Kathryn Martin and Vice Chancellor for Academic Affairs Vince Magnuson of the University of Minnesota–Duluth for their assistance.

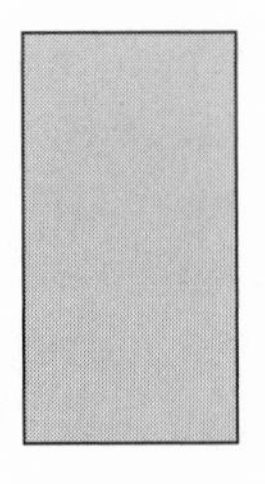

Introduction
Recognizing Darwin's Revolution

In scientific circles, the twentieth century is often termed the "century of physics" because of the remarkable progress made in that science: the theoretical work of Albert Einstein and Werner Heisenberg, the discoveries of the Curies, Ernest Rutherford's detection of the nucleus, the development of the atomic bomb by the Manhattan Project, and later the creation of fusion weapons. With its enormous potential, nuclear power also gave humans the ability to destroy much of life on the planet. These revelations and their consequences affected all people and states in international relations.

The twenty-first century has already been termed "the century of biology" by biologists and geneticists who are convinced its discoveries will rival those of physics in the last century and that those discoveries will have equally profound effects on people as well as on international relations. So close to the beginning of the century, such a description may seem precipitous, but not when one reflects on the exceptional recent advances in the science of evolutionary biology and the potential for even more progress in the near future. The progress of biological science is nothing short of revolutionary. It is as important for understanding human behavior as the great discoveries of Newton or Einstein are for the physical world. Charles Darwin is the Robespierre or Lenin of this revolution, and it should be termed Darwin's Revolution, as Darwin's work more than any other's produced dramatic changes in how we define our humanity and the human past. The year 1859, when Darwin published *On the Origin of Species,* is as important in the history of human thought as is 1543 for the Copernican Revolution or 1789 for the French Revolution. Darwin, crucially, forced us to recognize that *Homo sapiens* is an animal that has evolved through natural selection. He did not understand fully the force of his argument, but it is he more than any other who is responsible for the century of biology.

However, few outside the realm of biological science have recognized how the advances in evolutionary biology affect our understanding of human behavior. The social sciences in particular have largely not been aware of the progress evolutionary biology has made toward greater comprehension of human actions and motivations. An understanding of Darwin's Revolution allows evolutionary biologists and social scientists to grasp how human behavior evolved in radically different evolutionary conditions than what many (although not all) humans face today. This comprehension of human evolution provides major insights into human social and political behavior. Social scientists may build upon these insights to create better theories and have new understandings of social problems and issues.

At this point, it is important to stress that this is not a book about the triumph of "nature" over "nurture." That is a false dichotomy.[1] Human behavior cannot be neatly categorized into such a division. Rather than thinking in bifurcated concepts, we must think of human behavior as the product of the interaction of the genotype and the environment. When we consider a cake, it does not make much sense to examine a crumb and say: "this is the butter, but not the eggs" or this is "90 percent sugar and 3 percent flour" or "this is the result of cooking in a gas oven rather than an electric one."[2] The totality of the cake is what is important. Its ingredients and baking are equally significant for the final product. So it is with humans. Almost never can we examine a particular behavior and say: "this is 80 percent evolution and 20 percent environment." Both are necessary and intertwined.

This book is a first step in bringing evolutionary explanations into international relations. It is my hope that social scientists will become familiar with these explanations, and use them if they think them useful or necessary for their research into human behavior. Just as we need the ingredients and baking to make the cake, so let us use evolutionary and social explanations to explain human behavior.

Evolutionary biology consists of several related disciplines that together may be thought of as the life sciences. For my purposes, these are genetics, cognitive neuroscience, human ethology, human ecology, and evolutionary theory. This book uses ideas and concepts from these disciplines to study international relations. In essence, I use an evolutionary biological or life science approach to analyze international relations (although to advance my arguments I also employ concepts from social science disciplines, such as anthropology and psychology).

At its simplest level, genetics is the study of inheritance patterns of specific traits. These are contained in the genome, or all the genetic material in the chromosomes of an organism. Advances in genetics have led to three important scientific projects. The first, the Human Genome Project, generated much attention from the media. Its aim was to identify all of the approximately 30,000 to 40,000 genes in human DNA (deoxyribonucleic acid) and to determine the sequence of the 3.1 billion chemical bases that compose it.[3] In 2003, it completed its task and so provided a complete description of human life at its most fundamental level. Unraveling the mysteries of the human genome already has produced significant knowledge and promises to produce even greater discoveries. Scientists have discovered the genetic causes of proclivities to major health risks, such as certain cancers, hypertension, and obesity; of behaviors, such as alcoholism, depression, and eating disorders; as well as diseases like Alzheimer's, Tay-Sachs, and cystic fibrosis.[4] Understanding the human genome will produce advances in medicine, neuroscience, pharmaceuticals, psychiatry, and genetics that will rival the great scientific accomplishments in physics in the previous century.

The second project is cloning. Of all advances in the life sciences, this has received the greatest attention from the media—particularly since the cloning of the sheep "Dolly" in 1997—due to its fascinating, and some would say terrifying, implications for cloning human beings. Politicians, philosophers, and theologians now engage in serious debate about the ethics and morality of cloning human embryos to generate stem cells, or of cloning humans themselves, or animals with human genes, such as "Polly," a Poll Dorset lamb cloned five months after "Dolly."[5] The possibility of cloning humans has raised a host of equally intriguing and troubling issues: What does it mean to be human? Is it permissible to create life through cloning in order to harvest stem cells, thus ending the life of the clone? Yet, the benefits of cloning clearly are enormous. The cloning of embryos—so-called therapeutic cloning—makes possible regenerative medicine. Stem cells might be used to grow replacement cells or tissues, brain tissue for Alzheimer's patients, or pancreatic cells for diabetics, with much less likelihood of rejection by a patient's immune system. Doctors might replace cells and tissues as easily as a mechanic replaces a defective alternator. In a similar process called parthenogenesis, an egg cell is treated with chemicals that cause it to begin dividing into an embryo without fertilization; this might also generate stem cells. At some point cloning or parthenogenesis may permit human reproduction without the other sex.

The third project is in the field of human history. Genetic analysis of human DNA has proven useful for establishing paternity, for establishing the guilt or innocence of criminals, and for solving some historical mysteries. For example, we now know that the ten-year-old boy who died in a Parisian prison in 1795 was indeed Louis XVII. But it has also provided insights into the human past, notably human migration patterns and ancestry. DNA analysis is allowing scholars for the first time to understand the movements of our ancestors, *Homo sapiens,* from Africa into Asia and Europe, and finally into the Americas.[6] We now know about the first member of our genus *Homo: Homo habilis* lived in Africa approximately 2.5 million years ago, was bipedal, and could make tools. *Homo erectus* evolved from *habilis* and was the first member of our lineage to leave Africa for Asia and Europe, at least 1 million years ago. And 900,000 years before present, *Homo heidelbergensis* emerged from *erectus.* It was from *heidelbergensis* that *Homo sapiens* evolved about 500,000 years ago. The anatomically modern human (*Homo sapiens sapiens*) appeared approximately 100,000 years ago, most likely in southern and eastern Africa, and began to migrate from there some 70,000 to 75,000 years ago.[7]

In addition to determining broad human migration patterns, genetic analysis also tells us how closely we are related to other members of our genus. DNA extracted from the bones of Neanderthals (*Homo neanderthalensis*) has allowed scientists to determine that they are not directly related to modern humans, and are most probably a failed evolutionary branch of *heidelbergensis,* separating 500,000–400,000 years ago and dying out completely about 40,000 years ago.[8]

Cognitive or behavioral neuroscience is the study of the complex structure of the human brain and mind. It studies the way the brain develops, processes information, and produces other cognitive skills, as well as how it loses normal abilities due to age, injury, or toxicants. Equally important, it also helps define the mind. Steven Pinker, one of the leading scholars in the field, defines the mind as a "system of organs of computation, designed by natural selection to solve the kinds of problems our ancestors faced in their foraging way of life . . . understanding and outmaneuvering objects, animals, plants, and other people."[9] According to another major scholar in this field, Michael Gazzaniga, the process of human evolution explains the brain's workings, from how neurons talk to each other (synaptic relationships) to its complex circuits responsible for its highest functions, including problem solving and morality.[10] Cognitive neuroscience allows us to understand intelligence, imagination, language, mental illness, emotions (including love, hate, anger, humor, and fear), and other aspects of humanity, such as why we enjoy poems and melodies. Furthermore, this discipline allows us to grasp the homoplasy of intelligence, emotions, and feelings, and to create better social policies and principles to help people in mental or emotional distress.[11]

Table I. Geological Subdivisions of the Last 2.1 Million Years and Human Development

Epoch (The Pliocene, Pleistocene, and Holocene Epochs are of the Cenozoic Era)	*Estimated Number of Years Ago*	*Development of Humans (Estimated Time of Evolutionary Development)*
Very Late Pliocene	2,100,000–1,600,000	*Homo habilis* (2,100,000 years ago)
Pleistocene		*Homo habilis—H. erectus* (1,500,000 years ago)
-- Early (or Lower)	1,600,000–130,000	*H. erectus—H. heidelbergensis* (800,000 years ago) *H. heidelbergensis—H. sapiens* (250,000 years ago)
-- Middle	130,000–50,000	Anatomically Modern Humans (A.M.H.) in Africa and later Asia (*H. sapiens sapiens* 100,000 years ago)
--Late (or Upper)	50,000–10,000	A.M.H. in Australia, Europe (at first living with, then after Neanderthals), and finally America (radiating from Asia)
Holocene	10,000	Beginning of agriculture

On an evolutionary time scale, the human being is a recent creation. Placental mammals evolved around 100 million years ago, near the end of the Cretaceous, and primates diverged from other mammals about 75 million years ago. The hominids, of which humans are a part, split from the rest of the ape family (pongids) about eight million years ago, near the end of the Miocene Epoch.

Adapted from: Ernst Mayr, *What Evolution Is* (New York: Basic Books, 2001), p. 20; L. Luca Cavalli-Sforza, Paolo Menozzi, and Alberto Piazza, *The History and Geography of Human Genes*, abr. ed. (Princeton, N.J.: Princeton University Press, 1994), Table 2.1.1A, p. 61; Alan Strahler and Arthur Strahler, *Physical Geography: Science and Systems of the Human Environment* (New York: John Wiley and Sons, 1997), Table 11.1, p. 277; Ian Tattersall, *The Fossil Trail: How We Know What We Think We Know about Human Evolution* (New York: Oxford University Press, 1995), pp. 229–235; and Milford H. Wolpoff, *Paleoanthropology*, 2nd ed. (Boston: McGraw-Hill, 1999), chaps. 6–10.

Closely related to cognitive neuroscience is the relatively new discipline of evolutionary psychology. Evolutionary psychology is the application of the theory of evolution to human cognition and social behavior. It is essentially what Pinker calls an exercise of "reverse engineering," figuring out what the mind was designed to do.[12] It is not too far amiss to think of cognitive neuroscience as the hardware of the mind and evolutionary psychology as its software. Evolutionary psychology also has made important contributions for understanding human behavior. David Buss has studied the different mating strategies for men and women, jealousy, sexual aggression against women, and cooperation and conflict between the sexes.[13] Jerome Barkow, Leda Cosmides, and John Tooby have produced a major edited study that uses evolutionary psychology to explain the origins of human culture, social exchange, sex differences, and aesthetics, among other issues and behaviors.[14]

Closely allied to evolutionary psychology is human ethology. Ethology, also called behavioral biology, is the study of animal behavior. Human ethology—as the name suggests—is the application of ethological principles, arguments, and methodology to the human animal. Human ethologists search for the physiological causes of universal human behaviors. Ethologists Robert Hinde's, Konrad Lorenz's, Niko Tinbergen's, and Irenäus Eibl-Eibesfeldt's classic works reveal broad patterns of human behavior, including the need to identify with a community.[15] Other issues, such as homicide, infanticide, rape, human birth, and teen pregnancy have been studied through this perspective.[16] As one might expect, the discipline also has links to anthropology (due to a common interest in human universals and comparative cultural analysis) and evolutionary psychology (due to the study of human emotion and expressions).[17] For example, Paul Ekman's research reveals that smiles, frowns, grimaces, and other facial expressions are displayed and understood around the world.[18]

Human or behavioral ecology is sometimes considered a part of human ethology as well. Ecology is the study of the relationships between organisms and their environment, and human ecology studies the relationships between humans and a local or global environment, either at a specific date or over geological time. This is significant for this study because the scarcity of resources in the environments in which humans evolved significantly shaped behavior. Human ecology also informs the study of demographics, which of course is widely used by economists, political scientists, and sociologists.

Finally, evolutionary theory is the most important discipline involved, because it serves as the intellectual foundation of evolutionary biology, as well as of most of the arguments of this book. Unlike progress in genetics or the study of human mating strategies, advances in evolutionary theory receive comparatively little media attention. Nonetheless, over the years the work of major evolutionary theorists, such as Theodosius Dobzhansky, William Hamilton, John Maynard Smith, and Robert Trivers, is no less significant than the genome project, cloning, the tracing of human migration, or the study of the mind. Their explications of how evolution works, coupled with advances in genetics, have facilitated a deep understanding of human behavior that is as crucial for the social sciences as for the natural. Their scholarship serves as a foundation that may unite the natural and social sciences.

The dream of such a foundation is an ancient one. It dates to the classical Greeks, was present in the thought of the Rationalists, and flowered during the Enlightenment. Now, we are closer to realizing the goal of Aristotle, Descartes, and Condorcet than ever before due to great advances in genetics and evolutionary theory.[19]

Despite major advances in the field since his death, Darwin remains the preeminent theorist of evolution. The distinguished biologist Ernst Mayr argues that Darwin's theory of natural selection is the most revolutionary theory in history because of the ideas it refuted as well as those it advanced. Darwin dispatched the idea of creation: the belief that the diversity of life on earth was due to divine creation. He also defeated Lord Kelvin's argument that life on earth was new by estimating life on earth to be at least several thousand million years old. Finally, he overcame anthropocentrism by showing that humans are not a separate creation but the product of a common evolutionary process.[20]

Among the major ideas Darwin advanced was the seminal idea of evolution by natural selection. He used this process to explain the origins of life and human evolution.[21] Philosopher of science Daniel Dennett captures well the intellectual potency and elegance of the Darwinian Revolution: "If I were to give an award for the single best idea anyone has ever had, I'd give it to Darwin, ahead of Newton and Einstein and everyone else" because "in a single stroke, the idea of evolution by natural selection unifies the realm of life, meaning, and purpose with the realms of space and time, cause and effect, mechanism and physical law."[22]

Darwin's theory of evolution remains difficult for many to accept for theological reasons; even those with secular beliefs may think of humans as discrete from nature or natural processes.[23] In reference to Darwin's theory, Sigmund Freud once remarked that history is a series of blows to human narcissism. As Copernicus destroyed the geocentric model of the universe, and Freud himself the perfect rationality of the Cartesian ego, Darwin demonstrated that humans are not discrete from but rather a part of nature and subject to the same processes of evolution that affect fruit flies or *E. coli* and other bacteria.

While the cost of Darwin's Revolution is some damage to the amour propre of humans, the benefits for natural and social science far surpass it. The life sciences seized on Darwin's arguments to build the family of disciplines that make up evolutionary biology, and yet similar progress has not been made in social science—we do not yet have an evolutionary social science. Why this is the case is complicated. One reason certainly is the horrible abuse of Darwin's arguments by the social Darwinists. As I discuss below, their perversion of concepts such as "fitness" rightfully led to a powerful reaction against "Darwinism" in social science. They also made any effort to explain the value of Darwin's arguments to social scientists difficult because of the long half-life of the social Darwinist fallout. Social scientists treat evolutionary biology with some suspicion when humans are in the mix—and rightfully so, as I will stress time and again in this book.

But like all things, it seems, this suspicion has both positive and negative consequences. Of course it is good to be conscious of how arguments may be abused. Equally, evolutionary explanations have yet to prove themselves as valuable to the type of questions and issues studied by social scientists. Indeed, social scientists have contributed mightily to our understanding of human behavior without applying evolutionary theory or the other components of evolutionary biology to humans. For example, the causes of warfare have long been studied, well before Darwin, from multiple perspectives—the individual leader, state, or international system—and for many political scientists it is not self-evident what value the life sciences might bring to the discussion.

But the negative consequences are becoming increasingly onerous. The science of evolutionary biology continues to advance and generate important new understandings of human behavior. These are largely unknown or known only superficially in the social sciences, and this, in turn, leads to an artificially limited social science. Today we have a social science that has a few tools in its toolkit. It could have improved tools but does not because it is unaware of advances in the life sciences. The lack of awareness has a real cost. We want experts to be able to give policymakers the best solutions to problems such as preventing ethnic conflict. To do this, experts should be exposed to evolutionary and social explanations of human behavior in their professional education.

The gulf between evolutionary biology and social science has grown too great, and it is time to bridge this gap. Progress in evolutionary biology will allow us to realize Condorcet's dream of a unified natural and social science for the first time. That is, to understand human behavior at many different levels: the genes, the proteins, the brain, the mind, and the social actions. Each of these levels generates insights, but no one level of explanation is more important than the others. There is no simple equation of "one gene = one behavior."

Knowledge of the life sciences and their use in social science is not a threat to the social sciences, hindering understanding of human action, but the contrary. It enriches the social sciences. Irenäus Eibl-Eibesfeldt, Roger Masters, Edward O. Wilson, and perhaps most importantly Steven Pinker have led the way in synthesizing evolutionary theory and social science to explain important aspects of human behavior.[24] Eibl-Eibesfeldt might be considered the founder of human ethology. Masters shows the power of evolutionary theory for explaining such important topics as the origin of the state.[25] E.O. Wilson identifies aspects of human behavior that are universal.[26] Pinker has been instrumental in creating the new field of evolutionary psychology and for discovering the profound influence of natural selection not only on the ability of humans to speak a language but also on some of the rules that govern languages.[27] But perhaps most importantly, Pinker questions the denial of evolutionary causes of human behavior in social science.[28] He submits that the human mind is not a blank slate at birth, as John Locke suggested and many social theorists believe, but full of innate traits involving cognition, language, and social behaviors. Thus, as an empirical hypothesis about how the mind actually works, the conception of the mind as a blank slate fails. The sciences of genetics, cognitive neuroscience, psychology, and evolution have shown it to be false.[29]

Pinker's work is important for political science because it calls into question the epistemological and ontological foundations of liberalism, the dominant ideology of the Western world. No doubt liberal and other political theorists will reply to Pinker's arguments. For my purposes, his work is important not principally for the content of the argument but because of a more important point: Pinker's argument in his book *The Blank Slate* shows that social scientific theories that rely on assumptions of human behavior not informed by human evolution may be problematic, or even fundamentally flawed.

But intellectual exchange between evolutionary biology and social science will help to solve such problems. The synthesis largely developed by these scholars acknowledges that human behavior is simultaneously and inextricably a result of both evolutionary and environmental causes. The result of this synthesis should be an improved understanding of human behavior and better theories.[30]

At this time, there are few studies of international relations that use evolutionary biology. This should change in the future because this approach yields insights into some of the discipline's most important questions.[31] The time is right to recognize the Revolution of 1859, to bring Darwin into the study of international relations.

The Central Question and Argument of the Book

Bringing Darwin into the study of international relations means examining its major questions and issues through the lens of evolutionary biology. Of course, scholars of international relations have imported ideas from other disciplines before. They have used both psychological theories and formal modeling largely borrowed from economics to advance our understanding of important issues in the discipline. The application of evolutionary biology may generate equally important insights.[32] The central question of this book is to show how evolutionary biology and, particularly, evolutionary theory can contribute to some of the major theories and issues of international relations.

While the discipline of international relations has existed for many years without evolutionary biology, the latter should be incorporated into the discipline because it improves the understanding of warfare, ethnic conflict, decision making, and other issues. Evolution explains how humans evolved during the late-Pliocene, Pleistocene, and Holocene epochs, and how human evolution affects human behavior today. All students of human behavior must acknowledge that our species has spent over 99 percent of its evolutionary history largely as hunter-gatherers in those epochs. Darwin's natural selection argument (and its modifications) coupled with those conditions means that humans evolved behaviors well adapted to radically different evolutionary conditions than many humans—for example, those living in industrial democracies—face today.

We must keep in mind that the period most social scientists think of as human history or civilization, perhaps the last three thousand years, represents only the blink of an eye in human evolution. As evolutionary biologist Paul Ehrlich argues, evolution should be measured in terms of "generation time," rather than "clock time."[33] Looking at human history in this way, hunting and gathering was the basic hominid way of life for about 250,000 generations, agriculture has been in practice for about 400 generations, and modern industrial societies have only existed for about 8 generations. Thus Ehrlich finds it reasonable to assume "that to whatever degree humanity has been shaped by genetic evolution, it has largely been to adapt to hunting and gathering—to the lifestyles of our pre-agricultural ancestors."[34] Thus, to understand completely much of human behavior we must first comprehend how evolution affected humans in the past and continues to affect them in the present. The conditions of 250,000 generations do have an impact on the last 8. Unfortunately, social scientists, rarely recognizing this relationship, have explained human behavior with a limited repertoire of arguments. In this book I seek to expand the repertoire.

My central argument is that evolutionary biology contributes significantly to theories used in international relations and to the causes of war and ethnic conflict.[35] The benefits of such interdisciplinary scholarship are great, but to gain them requires a precise and ordered discussion of evolutionary theory, an explanation of when it is appropriate to apply evolutionary theory to issues and events studied by social scientists, as well as an analysis of the major—and misplaced—critiques of evolutionary theory. I discuss these issues in chapter 1.

In chapter 2, I explain how evolutionary theory contributes to the realist theory of international relations and to rational choice analysis. First, realism, like the Darwinian view of the natural world, submits that international relations is a competitive and dangerous realm, where statesmen must strive to protect the interests of their state through an almost constant appraisal of their state's power relative to others. In sum, they must behave egoistically, putting the interests of their state before the interests of others or international society. Traditional realist arguments rest principally on one of two discrete ultimate causes, or intellectual foundations of the theory.[36] The first is Reinhold Niebuhr's argument that humans are evil. The second, anchored in the thought of Thomas Hobbes and Hans Morgenthau, is that humans possess an innate *animus dominandi*—a drive to dominate. From these foundations, Niebuhr and Morgenthau argue that what is true for the individual is also true of the state: because individuals are evil or possess a drive to dominate, so too do states because their leaders are individuals who have these motivations.

I argue that realists have a much stronger foundation for the realist argument than that used by either Morgenthau or Niebuhr. My intent is to present an alternative ultimate cause of classical realism: evolutionary theory. The use of evolutionary theory allows realism to be scientifically grounded for the first time, because evolution explains egoism. Thus a scientific explanation provides a better foundation for their arguments than either theology or metaphysics. Moreover, evolutionary theory can anchor the branch of realism termed offensive realism and advanced most forcefully by John Mearsheimer. He argues that the anarchy of the international system, the fact that there is no world government, forces leaders of states to strive to maximize their relative power in order to be secure.[37] I argue that theorists of international relations must recognize that human evolution occurred in an anarchic environment and that this explains why leaders act as offensive realism predicts. Humans evolved in anarchic conditions, and the implications of this are profound for theories of human behavior. It is also important to note at this point that my argument does not depend upon "anarchy" as it is traditionally used in the discipline—as the ordering principle of the post-1648 Westphalian state system.

When human evolution is used to ground offensive realism, it immediately becomes a more powerful theory than is currently recognized. It explains more than just state behavior; it begins to explain human behavior. It applies equally to non-state actors, be they individuals, tribes, or organizations. Moreover, it explains this behavior before the creation of the modern state system. Offensive realists do not need an anarchic state system to advance their argument. They only need humans. Thus, their argument applies equally well before or after 1648, whenever humans form groups, be they tribes in Papua New Guinea, conflicting city-states in ancient Greece, organizations like the Catholic Church, or contemporary states in international relations.

Like realists, rational choice theorists also depend on egoism as a foundation of their theory; thus, evolutionary theory provides them the same benefit. Rational choice is a diverse theory, but its essence is captured by Jon Elster, one of its major theorists: "when faced with several courses of action, people usually do what they believe is likely to have the best overall outcome."[38] So we choose actions as a means to an end, usually our self-interest: "to act rationally is to do as well for oneself as one can."[39] This is an important assumption of classical and neoclassical economics, so important that the distinguished economist George Stigler calls "self-interest" the "granite" upon which Adam Smith built the "palace of economics."[40] Rational choice is also widely used in other disciplines, particularly political science and sociology. Nonetheless, rational choice theorists are often criticized for making this assumption.[41] Evolutionary theory gives rational choice theorists the first scientific foundation for their theory; it also affords a better understanding of human preferences and decision making.

In addition to the theoretical contributions, evolutionary theory allows international relations scholars to generate insights into specific issues in the discipline, the origins of war and ethnic conflict. In chapter 3, I show how evolutionary theory and human ecology allows us to comprehend why our ancestors would wage war for offensive and defensive reasons—to gain and to protect resources from attack. I then test this argument by drawing from both ethnological studies and statistical studies of warfare among tribal societies to determine whether these societies do indeed wage war for the reasons evolutionary theory would predict. Based on this evidence, I will argue that they indeed do so and submit that an evolutionary explanation for warfare allows scholars of international relations to comprehend better its origins by recognizing the importance of resources as causes of war. Present scholarship on the origins of war emphasizes the importance of political factors, such as the security dilemma, the structure of the international system, or the type of regime. Of course, all of these are important and deserve careful study, but wars over resources were centrally important in human evolution and remain so today, especially in developing countries, where struggles over water or suitable farmland can lead to conflict. Indeed, if the earth's climate continues to grow warmer, warfare over resources will become a more important casus belli.

Chapter 4 builds directly on chapter 3. Having explained the origins of warfare among humans, I discuss the implications of warfare for human evolution, in particular the growth of human intelligence and human society. I find that the threat of external attack, from both predators and other humans, provides a strong ultimate cause of the rapid growth of both intelligence and social living in increasingly larger groups. Second, using ethological studies, I explore warfare among other animals, particularly ants and chimpanzees, and argue that Carl von Clausewitz's famous observation that war is politics conducted by other means is correct, but in a sense has become too influential. Wars certainly are conducted for political ends, but this definition of warfare obscures as well, precisely because it emphasizes the political aspects of warfare. The understanding of the origins of warfare provided by the life sciences lets us recognize that other animals fight wars, and that war evolved in humans because it is an effective way to gain and defend resources. Third, I examine how evolutionary theory helps explain the physically and emotionally stimulating effects of warfare on combatants. I suggest that the stimulation often found among warriors may be the residue of similar experiences among groups that hunted and fought in the conditions of the Pliocene, Pleistocene, and Holocene. Lastly, I argue that a life science perspective allows scholars to understand why disease will become increasingly important in international relations as new diseases and new strains of existing diseases emerge and make biological warfare a progressively more effective weapon of war. In addition, this perspective permits us to understand that disease has played a significant but neglected role in imperialism, particularly in the European conquest of North and South America. I argue that a critically important factor in explaining the cost of European imperialism is the epidemiological "balance of power" between conquering people and those conquered.

In chapter 5, I use evolutionary theory to explain why the in-group/out-group distinctions, xenophobia, and ethnocentrism evolved in humans, and in turn why ethnic conflict occurs and reoccurs in international relations. Moreover, the chapter provides important insights for scholars and policymakers who seek to mitigate or prevent ethnic conflict. In the final section of this book, I provide conclusions to the study, discuss an agenda for further research, and offer thoughts on the underdeveloped relationship between Darwin's Revolution and the social sciences.

The Significance of the Study

The central argument of this book is important for four reasons. First, it builds on the foundation laid by Eibl-Eibesfeldt, Masters, Wilson, and Pinker for a new type of social science, one that is significantly informed by evolutionary biology, and that recognizes human behavior as the product of the interaction of the genotype and the environment.[42] For much of the twentieth century, most social scientists explained social events only in terms of their social causes. This occurred for two reasons. The first is the legacy of the renowned sociologist Émile Durkheim. Durkheim is often said to have established this belief when he argued that social facts may only be explained by other social facts. In 1895, he advanced this influential argument: "the general characteristics of human nature participate in the work of elaboration from which social life results. But they are not the cause of it, nor do they give it special form; they only make it possible."[43] Rather, they are "the indeterminate material that the social factor molds and transforms," whose "contribution consists exclusively in very general attitudes, in vague and consequently plastic predispositions which . . . could not take on the definite and complex forms which characterize social phenomena."[44]

Durkheim's argument has been significantly refined over the years, but its essence remains the same: social causes can best explain differences among people because individuals are born essentially the same, as blank slates or empty vessels waiting to be filled, and they vary only in beliefs, norms, and cultures.[45] So the effects of social phenomena like culture, the politics of the state, and other social factors must explain the differences found across societies. This is the heart of the standard social science model used to explain behavior across the social sciences.[46]

No doubt humans are born almost entirely the same, and a multitude of social factors such as culture, ideology, the political system, history, and norms do have an enormous impact on individual lives. Yet it is not the case that culture rules once humans have passed through infancy, childhood, and adolescence. Here is the flaw in the standard social science model: it does not acknowledge Darwin's Revolution. It offers, if you will, only half of the pieces of the puzzle that is human behavior—the baking but not the ingredients to the cake. Human behavior is simultaneously affected by the environment and by the genotype through evolution by natural selection. Again, I hasten to add that this book is not a debate about nature versus nurture. Human behavior must be understood as the synthesis of these forces, the interaction of the environment and evolution by natural selection working largely through the genotype. In explaining human behavior we cannot slight either force.[47] Social scientists and natural scientists must have both to understand and explain fully human behavior.

Evolutionary biology is beneficial for social scientists because it allows us to acknowledge that evolution's effects do not disappear, but are present throughout our lives irrespective of our culture, class, or social standing. This is true for much human behavior, but at the same time is only a partial explanation of the human behavior I examine.[48] Human behavior, like former mayor David Dinkins's description of New York City, is a "gorgeous mosaic" of evolutionary and social causes.

A second reason why social scientists have preferred to explain social phenomena without referencing evolutionary biology is the abuse of the concept of "fitness" and other evolutionary ideas. Richard Hofstadter documents how social Darwinists writing in the late nineteenth and early twentieth centuries used Darwinian rhetoric to support their ideologies; among them were British sociologist Herbert Spencer and important intellectual figures in American politics, such as William Graham Sumner.[49] These writings provided the aura of scientific certainty for racist immigration, education, eugenics and forced sterilization, and other policies of the American government as well as some state governments, such as Vermont and Virginia.[50]

This alone is sufficient to make many social scientists wary of Darwinian thought. However, since social Darwinian ideas contributed to Nazi ideology and thus to its horrible eugenics policies and ultimately to the Holocaust, social Darwinism has been universally condemned.[51] Social Darwinists abused and contorted Darwin's ideas to suit their favored policies and to support their prejudices. Nonetheless, the fact that the grotesque caricature of Darwinism that is social Darwinism had such a negative impact explains the considerable suspicion that many social scientists still feel when evolutionary arguments are applied to *Homo sapiens.*

Beyond being understandable, this suspicion was appropriate. Fortunately, conditions have changed as knowledge has advanced due to the scholarship of evolutionary theorists, ethologists, geneticists, and social scientists that are sensitive to the insights of evolutionary biology. Indeed, this knowledge is changing our very conception of what it is to be human and what is unique about our species. It is also allowing social scientists and philosophers to ask, and indeed forcing them to confront, a host of questions that would have seemed incongruous or ridiculous even a generation ago: Are we the only animals that possess self-consciousness, or culture, or even language? Do other animals fight wars? Do other animals have complex emotions? Can they deceive? Certainly, the animal rights movement has its intellectual origins in Darwin's recognition of the continuity between animals and humans.[52] As philosopher Mary Midgley wrote in 1978, "Drawing analogies 'between people and animals' is, on the face of it, rather like drawing them 'between foreigners and people.'"[53]

Building on some of these questions and the scholarship of Eibl-Eibesfeldt, Masters, Wilson, and Pinker are scholars such as Donald Brown, Leda Cosmides and John Tooby, Francis Fukuyama, Sarah Blaffer Hrdy, Arnold Ludwig, Stephen Sanderson, Albert Somit, and Steven Peterson.[54] Brown demonstrates how anthropology may be enriched and broadened when anthropologists use evolution to determine universal human behaviors.[55] Cosmides and Tooby, using this approach, reveal important insights in psychology.[56] Fukuyama finds it relevant for solving collective action problems that bedevil economics and political science and for the invention of moral rules that constrain individual choice.[57] Hrdy's analysis is important for sociology because she examines the universal elements of motherhood and its complications in modern society.[58] Ludwig applies the argument to explain political leadership, why people want to become rulers, and their psychological similarities.[59] Sanderson uses the approach to improve theories of social conflict in sociology. He develops a Darwinian conflict theory that surpasses functionalist, Marxist, and Weberian explanations of social conflict.[60] Somit and Peterson argue that it provides a basis for specific forms of government.[61]

My argument should be seen as a contribution to the project of applying evolutionary biology to social science in order to produce a new synthesis to explain human behavior.[62] Again, human behavior in general is determined not by evolutionary forces or culture alone but by the interaction of both forces. As Masters writes, "the first requisite for a rigorously scientific approach to human nature is . . . willingness to abandon the belief that answers are either/or: our behavior can be both innate and acquired; both selfish and cooperative; both similar to other species and uniquely human."[63] My aim is precisely to generate scholarship that is informed by both forces and to move away from strict and inappropriate "hard" models of either a biological or social determinism for the topics I address in this book. These "hard" models of human behavior may be appropriate to explain some aspects of human behavior. However, speaking broadly, human behavior can be studied well through a "soft" version of either position, that is, a position that recognizes the value of the other. The evolutionary anthropologist Lionel Tiger argues: "Biology is not destiny, but it's good statistical probability."[64] I amend this statement: Neither biological nor social explanations of human behavior are "destiny," but most often they are equally important and necessary steps toward establishing a complete and scientific explanation of a particular human behavior. In this spirit I offer this book.

This book's second contribution is in the realm of what methodologists term "theory improvement." Evolutionary theory may be used to place some social scientific theories on a scientific foundation. Specifically, evolutionary theory provides a scientific foundation for realism and rational choice. The basic argument of evolutionary theory is widely accepted by evolutionary biologists. Its use by realists will help scholars construct verifiable scientific explanations that will improve the theory according to the standard set by Gary King, Robert Keohane, and Sidney Verba in their influential study of social science methodology.[65] Improving realism in this way can invigorate realist scholarship by anchoring it in a scientific theory for the first time in its long history. The egoistic and dominating behavior that realism expects of individuals and states is a result of human evolution. Scholars may be attracted to realism and yet reluctant to anchor scholarly arguments on a theological force like evil, a metaphysical precept such as *animus dominandi,* or anarchy. They will find a sound scientific substructure in evolutionary theory. The same is true for scholars who use rational choice analysis: evolutionary theory will let them ground the assumptions made by their theory using evolutionary science and will in turn improve the theory.

My third contribution concerns specific issues in international relations. I use concepts from evolutionary biology to explain the origins of warfare and ethnic conflict. The perspective offered by evolutionary biology allows us to comprehend that warfare can help people acquire the resources they have needed to survive throughout our evolutionary history. These may be food, a mate, shelter, or territory, especially land with high-quality soil or wild foods, or abundant firewood, or territory free of dangerous animals, insect infestation, or disease. What evolutionary theory and human ecology allow students of warfare is a grasp of the interconnections between the origins of warfare and the functions of warfare. That is, warfare originates in the need to gain and defend resources against competitors. In addition, evolutionary theory may be used to explain in-group/out-group distinctions, xenophobia, and ethnocentrism, all contributing causes of ethnic conflict. Scholars of ethnic conflict who argue that "ancient hatreds" cause such clashes can now use evolutionary theory to explain scientifically the origin of these hatreds, and thus eliminate the need to refer to the nebulous term "ancient hatreds."[66]

Fourth, incorporating evolutionary biology into the discipline of international relations can benefit each of the levels of analysis traditionally used in the discipline, the first (the individual), the second or unit level (the state), and the third or systemic level (the international system).[67] For instance, evolutionary theory can directly inform the individual level of analysis because it explains individual behavior, for example, showing why individuals are egoistic.[68] Moreover, evolutionary theory, and related disciplines, such as cognitive neuroscience, ethology, and genetics, will generate novel understandings of some of the central issues studied in international relations, such as human decision making.[69]

Evolutionary theory also allows scholars to use evolutionary models or metaphors to explain how the environment (the international system) conditions the behavior of individual leaders of states or even the actions of the states themselves. For example, William Thompson suggests that the metaphor of evolution may remind scholars of international relations of the ubiquity of change, the complexity of international relations, and the difficulties of predicting the future, or it may let them be more sensitive to trends in conflict or cooperation.[70]

Perhaps most importantly, Kenneth Waltz's conception of socialization in international relations is very similar to Darwin's concept of natural selection as a process of evolution, making it an example of how evolutionary theory may provide metaphors that better explain state behavior in a competitive environment. In his magnum opus, Waltz explains how the international system forces states to socialize, adapt, and emulate successful states or else fall prey to the depredations of other states.[71] The Darwinian model of evolution strongly reinforces Waltz's argument and also allows us to understand how states, like animals in the natural world, adapt to an external environment—at times consciously, but also through unconscious forces like natural selection.

Moreover, the metaphor of evolution through natural selection provides insights for scholars who use the systemic analysis of international relations. As I explain in the next chapter, natural selection works through many forces of nature, including external selection pressures, climate, predators, and parasites, but similarly important are individual changes, differences within the genotype, that allow a particular animal with a positive or better adaptation to flourish in a slightly changed environment. Thus, when applied to international relations, we can understand that aspects of the particular state, that is, either positive or negative adaptations, may have significant effects on the international system. We may think of the command economies of the Soviet Union, Cuba, or North Korea as being evolutionary dead ends in the environment of globalization, or as nuclear weapons as contributing to or detracting from the "fitness" of a state in international relations.

The condition of the natural world may also serve as a metaphor for international relations. Indeed, probably the most famous metaphor in the history of international relations is Thomas Hobbes's famous description in *Leviathan* of the "state of nature." As an ecologist and ethologist, Hobbes was not bad. His conception of the natural world captured much of what Darwin would recognize two hundred years later. In the natural world, animals must provide resources—food, water, and shelter—for themselves and perhaps for their genetic relatives and others in very occasional and specific circumstances, such as to attract a mate. They also live in a dangerous environment, constantly threatened by predators as well as their own kind, and by forces of nature, such as disease, parasites, and weather. Animals act the way they do—they are egoistic, fearful of others, and react to threats by fighting, freezing, or through flight—because natural selection has favored these behaviors. They help animals survive in a dangerous world.

Similarly, the realist theory of international relations argues that states face threats and should behave in similar ways to maximize their chances of survival in a dangerous and competitive environment. States are in this environment, the state of nature as described by Hobbes, due to anarchy and the desire to dominate to ensure adequate resources for survival. I submit that this desire to dominate is a stronger force for determining behavior than anarchy: even when the problem of anarchy is solved, people still compete for dominance. Replacing anarchy with hierarchy—or government—does not eliminate evolution as a cause of human behavior.

Anarchy and the desire to dominate create similar consequences: states, like animals, are egoistic, and will place their needs, for food or for security, before the needs of conspecifics (their own kind). The use of force is always possible. External threats are always present, as are natural threats. In addition, they may encounter threats from within the group. Cooperation will be difficult and occasional. States may cooperate, but only when they share a mutual interest. When a state or coalition becomes dominant, it will worry greatly about challenges from conspecifics, just as the alpha male gorilla or lion worries about threats to his resources. Just as he may kill other males, or drive them off, or forge a coalition with them to maintain his power, the hegemonic state does the same with respect to its challengers. Just as alpha males living in what ethologists term "dominance hierarchies" bring stability and other important social benefits to the pack, herd, or pride, so too does the hegemon bring stability and the benefits of other collective goods to international relations. Evolutionary theory allows us to understand animal behavior; in turn we can see that Hobbes's description was more accurate than he or anyone could have realized before Darwin's Revolution.

The use of evolution as a metaphor may generate insights, but like all metaphors, it must be sensitively used. A metaphor is a useful teaching tool. It may capture concisely a similarity in causation or allow greater sensitivity or identification of parallel conditions. Of course, when we move beyond metaphors to study specific empirical issues or to theorize, then we must focus on the pressures of the anarchic international system or unit level causes, such as the actions of bureaucracies within a state, as they are more important to explain or predict behavior.

Finally, although the use of evolutionary theory in international relations will help its scholars improve their theories and study of major issues, it does not require that students of international relations adopt or subordinate themselves to the agenda or priorities of evolutionary theory or evolutionary biology. Using the ideas developed in another discipline does not entail acceptance of all of its concepts. In this book, I use ideas and concepts from evolutionary theory, ethology, and genetics, as well as from other disciplines like anthropology, archaeology, and psychology. As more scholars adopt this perspective in the discipline of international relations, they will not be concerned with the problems that interest evolutionary theorists, ethologists, biologists, or archaeologists. They will still study international relations, and will focus on the traditional concerns or major questions of the discipline. What will be different are the ways they think of these problems and questions, and as a result, expand our understanding of international relations.

1 Evolutionary Theory and Its Application to Social Science

In this chapter I first explain what evolutionary theory is and the importance of Darwin's idea of natural selection. Second, I discuss major conceptual issues that social scientists should consider when they apply evolutionary theory to their work. Finally, I address the major criticisms of evolutionary theory. A thorough discussion of evolutionary theory is required because this is the theoretical foundation upon which the subsequent chapters build.

What Is Evolutionary Theory?

Evolutionary theory explains why and how life changes over time. It explains the great diversity of life that now exists on this planet and all that has lived in the past, from single-celled organisms and trilobites to dinosaurs and primates. The fundamental question is: How did species evolve over time, causing the broad scope and great diversity of life found today? In 1858, Alfred Russel Wallace, who developed aspects of what we now term "Darwinism" independently of Charles Darwin, succinctly captured a major part of the answer, natural selection: "There is a general principle in nature which will cause many varieties to survive the parent species, and to give rise to successive variations, departing further and further from the original type."[1] However, a complete answer to the question requires a discussion of Darwin's thought, particularly his idea of natural selection, and how genetics affects evolution.

Modern evolutionary theory rightfully begins with Darwin because in the totality of his work he advanced five theories that form the core of Darwinism: evolution, the common descent of life, the gradualness of evolution, the multiplication of species, and natural selection.[2] I will discuss each in turn. Table 1.1 summarizes Darwin's ideas.

Table 1.1. Darwinism

Core Idea of Darwinism	*Mid-Nineteenth-Century Intellectual Thought*
1. Evolution: The world is neither constant nor perpetually cycling, and organisms are transformed steadily over time. Mayr describes this as the first Darwinian Revolution.	1. Belief that the world is constant, ordered, and young.
2. Common Descent: Every group of organisms descends from an ancestral species.	2. Belief in the essentially unchanging nature of species perfectly adapted for the environment.
3. Gradualism: Evolution proceeds relatively slowly, gradually rather than in jumps.	3. Belief in no change or change by saltations, Lamarckism, or orthogenesis.
4. Multiplication of Species: Great diversity of life expected at present and in the past.	4. Belief that only the known extant species have ever existed. Fossil evidence is proof of the Great Flood.
5. Natural Selection: The mechanism of evolutionary change. Heritable modifications that assist its ability to survive and reproduce are passed to subsequent generations. Mayr terms this the second Darwinian Revolution.	5. Once Darwin's idea was advanced in 1859, it met great popular and academic resistance.

Sources: Ernst Mayr, *The Growth of Biological Thought: Diversity, Evolution, and Inheritance* (Cambridge, Mass.: Belknap Press of Harvard University Press, 1982), pp. 426–525; Mayr, *Toward a New Philosophy of Biology: Observations of an Evolutionist* (Cambridge, Mass.: Harvard University Press, 1988), pp. 198–212; and Mayr, *One Long Argument: Charles Darwin and the Genesis of Modern Evolutionary Thought* (Cambridge, Mass.: Harvard University Press, 1991), pp. 68–107.

The central idea of evolution is that the world is "neither constant nor perpetually cycling but rather is steadily and perhaps directionally changing, and that organisms are being transformed in time."[3] Today this is understood as a scientific fact, but when Darwin was writing it was not widely accepted. The common belief was that the earth was unchanging and new.[4] Moreover, it was widely held that each species passed on to the next generation an essence that did not change or evolve. But scholarship by Darwin and others brought about a paradigm shift. Now, of course, the evolution of life is accepted in all but a few circles. Ernst Mayr terms this the "first Darwinian revolution" because it replaced the view that the earth's natural life was unchanging; it also deprived "man of his unique position in the universe" and placed "him into the stream of animal evolution."[5]

Second, Darwin's famous excursion to the Galápagos archipelago led him to conclude that the many species of finches, for example, on those islands descended from a common ancestor.[6] As he writes in his classic, *On the Origin of Species,* "a number of species . . . keeping in a body might remain for a long period unchanged, whilst within this same period, several of these species, by migrating into new countries and coming into competition with foreign associates might become modified."[7] He realized that by tracing an animal's ancestry into the higher taxa he could determine a common ancestor. Darwin wrote that he viewed all beings "as the lineal descendants of some few beings which lived long before the first bed of the Silurian system was deposited. . . . Judging from the past, we may safely infer that not one living species will transmit its unaltered likeness to a distant futurity."[8]

Third, Darwin proposed that evolution was a gradual process. As Mayr notes, this was a "drastic departure from tradition" which argued that new species could be created only by new origins or saltations.[9] Darwin believed science faced a choice: either new species arose through a Creator or else the new species arose from extant ones, changing slowly over time. He maintained his belief in gradual evolution despite great criticism from even his close supporters. One such supporter was Thomas H. Huxley, who wrote to him just before *On the Origin of Species* was published in 1859: "You have loaded yourself with an unnecessary difficulty adopting *Natura non facit saltum* [nature does not make jumps] so unreservedly."[10] Notwithstanding this criticism, evolutionary theorists now accept Darwin's argument that evolution is gradual.

However, Darwin's theory of gradualism is the source of some debate among evolutionary theorists. This debate pivots on the question: What is "gradual"? In 1972, Niles Eldredge and Stephen Jay Gould advanced an argument for "punctuated equilibria," the idea that species may undergo relatively long periods with little evolutionary change, but these periods of stasis or equilibrium may be broken or punctuated by relatively rapid change.[11] While their work received attention, many evolutionary theorists argued that this was not a new argument.[12] Mayr argues forcefully that Darwin was aware that evolution may not occur for long periods of time and also that it may occur rapidly.[13] Nonetheless, the work of Eldredge and Gould is useful because it reminds us that there are degrees of evolutionary gradualness.

Fourth, the work of Darwin and his contemporaries answered the question of why there are so many species in the world.[14] This question might seem odd today, when we know much more about the variation of life on the planet and are more sensitive to human impact on it.[15] Darwin's recognition of evolution, its gradualness, and the mechanism of natural selection allowed him to explain the great number of species found in the world. Even more importantly, however, he understood that the extant species were only a portion of the life that must have existed on earth. This insight provided a valid, scientific explanation for the origins of the many and widely scattered dinosaur fossils that were often attributed to the biblical Great Flood.[16]

Fifth, natural selection is perhaps his greatest contribution, because it describes the most important mechanism of evolutionary change. Darwin recognized that animals struggle for life in a dangerous and competitive world. As he writes in *On the Origin of Species,* "owing to this struggle for life, any variation, however slight and from whatever causes proceeding, if it be in any degree profitable to an individual of any species . . . will tend to the preservation of that individual, and will generally be inherited by its offspring. The offspring, also, will thus have a better chance of surviving."[17] Describing this principle as "natural selection," Darwin continues, "for, of the many individuals of any species which are periodically born, but a small number can survive. I have called this principle, by which each slight variation, if useful, is preserved, by the term of Natural Selection, in order to mark its relation to man's power of selection [artificial selection]."[18] Darwin's profound argument is that heritable modifications in individual species that help it reproduce and survive are passed to subsequent generations. Mayr terms this the "second Darwinian revolution" because it provides humans with an entirely new way of explaining the natural world.[19]

Building on Darwin's idea of natural selection, evolutionary theorists still agree with the core of Darwin's elegant idea but have now refined the concept in two ways. First, natural selection at the level of the genotype (the total genetic constitution of an organism or population) affects the phenotype (the totality of characteristics of an individual or population).[20] Described more precisely, natural selection requires both replicators and interactors. It depends on the replication of the organism but also on some interaction with the environment so that replication is slightly, perhaps ever so slightly, different.

Genes are the archetypical example of replicators, but there are others.[21] For example, in asexual species, the entire genome would be the replicator. Several qualities make for a good replicator: longevity, fecundity, and fidelity. Longevity is the endurance and durability of the copies the organism may pass through reproduction to subsequent generations. In general, "the more copies a replicator produces (fecundity) and the more accurately it produces them (fidelity), the greater its longevity and evolutionary success."[22] Replication is a necessary but not sufficient process for natural selection to occur.

Interactors are the other component necessary for natural selection. David Hull, a philosopher of biology, coined the term; he defines interactors as "those entities that interact as cohesive wholes with their environments in such a way as to make replication differential."[23] Organisms (phenotypes) are examples of interactors. But thanks to Hull, interactors are seen to be more than phenotypes. Genes are also affected or interact with the environment through genetic mutation, and in addition chromosomes and gametes (reproductive cells such as the egg in the female or spermatozoon in males) interact with the environment as surely as organisms do.

In the second refinement of Darwin's thought, evolutionary theorists now recognize that natural selection is a two-step process. The first step is the production of genetically differing individuals, which is essentially what Darwin recognized; the second, equally important step concerns "the survival and reproductive success of these individuals."[24] If an individual animal does not survive long enough to reproduce, then its genotype—perhaps with its slight improvements that would benefit the species over time—may be lost.

Modern Evolutionary Theory: Incorporating Natural Selection and Mendelian Genetics

In 1865, six years after Darwin published *On the Origins of Species,* Gregor Mendel published a paper on reproduction in pea plants in which he identified a previously unsuspected mode of heredity, which we now call genes. This discovery was the beginning of modern genetics.[25] He also argued that some of these genes come from the male parent and some from the female, and that they are transmitted intact and in a random manner.[26] Modern evolutionary theory synthesizes Darwin's basic ideas and Mendelian genetics to produce a theory sometimes termed the "synthetic" theory; evolutionary theorist Julian Huxley described it as "the evolutionary synthesis."[27] The great value of adding genetics to Darwinism is that it introduced several new media through which evolution can occur, principally migration, mutation, and drift. Darwin knew little or nothing of these sources of genetic variation, which were largely developed by evolutionary theorists and geneticists after his death. I will explain each in turn.

First, migration is the most obvious mechanism of change. It occurs as breeding migrants move into and out of a population. These visitors may thus affect the rates at which genes move into and out of a population, or its gene frequency.[28] Migration increases the genetically effective population size for a species. Rarely, a species may become segregated into colonies that are completely isolated from each other, preventing migration between populations, and possibly over time leading to a deleterious impact, such as inbreeding or even extinction, for a small, remote colony.

Second, mutation facilitates spontaneous novelty in DNA, a very long molecule comprising two strands. Each strand links a sequence of molecular subunits called bases, of which there are four types, adenine, cytosine, guanine, and thymine, or A, C, G, and T. These bases may be conceived of as letters strung together in a particular order, just as letters are ordered in every word. They contain the code necessary to create and reproduce life. Most cells have several DNA molecules of various lengths; each individual DNA molecule and its associated proteins is called a chromosome. The numbers and lengths of chromosomes almost always differ between species; for example, humans have twenty-three pairs of chromosomes in the nucleus of each cell, for a total of forty-six, whereas fruit flies have four pairs of chromosomes per cell.[29] DNA replication normally occurs every time a cell divides, so that each daughter cell inherits one complete set of DNA molecules. This division is called mitosis. Another form of division, meiosis, is less common but equally important and is essential for human sexual reproduction. It occurs in some of the cells of the testicles and ovaries. As these cells divide, the two strands of each chromosome pair separate so that the progeny cells have only twenty-three rather than forty-six chromosomes in their nuclei. Consequently, each parent only puts half of his or her chromosomes into each sexual cell, resulting in every individual's receiving twenty-three chromosomes from the father and a like number from the mother.[30] However, mutations can occur during this process of DNA replication through either mitosis or meiosis.

Mutation is a theoretically more complicated process than migration, and may have one of several causes. It may be the result of a natural process, or event, such as nuclear radiation. Most often, however, mutation takes one of three forms.[31] First, a gene may mutate due to a "copying error," the substitution, addition, or deletion of nucleotides within a section of the gene's DNA or RNA (ribonucleic acid) during mitosis or meiosis. Second, structural chromosomal changes may affect the arrangement of genes on the chromosome. Third, there can be changes in the number of chromosomes in the gene. While the origins of mutation are important, population size also must be considered. A single new mutation, no matter how adaptive, is likely to be overwhelmed in its effects if it appears in a large population.

To illustrate this point, consider a case where one genetic mutation makes a swifter cheetah. One genetically modified cheetah in a million is not likely to make a difference: though it may be better able to catch food, it could also die before it breeds. Even if it does breed, the one gene (we assume one for the sake of simplicity) that accounts for swifter speed may combine with any of the many genes from its mate that account for a relatively slower speed, so the offspring will not be swifter. As a result, the mutation is likely to be lost. The mutation is more likely to make a difference if the population is small, so that replications of the gene in future generations can combine to produce faster speed for that specific population of cheetahs, and perhaps later as out-breeding occurs, for the larger population of cheetahs. Population size does matter for the evolution of an adaptive mutation. If a better-adapted population is too small, it could be wiped out by predators, disease, or other natural forces—such as flood, drought, fire, earthquake, or volcanic eruption—before it has a chance to spread to a general population.

Third, genetic drift, or the "Sewall Wright effect," also causes change. The renowned geneticist Theodosius Dobzhansky explains how this may occur. In different human populations the frequencies of genes may vary. They "need not necessarily be caused by natural selection. They may . . . be explained . . . by random fluctuations in gene frequencies in effectively small populations."[32] Small populations are more likely to change in this way than large ones because certain alleles (alternative forms of a gene occupying the same chromosomal locus) may be lost over generations, and others may reach fixation, perhaps reaching a frequency of 100 percent.[33] This process is likely to occur over time because the loss of alleles or fixation of specific genes is likely to be incremental. In different populations, genetic drift will occur independently, with certain alleles spreading in some populations and different alleles in others.[34] For large populations, such as six billion humans, Dobzhansky says it is less likely to be an important cause of evolution.[35]

Modern evolutionary theory recognizes that evolution occurs due to natural selection, genetic migration, mutation, and drift. Thus, it is the result of both natural processes and chance.[36] In addition, two other forces of evolution, artificial selection and sexual selection, can be significant and so deserve to be mentioned. Artificial selection, long practiced in agricultural and animal breeding, was familiar to Darwin, who used it as a model for his idea of natural selection.[37] The animal breeder changes the properties of his herd by deliberately selecting certain sires and dams to breed the next generation, in which he hopes those desired properties will be present. Thus it is a directed and not a natural process.

Darwin coined the term sexual selection for the process by which an individual gains reproductive advantage by being more attractive to the opposite sex. He briefly described this process in *On the Origin of Species* and then thoroughly presented his ideas in *The Descent of Man*; indeed, most of that work concerns sexual selection.[38] Darwin argued that an individual might contribute more genes to subsequent generations by being more successful at reproduction rather than through physiological efficiency. Darwin stated that sexual selection might be another evolutionary force, but one that "acts in a less rigorous manner than natural selection" because natural selection "produces its effects by the life or death at all ages of the more or less successful individuals" and controls the number of resources that may be devoted to attracting females.[39] In the case of sexual selection, adaptations arise through competition for reproduction; they then spread in a population because they increase the chance of mating. Darwin's most famous example was the peacock's tail, which is not useful for finding food or for defense but is useful for attracting peahens.[40]

More technically, sexual selection is competition among gametes for fusion partners. As the geneticist Graham Bell describes it, "in multicellular animals and plants, gametes may fuse inside or outside the bodies of gamonts [individuals that generate gametes], and it is very often the behavior of the gamonts (such as peacocks) that determines the success of gametes, rather than the behavior of gametes themselves."[41]

Obviously there must be limits on such extravagances as peacocks' tails since these adaptations may quickly hinder the bird in gathering food, leading to its poor health and even death. Also, such an adaptation might make the animal easier to detect or impede its flight from predators. In this manner, natural selection serves as a check on sexual selection.

Sexual selection remains a controversial idea and is not wholly accepted by evolutionary theorists. Many evolutionary theorists and population geneticists reject sexual selection because they focus on gene frequencies and thus see the critical component of evolution as the phenotype's ability to reproduce genes successfully. Sexual selection, if it is operative at all, does not function at the level of the gene. And clearly, it does not obtain for asexual species. However, if evolution operates at the level of the individual for sexual species, then it is conceivable that sexual selection might obtain. Despite the controversy, Mayr says "there is now a justified tendency to interpret sexual selection rather broadly as any morphological or behavioral characteristic that gives a reproductive advantage" and thus is a process of evolution.[42]

Table 1.2 summarizes the mechanisms of evolution. While all of these causes of evolution are important, evolutionary theorists consider natural selection to be significant for studying animals like *Homo sapiens.* Mayr succinctly captures the importance of natural selection: "long-term evolution without natural selection is inconceivable."[43] In this study, I will most often refer to evolution by natural selection, but the reader should always remember the other mechanisms of evolution, artificial and sexual selection, and the other components of the evolutionary synthesis: genetic migration, mutation, and drift.

Fitness and the Processes of Evolution by Natural Selection

The essence of the evolutionary argument is that most of the behavioral characteristics of a species evolve because they help it survive and reproduce: "Those whose genes promote characteristics that are advantageous in the struggle to survive and reproduce are rewarded through the transmission of their genes to the next generation."[44] This process has three constituents.[45] First, some genetic variation must be present in the species. If all individuals are the same, then there is no basis for change. Fortunately, gene frequencies change regularly due to genetic drift, migration, mutation, and natural selection, and for some animals through sexual and artificial selection.[46] Thus, for sexually reproducing species, only identical twins (or other monozygotic multiple births) are truly identical; all others possess differences. Gould calls these differences the "raw material" of evolutionary change.[47] Second, this variation must affect and improve what evolutionary theorists term "fitness"—a member of a species is fit if it is better able than others to survive and reproduce.[48] Survival of the fittest means that those individuals possessing characteristics that cause them to be better able to survive and reproduce will do so.[49] They will be better represented in the next generation than those who are less fit. Finally, there must be heritable variation in fitness: the characteristic must be passed from parent to offspring. The ability of zebras to run quickly is one example; on average, fast zebras tend to have faster offspring.[50]

Table 1.2. The Processes of Evolution	
Process	*Comments*
1. Natural selection	Darwin's greatest contribution to evolutionary theory. At its root, it is a simple idea: genes that promote characteristics that help in the struggle to survive and reproduce are extant in the next generation.
2. Genetic migration	Genes move in and out of a given population as a phenotype does.
3. Genetic mutation	Mutation occurs via copying error, structural chromosomal changes, or numerical changes affecting the arrangement of genes in a chromosome.
4. Genetic drift	Caused by either natural selection or random fluctuations in gene frequencies, it mostly affects small populations.
5. Artificial selection	Darwin's observation of this common practice among contemporary breeders influenced his thinking.
6. Sexual selection	Adaptations arise as a result of competition over reproduction. Controversial as a discrete process of evolution.

The concept of fitness is critical because it is the major causal mechanism used in evolutionary theory to explain how life adapts to new environments.[51] According to evolutionary theorist John Maynard Smith, "evolution is concerned with how, during the long history of life on this planet, different animals and plants have become adapted to different conditions."[52] He notes: "some adaptations, such as the process whereby energy is obtained by fermentation, are common to almost all living things. For this very reason it is difficult to discover their evolutionary history."[53] In contrast, other adaptations "enable some animals or plants to live in a special way in a particular environment. In such cases we can observe differences between animals, and can often see how these differences are suited to the ways in which animals live; we then have a chance to study how and why they have evolved."[54]

Natural selection results in changes in traits that are directional and not random; this is adaptation. Adaptations are phenotypic features or traits such as morphological structures, physiological mechanisms, and inherited behaviors that are present in individual organisms because the process of natural selection favored them in the past. As Mayr explains, an adaptation is "a property of an organism, whether a structure, a physiological trait, a behavior, or anything else that the organism possesses, that is favored by selection over alternative traits."[55] Because these features increase fitness, they spread though the population. By possessing these adaptations, an individual stands a better chance of surviving long enough to reproduce and of having his genes represented in the next generation.[56] That is, adaptations that enhance survival and reproduction (fitness) will increase in frequency in relation to those that do not.[57] This is the basic model of evolutionary theory.

It is important to note that evolutionary theory also applies to *Homo sapiens.* Like all animals, *Homo sapiens* behaves as he does largely due to evolution by natural selection.[58] Evolution shows us that human behavioral traits—the inherited, or genetic, causes of human behavior—evolved due to natural selection and those genes that increased fitness spread though the population. It is important here to stress what is meant by fitness because the concept may be misunderstood: By displaying these traits, an individual stands a better chance of surviving long enough to reproduce and of having his genes represented in the next generation. Thus all organisms, including humans, undergo constant evolution, however slow the process may be.[59]

Now this is not an argument that individuals are perfectly adapted to an environment. As E.O. Wilson argues: "No organism is ever perfectly adapted" to a given environment and not all phenotypic characteristics are necessarily adaptive.[60] For example, it was suggested that the flamingo's pink coloration is an adaptation that makes it less visible to predators at dawn and dusk. However, ethologists now know that the coloration is largely a consequence of the flamingo's diet of shrimp.[61] Moreover, natural selection must operate within strong constraints, at least in the evolutionary near-term, such as physical and chemical processes that restrict what is possible for the phenotype.[62]

In addition, no species, including *Homo sapiens,* can consciously strive to maximize fitness. Evolution selects for fitness only indirectly, by seeking adaptive phenotypes. This point is well captured by philosopher of science Alexander Rosenberg: "among giraffes it selects for long necks, among cheetahs for great foot speed, among chameleons for mimicry, and among eagles for eyesight. But each of these is selected because it makes for the survival . . . [and reproductive success] of the organisms endowed with it."[63] As a result of this process, "all organisms are at best approximate . . . [and] jury-rigged," Rosenberg writes; the best or most fit among them "out-reproduce the others. But the best may not be very good on any absolute scale. It need only be good enough to survive and outlive the other variant among which it emerges."[64]

Matt Ridley also captures this point about needing only to be "good enough" to survive and outlast others—what may be thought of as relative fitness—with his anecdote of the sage and his disciple who live in a remote forest. One day, as they were gathering berries, a ferocious bear suddenly attacked them. As they fled through the forest, the bear was gaining on them. The disciple panted despairingly to the sage, "I am afraid sire, that it is no use: we cannot outrun the bear." The sage replied, "It is not necessary that I run faster than the bear. It will be enough if only I can run faster than *you*."[65]

The word game presented in table 1.3 illustrates how evolution by natural selection occurs. This table shows how one may begin with the word "word," which may be transformed letter by letter into the word "GENE." Given that we are changing one letter at a time and the alphabet has 26 different letters, Bell argues that the transformation would take about 26^4, or 5×10^5, words, each four letters long. "If we change one letter at random every second, we can expect with average luck to spend about 200,000 s[econds], between two and three days, before getting GENE."[66] Following Bell, table 1.3 provides an even better analogy of how evolution works: first, it allows only alterations that produce meaningful English words, and second, any variants that are more similar to GENE, because they have the right letter in the right place, will replace any less similar.

Bell submits that these rules are analogous to assuming that different sequences of nucleotides in nucleic acids or of amino acids in proteins have different properties because some are nonfunctional and cannot produce a lineage, whereas some of the functional lineages are superior to others and thus replace them fairly quickly.[67] The successive selection of slight improvements facilitates rapid and apparently improbable transformations. This process does not depend upon the right variant arising fully formed, like Athena from the head of Zeus. It is most often a gradual and iterative process, and

Table 1.3. A Word Game Illustrating the Darwinian Theory of Evolution

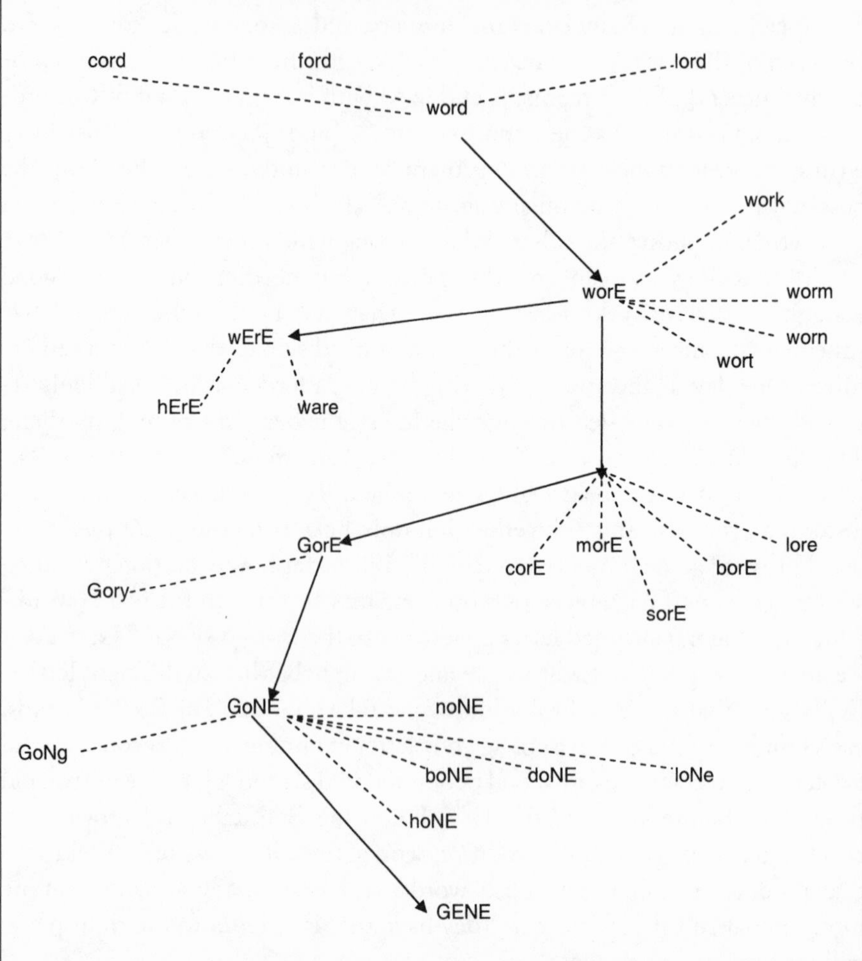

Single letter variants still arise at random, but only meaningful English words are viable sequences, and any variant that more closely resembles GENE is selected, thus rapidly replacing competing sequences. The solid line traces the ancestry of the successful linage. Unsuccessful lineages are shown as dashed lines. *Adapted from:* Graham Bell, *Selection: The Mechanism of Evolution* (London: Chapman and Hall, 1997), p. 22.

may be slight. As Paul Ehrlich argues, even a 1 percent "differential in the reproductive performance of genotypes is quite strong selection"; indeed it "probably would be enough to account for all of evolution between the origin of life (at least 3.5 billion years ago) and the appearance of upper-class Englishmen (assumed by upper-class Englishmen of Darwin's day to be the pinnacle of evolution)."[68]

Having explained the core ideas of the Darwinian paradigm and natural selection, I turn next to the processes of evolution by natural selection. Evolution works largely through natural selection, but this concept may be divided into three major processes, or subdivisions: neoclassical Darwinian theory, inclusive fitness, and group selection. Both inclusive fitness and group selection build upon the insights of neoclassical Darwinian evolutionary theory; it is important to stress that they are subdivisions of, and not alternatives to, natural selection.

Neoclassical Darwinian evolutionary theory submits that natural selection occurs at the level of the gene and will produce behavior that is egoistic: the individual animal will be concerned with meeting its own needs. Inclusive fitness shares much with neoclassical Darwinism, but submits that one also tries to protect one's genetic relatives. Finally, group selection argues that natural selection can occur at the level of the group rather than just at the individual level of analysis. These processes of natural selection are important for the theories and issues of international relations I examine in this book, and so I will elaborate on each.

Neoclassical Darwinian Theory and Individual Selection

Darwinian evolutionary theory begins with the recognition that evolution occurs by natural selection. For Darwin, natural selection operates at the individual level of analysis; that is, it worked on the individual animal (phenotype). Darwin did not know of the existence of genes, but it is a testament to the strength of his argument that, once genes were discovered, his theoretical framework was strengthened. Building upon the Darwinian paradigm, neoclassical Darwinists propose that natural selection occurs at the level of the individual gene. A conceptual shift is required here because the level of analysis is still the individual, but the individual gene and not the organism, although none would dispute that natural selection may operate on the entire phenotype. The gene is probably the most appropriate level of analysis for selection, an argument made most famously by the celebrated evolutionary theorist Richard Dawkins.

As Dawkins explains, at one time there were no organisms, just chemicals in the primordial "soup."[69] At first, different types of molecules started to form, by accident. Some, it turned out, could reproduce themselves. They used the constituents of the soup—carbon, nitrogen, hydrogen, and oxygen—to make copies of themselves.[70] These various components were not in limitless supply, so the different molecules competed for them as they went about making copies. From this competition, the most efficient copy makers emerged. But the process was never perfect; sometimes the molecules made mistakes, and occasionally these accidents were better at replication or made some other contribution to fitness. One such mistake might have been the formation of a thin membrane that held the contents of the molecule together—a primitive cell. A second might have involved dividing the cell into ever larger components, organs, and so on to create what Dawkins calls "survival machines." He explains, "the first survival machines probably consisted of nothing more than a protective coat. But making a living got steadily harder as new rivals arose with better and more effective survival machines. Survival machines got bigger and more elaborate, and the process was cumulative and progressive."[71]

From the perspective of genes, there is no intentionality. They did not want to create or inhabit people, but due to evolution the process continued nonetheless. "They created us, body and mind; and their preservation is the ultimate rationale for our existence."[72] However, Dawkins makes it clear that the interests of the gene and the organism need not coincide at different stages of an organism's life, particularly after reproduction.[73] In the cold calculus of evolution, the phenotype is just the ephemeral vehicle of the gene. In general, however, the actions of the gene—its "selfishness" if you will—increase its fitness, and so the behavior spreads.[74]

The use of the word "selfish" has to be considered carefully, as prominent evolutionary theorist George Williams notes: "Dawkins had no wish to imply that a tiny trace of DNA could be an entity that could have a mental experience of malice or guilt."[75] Rather, Williams comments, the gene is the bearer of a selfish message: "Exploit your environment, including your friends and relatives, in ways that maximize my proliferation."[76] Moreover, Williams argues, selection at the level of the gene is an important contribution to evolutionary theory but is usually slightly misleading because, strictly speaking, genotypes "do not replicate themselves in sexual reproduction . . . [and so] they can not be units of selection."[77]

Williams finds it more accurate to speak of individual selection, the "primary mechanism of selection at the genic level."[78] As he explains in his exceptional book *Natural Selection:* "The individual arises as a unique temporary genotype formed in sexual reproduction by a sampling of gametes from the gene pool. The genotype thus formed is a set of instructions for producing an interactor, the fitness of which will be determined mainly by how favorable an environment it encounters during development."[79] He continues, "for most individuals formed from a gene pool, fitness will soon be zero, because environmental stresses will destroy them before maturity."[80] Despite those losses, however, "of those that mature, fitness may vary widely, and individuals of similar fitness may have greatly dissimilar reproductive success because of chance environmental events."[81] Williams concludes, "Reproductive success is measured by the magnitude of the interactor's lifetime total of genes put back into the gene pool, relative to the performance of others in the population."[82]

Neoclassical Darwinian theory is still regarded as profound and insightful by contemporary evolutionary theorists. But inclusive fitness and group selection are also significant for examining issues both in the natural and social sciences. For the arguments of the subsequent chapters, inclusive fitness is particularly important.

Inclusive Fitness

The theory of inclusive fitness, also called kin selection, builds upon and improves neoclassical Darwinian theory. It was principally developed by evolutionary theorist William Hamilton, who sees reproductive success as involving the success of one's relatives as much as one's own.[83] Roger Masters summarizes how inclusive fitness modifies traditional Darwinian evolutionary theory: "natural selection favors the ability of individuals to transmit their genes to posterity," but it is true that "an organism's reproductive success can sometimes be furthered by assisting others, instead of by mating."[84] Thus understood, an individual's self interest can be served by assisting genetically related individuals.[85]

Of course, this process may often be instinctual, rather than a matter of cognitive awareness.[86] For example, why do birds sound the alarm when doing so calls the predator's attention to them? Hamilton answers that a bird in this situation is acting to save genetically related birds—its relatives.[87] Thus it advances its own reproductive success by assisting those who are related to it. This behavior has profound implications for the way we measure reproductive success and understand nepotism and altruism. From this perspective, what is termed altruism—caring for one's relatives, for example—is actually selfish. Mayr explains how altruistic behavior could evolve: "if there is 1 chance in 10 that an altruistic act would cost the life of the altruist, but the beneficiaries were the children, siblings, or grandchildren of the altruist, with all of whom he shares more than 10 percent of his genes, selection would favor the development of altruism."[88]

The force of Hamilton's argument is found not only in human conceptions of nepotism and altruism, but also among social insects. Explaining the behavior of these insects was a major problem for Darwin, who recognized the difficulty that nonreproductive social castes posed for the theory of natural selection. But Hamilton's improvement on Darwin's argument allows their behavior to be explained. Social insects (*Hymenoptera*) such as ants, bees, and wasps provide many examples of altruistic behavior that demonstrate the force of Hamilton's argument.[89] *Hymenoptera* are eusocial: female workers forego reproduction to assist the queen in raising more of the worker's sisters rather than producing their own offspring.[90] The genetics of *Hymenoptera* is called haplodiploidy. The males are haploid: they contain one set of chromosomes and develop from unfertilized eggs. The females are diploid: they have two sets of chromosomes and develop from haploid eggs laid by the queen and fertilized by a haploid male. The diploid females create haploid sexual eggs by meiosis, and haploid males make haploid sperm by mitosis. Consequently, a fertilized egg receives half of its mother's chromosomes, but all of its father's chromosomes.

Table 1.4 compares the relatedness of eusocial (haplodiploidy) insects with that of most diploid animals, such as humans. To describe relatedness geneticists use the term *r*, the coefficient of relatedness, which is the probability that a gene sampled at random in one individual will also be present in another.[91] The queen ant and her daughter share half of their genes, as in more conventional species, but full sisters share half of their mother's genes and all of their father's, so the *r* value is 0.75. Thus, a female ant is more closely related to her sisters than she is to a potential daughter; that relationship would have an *r* value of 0.50 for a potential daughter, the same as a mother and daughter in diploid species like humans.[92] Consequently, Hamilton suggests that this closer relationship among eusocial sisters causes them to help one another rather than reproducing. In fact, aiding sisters rather than daughters is a distinctive feature of hymenopteran societies.

Table 1.4. Coefficients of Relatedness (*r*) between Kin Pairs in Humans and *Hymenoptera*		
Actor	*Recipient*	r *value* (*The probability of recipient holding any one gene identified in the actor)*
	Diploid For example, humans	
Parent	Offspring	0.5
Individual	Full sibling	0.5
Individual	Half-sibling	0.25
Individual	Identical twins (monozygotic)	1.0
Grandparent	Grandchild	0.25
	Haplodiploid *Hymenoptera,* assuming queen mates with only one male	
Queen	Daughter	0.5
Queen	Son	0.5
Son	Queen	1
Sister	Sister	0.75
Brother	Sister	0.5
Sister	Brother	0.25

Source: John Cartwright, *Evolution and Human Behavior: Darwinian Perspectives on Human Nature* (Cambridge, Mass.: MIT Press, 2000), p. 78.

Group Selection

The third subdivision of natural selection, group selection, operates on local populations of a particular species within a global population. Individual selection favors traits that maximize fitness for an individual or even a subset of a population, but group selection favors traits that maximize the relative fitness of entire groups.[93] Group selection is more controversial than neo-Darwinism or inclusive fitness, and many evolutionary theorists submit that natural selection only operates at the individual level—individual organisms or genes—not at a group level.[94] I will touch on criticisms below, but this controversy is not critical for my purposes. For the purposes of this book it is important that I explain its core ideas and acknowledge that prominent evolutionary theorists and philosophers of biology advance group selection as a type of natural selection.

One of group selection's foremost proponents is evolutionary theorist David Sloan Wilson. He argues that even briefly interacting individuals constitute trait-groups that are at a higher level of selection (a group) than the individual selection of neo-Darwinism or inclusive fitness. According to Wilson, group selection is the component of natural selection that operates on the differential productivity of local populations within a global population.[95] He provides an example of how group selection may occur. Wilson examines winter flocks of a song bird, the Harris sparrow (*Zonotrichia querula*), among whom kinship is likely to be low and so kin selection is unlikely to have much effect.[96] Flocks do vary in composition, often more than one would expect due to simple random association. Different flocks may have different average frequencies of aggression, use of alarm calls, and other characteristics likely to affect the fitness of the flock as a whole. In the spring, when the birds go their separate ways and establish breeding territories, those from the more socially benign flocks can be expected to be more numerous, better fed, and better able to reproduce. Thus, in general, the more successful flocks will tend to preserve more cooperative individuals, and lead to an increase in those social traits that contribute to the flock's fitness. So his example shows that altruism as well as other behaviors are anchored in the evolutionary process.[97]

In response to criticism from evolutionary theorists who support individual selection, Sober and Wilson argue that evolution no doubt does occur at an individual level as their critics submit. They say Williams was correct when he famously argued in response to advocates of group selection that "a fleet herd of deer is just a herd of fleet deer—the group runs fast not because this benefits the group but because it benefits each individual."[98] Natural selection occurs without question at the individual level. All evolutionary theorists accept this argument.

Nor would they argue with Darwin's important observation concerning his own theory: "If it could be proved that any part of the structure of any one species had been formed for the exclusive good of another species, it would annihilate my theory, for such could not have been produced through natural selection."[99] No, Sober and Wilson are not annihilating Darwin. Their argument is not that the behavior of a species develops mostly for the good of others. Their primary message is that: "groups, too, can be functional units and that individuals sometimes behave more like organs than like organisms."[100]

In an elaborate discussion of group selection, Sober and Wilson argue that group selection requires four conditions to occur.[101] First, a population of groups must exist, instead of just one group; second, the groups must vary in the population of the particular trait, such as altruism; third (and significantly), a direct relationship must exist between the proportion of individuals possessing a certain trait in the group and the fitness, in terms of offspring, of the group; and finally, they must both reproduce and mix with other groups to create new groups. According to this argument, some individuals may possess traits that contribute to the group's fitness rather than their own. Thus, "individual selection favors traits that maximize relative fitness within single groups. Group selection favors traits that maximize the relative fitness of groups."[102]

Williams agrees that this type of selection could happen. It is a possibility. But he also argues convincingly that there is no persistent codex that could reliably cause this to occur. For example, in Wilson's flock example, Williams submits that the membership of the flock is likely to be unstable as individuals disappear, "disperse and mate exogamously."[103] Bell agrees with Williams and submits: "selection will favor lineages in which social acts tend to benefit members of the same lineage; it cannot favor lineages that promote the reproduction of other lineages."[104] Moreover, "group selection is a fallacious concept when it is held to cause the evolution of characteristics that benefit populations, or species, or communities, as a whole, without any distinction of ancestry."[105] Similarly, evolutionary theorist Raghavendra Gadagkar submits that: "Natural selection almost always acts at the level of the individual organisms and selects those that are best adapted to their environment, even if that hurts the group or species as a whole."[106] He continues, most of the "natural phenomena" group selection advocates "imagined could be explained only by *group selection* are better explained by *individual selection*."[107] In the same vein, evolutionary anthropologist Donald Symons maintains that "because the only known evolutionary process that produces adaptations is natural selection, Darwinian theory directs attention to properties of individual human beings rather than the properties of groups," and so "there is no reason to suppose that larger groupings of organisms . . . are ever the beneficiaries of adaptations."[108] Other evolutionary theorists and ethologists have strongly criticized group selection, and have argued that group selection may occur at best in only very precise and narrow conditions.[109]

Despite debate over the degree or conditions in which group selection may occur, I stress that it is sufficient for the purposes of this book that some major evolutionary theorists agree that groups can be functional units of selection. Nonetheless, please note that the arguments of the following chapters do not depend upon the accuracy of group selection as an evolutionary theory. I will discuss group selection again only in chapter 3 because it can explain the origins of warfare.

Having discussed evolutionary theory and the processes of natural selection, I now move to a discussion of how evolutionary theory can be applied to social science. In particular, I address the type of causation used in evolutionary theory, and consider issues that may be studied and the level of analysis appropriate to study them. After this section I address criticisms of evolutionary theory.

Evolutionary Theory and Social Science: Three Considerations

To apply evolutionary theory usefully to social science, we must address three issues. The first is the type of causal argument most often used in evolutionary theory, which is concerned with ultimate rather than proximate causation. Second, which particular issues or topics can be studied usefully through evolutionary theory, and are any explanations also sensitive to the impact of human culture on the issue or event studied? Third, what level of analysis does the scholar use? Explanations of events generate insights at different levels of analysis, and it is often critical for social scientists to be explicit about the level of analysis of their explanation so that others know the benefits and liabilities of the analysis.

Ultimate versus Proximate Causation: Evolutionary Theory Is Ultimate Causation

In the philosophy of science, as well as the natural and social sciences, scholars making causal arguments often distinguish between ultimate and proximate causation. Evolutionary theorists do so, too. Ultimate causes are universal statements that explain proximate causes.[110] Proximate causes can be derived or deduced from ultimate causes and focus on explanations of immediate occurrences. In general, a theory is better to the extent that its ultimate and proximate causes can be tested. Evolutionary theory is concerned with ultimate causes of behavior rather than proximate causes.[111] Ultimate causal analysis, such as natural selection, explains why proximate mechanisms occur and why animals respond to them as they do. It does not directly describe behavior but rather frames the parameters of a proximate causal explanation, explaining for example how a particular behavior permits or facilitates the survival of the animal.[112] As E.O. Wilson explains: "Proximate explanations answer the question of *how* biological phenomena work, usually at the cellular and molecular levels," whereas ultimate causes explain "*why* they work . . . the advantages the organism enjoys as a result of evolution that created the mechanisms in the first place."[113] Put differently, every phenomenon or process in a living organism is the result of both the proximate, or functional, cause and the ultimate, or evolutionary, cause.

Proximate causal analysis explains many facts about an animal: why or how hormonal or stimulus-specific factors operate within it, its particular features or physiology, and the specific situations in which it demonstrates the trait. If we want to understand why birds fly south for the winter, an ultimate causal explanation of bird migration could consider factors that contribute to fitness, such as the availability of food, mates, and predators at both the indigenous and wintering areas. Ultimate or evolutionary causations are those that "lead to the origin of new genetic programs or the modification of existing ones. . . . They are past events or processes that changed the genotype."[114] A proximate explanation would address high sex hormone levels that are correlated with spring migrations, or changing environmental conditions to which birds are sensitive, such as temperature, rainfall, and barometric pressure.[115] Symons summarizes the distinction nicely when he writes, "proximate causes deal with matters such as development, physiology, and present circumstances," whereas "ultimate causes are particular circumstances in the ancestral populations that led to selection for the trait in question."[116] Finally, as biologist John Alcock notes, the use of the term "ultimate" causation should not suggest that these explanations are more important than proximate explanations, only that they are different types of explanations: "'ultimate' does not mean 'the last word' or 'truly important' but merely 'evolutionary.'"[117] The use of the two terms, however, "helps us acknowledge the fundamental difference between the immediate causes of something and the evolutionary causes of that something."[118]

While evolutionary theorists are usually concerned with ultimate causes, they do universally recognize that among animals—and humans in particular—ultimate causes are affected by the animal's psychology and, for humans and perhaps chimpanzees, culture.[119] For example, individual animals may not consciously strive to maximize fitness and they behave in ways greatly influenced by proximate causes, such as their environment and culture. This is particularly important because, as I discuss below, evolutionary theorists have been accused of genetic determinism, or submitting that there is a gene "for" a specific social behavior as if there were an exact one-to-one correlation between one gene and a particular social action. It is no exaggeration to state that no evolutionary theorist believes this. Evolutionary theorists subscribe to the standard interactionist model of human behavior: genes, or specifically an individual's genotype, may greatly inform why an individual undertakes an action, but environment is almost always important. Patrick Bateson introduced the metaphor of a cake I used in the introduction to illustrate the conventional interactionist model of human behavior. There is no one-to-one correlation between a recipe and the cake; the whole recipe is needed to produce the cake, and the raw ingredients and the way they are combined are also important, as are timing and baking.[120] The raw ingredients of the cake are the genetic and environmental causes, while the baking might be thought of as the biological and psychological processes of development. All are necessary to produce the final product.

The Issue Studied: Ruler or Tape Measure

The second critical question is that of issue or topic: Is a social topic usefully studied through evolutionary theory? Equally important, is the explanation sensitive to the influence of human culture on the event or issue? In general, evolutionary theory can provide an ultimate cause for most human behavior, but whether it will usefully explain a particular question or issue is an empirical question and largely depends on the social question addressed and the particular level of analysis the scholar uses for explanation, as discussed below. The metaphor of tools of measurement—a ruler or tape measure—is helpful here. Just as one decides whether to use a ruler or tape measure depending upon what one wants to measure (a ruler is better for a piece of paper while a tape measure can better measure the dimensions of a house), one should let the issue being studied determine one's choice of measuring tool.

Recognizing the importance of the issue studied, I suggest two major reasons why evolutionary theory is particularly valuable for studying issues in social science.[121] First, it allows us to discover the biological or evolutionary causes of social phenomena. Contemporary evolutionary theorists offer excellent arguments to explain much of human behavior.[122] For example, the scholarship of evolutionary psychologists Leda Cosmides and John Tooby shows that the human brain contains a mechanism designed to detect cheating in social exchanges.[123] This insight has important implications for psychological and economic models of human behavior and was made possible by the scholarship of evolutionary theorist Robert Trivers working at the level of ultimate causation.[124] Nonetheless, the usefulness of evolutionary theory has been recognized by too few social scientists. This perspective is certain to change as more scholars recognize that it allows a firmer foundation for some social science theories and a fecund new level of analysis for studying human behavior.[125]

Second, ultimate causation also may suggest where to find new or alternative proximate causes of human behavior. Randy Thornhill and Craig Palmer use evolutionary theory to generate new understandings of human sexual coercion that help to reduce its occurrence.[126] It has also been useful to improve understanding of other important issues, such as child abuse, homicide, and the relationship between creativity and birth order.[127] The scholarship of these authors helps to demonstrate the usefulness of the interaction between evolutionary theory and social science while remaining attuned to the influence of human culture on social issues and events.

The Level of Analysis: Each Has Costs and Benefits

Finally, the level of analysis used to explain social phenomena is also important. Almost any event may be explained at multiple levels or organized for methodical analysis in many ways.[128] The benefits of focusing on any particular level of analysis are simplification and abstraction for methodological and conceptual coherence. By abstracting, a scholar can discern the power of one or multiple causes of the event studied, and thus assist in a thorough and analytic explanation of a phenomenon.

For example, the origins of World War I may be studied in several ways: by focusing on the interaction of statesmen, or the characteristics of the states (were they militaristic or democratic?), or the pressures that the international system placed on the major European states. The causes of immigration may be studied through the motivations of particular immigrants, the economic and social conditions of their homeland, or the new state's attractive economic conditions, cultural receptivity, and political attitudes.

Evolutionary theory provides a new level of analysis for social science explanations that generate important insights into major issues in social science, such as those listed above and those examined in this study. Nonetheless, evolutionary theory must prove its worth in social science to a somewhat skeptical audience, and this will take time as more scholars find that this level of analysis supports research in specific issue areas. The appropriateness of evolutionary theory is linked to the issue studied. It is relevant for understanding the origins of warfare, where one needs the tape measure, rather than understanding the origins of a specific war, where one needs a ruler. Finally, the use of evolutionary theory in social science is not intended to supplant any existing social science theory or paradigm; rather it is intended to augment and strengthen the study of social issues.

Of course, costs are attached to focusing on particular levels of analysis; the greatest may be the potential to impose a false uniformity or determinism on a complicated and complex reality. As both David Singer and Kenneth Waltz have noted, this is an inevitable cost of theorizing, and scholarship using nearly any theory will face this problem.[129] Evolutionary theory's level of analysis, ultimate causation, must be pursued as it generates insights concerning human behavior. But all scholarship done in this tradition must recognize that other explanations can also solidly inform a phenomenon as complicated as human behavior. Good scholarship knows the value and limitations of a particular level of analysis.

Kenneth Waltz's application of structuralism to the study of international relations is one example. His *Theory of International Politics* produced important insights and generated a productive research agenda but could not address many significant issues that were better studied by other levels of analyses, as he and many other scholars have stressed over the years. Like Waltz's use of structuralism, the use of evolutionary theory in social science will make major contributions and generate a productive research agenda that both augment and critique mainstream social scientific theories and thus advance our knowledge of human social behavior.

Of course, no theory will be able to explain all aspects of warfare or every particular war. However, this fact does not diminish the power of Waltz's explanation or the usefulness of systemic theories. Indeed, to even expect it to apply so universally is to misunderstand the role of theories in studying international relations. So it is with the use of evolutionary theory to study specific topics in international relations: we will gain knowledge of the origins of warfare, but it will not tell us why the Thirty Years' War lasted thirty years or why the invasion of Europe occurred in 1944 rather than 1942, 1943, or 1945.

Major Criticisms of Evolutionary Theory

In this section I present and discuss some of the major and more common criticisms made of evolutionary theory. Addressing these criticisms is important so that social scientists are able to use evolutionary theory with confidence. The major criticisms are that evolutionary theory does not apply appropriately to humans because it is deterministic; that it is reductionist; and finally, that evolution by natural selection is tautological and Panglossian.

Is Evolutionary Theory Deterministic?

The most cogent argument that the use of evolutionary theory to study humans constitutes biological determinism is advanced by Richard Lewontin, Steven Rose, and Leon Kamin.[130] They formulate their critique succinctly: "Biological determinists ask, in essence, Why are individuals as they are? Why do they do what they do? And they answer that human lives and actions are inevitable consequences of the biochemical properties of the cells that make up the individual."[131] In turn, "these characteristics are . . . uniquely determined by the constituents of the genes possessed by each individual. Ultimately, all human behavior—hence all human society—is governed by a chain of determinants that runs from the gene to the individual to the sum of the behaviors of all individuals."[132]

These authors advance a second argument, that politically conservative forces use biological determinism to legitimize social inequalities. It "has been a power mode of explaining the observed inequalities of status, wealth, and power in contemporary industrial capitalist societies, and of defining human 'universals' of behavior as natural characteristics of these societies."[133] As a result, "it has been gratefully seized upon as a political legitimator by the New Right, which finds its social nostrums so neatly mirrored in nature; for if these inequalities are biologically determined, they are therefore inevitable and immutable."[134]

These arguments are grave and compel a response. If they are true, then social scientists—appropriately—would be quite reluctant to use evolutionary theory to inform their analyses of human behavior, and most academics would not support the use of "biological determinism" to support inequality in society. Fortunately, however, these arguments are misplaced.

My first response to these arguments is a matter of precise terminology. As Sober argues, "biological determinism" is too imprecise a term to support the argument because it excludes environmental factors like nutrition that can greatly affect biology.[135] Therefore, it is more accurate to make the charge of genetic determinism. Second, the impact of genes on behavior most often are probabilistic, not determinative, as Steven Pinker reminds us.[136] Third, all contemporary evolutionary theorists, including Hamilton, Maynard Smith, Mayr, Trivers, Williams, D.S. Wilson, and E.O. Wilson, are not arguing that human behavior is solely the product of genetic determinism. As Maynard Smith argues, the charge of genetic determinism "is largely irrelevant, because it is not held by anyone, or at least not by any competent evolutionary biologist."[137] As I stressed above, all contemporary evolutionary theorists recognize that human behavior results from interaction between the genotype and the phenotype's environment. The result is neither wholly random nor deterministic. The great biologist Sewall Wright captured this point: "The Darwinian process of continued interplay of a random and a selective process is not intermediate between pure chance and pure determination, but in its consequences qualitatively utterly different from either."[138]

To demonstrate this point, Sober provides the example of parental care.[139] Empirically, on average women have taken greater care of the children than have men in each environment that humans have thus far inhabited. To explain this, evolutionary theorists hypothesize that natural selection has favored different behaviors between the two sexes—one set of behavior within men and another within women. "Of course, this hypothesis does not exclude what is obviously true—that some men provide more parental care than others and that there is variation among women for the trait as well. The hypothesis attempts to account for variation *between* the sexes, not *within* them."[140] The differences within the sexes may be largely environmental, and the differences between them mostly genetic. Thus genes and environment are responsible for human behavior to different degrees, but both are responsible.

Nonetheless, proponents of "determinism" appear to have a hard time accepting this, as Pinker recounts: "Among the radical scientists and the many intellectuals they have influenced, 'determinism' has taken on a meaning that is diametrically opposed to its true meaning," now it is "used to refer to any claim that people have a *tendency* to act in certain ways in certain circumstances."[141] For those who make the charge of "genetic determinism" when we analyze human behavior, it seems that any "probability greater than zero is equated with a probability of 100 percent. Zero innateness is the only acceptable belief, and all departures from it are treated as equivalent."[142]

My response to the second argument of Lewontin, Rose, and Kamin is to repeat that evolutionary theory recognizes the impact of the environment on the phenotype, and it is not prescriptive. While evolutionary theory does provide insights into human behavior, it does not prescribe what humans should do. It may be used to explain the origin of war; it does not imply that war is good. To demonstrate this point it is helpful to recall the argument David Hume developed in *A Treatise of Human Nature.*[143] Though the argument is usually termed the "naturalistic fallacy," that term was first used by G.E. Moore over a century after Hume.[144]

Hume's argument is often summarized as: You cannot deduce an "*ought*" from an "*is.*" What this means is that a distinction must be made in augmentation between "is-statements" and "ought-statements." An "is-statement" is a factual statement that makes no moral judgment. An "ought-statement" makes a moral judgment about the moral characteristics of some action or class of actions.

For Hume, factual claims and value judgments are different kinds of statements; ought-statements cannot be deduced from exclusively is-statements. Thus, he would regard the following argument as invalid:

> Ethnic cleansing causes suffering.
>
> ---
>
> Ethnic cleansing is wrong.

In this argument the conclusion does not follow deductively from the premise. A deductively valid argument for an ought-conclusion must have at least one ought-premise. So if an additional premise is supplied, it will be deductively valid:

> Ethnic cleansing causes great suffering.
>
> It is wrong to cause suffering.
>
> ---
>
> Ethnic cleansing is wrong.

The genetic determinism argument commits the naturalistic fallacy if its authors claim that evolutionary theory does not merely make factual statements about the world, but also makes moral judgments. Evolutionary theorists do not argue that what is "natural" is thus "good."[145] The argument that evolutionary theory masks a political agenda is not supported. Knowing facts about the world does not prove any moral conclusion on the basis of those facts alone.

Finally, even setting the naturalistic fallacy aside, it is rather curious that these authors would claim that evolutionary theorists argue human behavior is immutable. Even a cursory understanding of evolutionary theory supports the opposite conclusion. Humans are evolving over time now as they have in the past—for the human genotype almost nothing is inevitable or unchangeable. As the earth's environment changes in the future, and perhaps as humans encounter new environments on other planets, so too will the human genotype and phenotype change. Evolutionary theory recognizes that humans are evolving gradually and provides no moral justification or Archimedean point for any political philosophy. Again, evolutionary theory describes human behavior; it does not prescribe what humans should do.

Is Evolutionary Theory Reductionist?

A second major argument made against evolutionary theory is that it is reductionist. Reductionism occurs when a theorist explains the properties of the whole through the units that constitute it or the fundamental forces that act upon it. For example, Marx and Engels explained the condition of any given society through class interests and conflict, and Freud analyzed diverse mental illnesses and numerous emotional problems through his theory of childhood sexual development and trauma. Lewontin, Rose, and Kamin argue that the application of evolutionary theory to the study of humans is reductionist. They argue that social phenomena are not the sums of individual behavior and cannot be reified into properties located in the human brain or elsewhere.[146] They also argue that human behavior is too complicated and multifaceted to be reduced to a single principle or principles; the whole cannot be reduced to any particular component or principle.

In response, I explain why reductionism is a basic procedure in all sciences. Next, I introduce the concept of hierarchical reductionism provided by Dawkins that explains why reductionism is both right and proper in scientific analysis. Reductionism is like cholesterol—it comes in good and bad forms. I conclude by arguing that the claim that evolutionary theory is reductionist is as inappropriate as the argument that it is deterministic.

Reductionism is commonly used in the natural and social sciences because it is often necessary to explain complicated or apparently distinct phenomena through fundamental principles or conditions. Very few scientists would argue that anything is wrong with abstracting or simplifying in order to understand underlying principles, even if those principles are the purview of another discipline. Clearly, these actions are often necessary for theorizing, and interdisciplinary scholarship can obviously be fruitful for all concerned. Nonetheless, reductionism would be a problem in a particular subject or issue if the actions of abstracting and simplifying led away from the discovery of knowledge. And this is the central issue at stake here: Does reductionism hinder the discovery of knowledge?

Above, I raised questions concerning the application of evolutionary theory to social science; they must be addressed here as well. The particular issue to be studied and the level of analysis a scholar chooses are important considerations when scholars charge an explanation is reductionist. For example, it is possible to study the Cuban Missile Crisis from a molecular level, but it is not likely to be rewarding. On the other hand, it is equally true that studying certain issues in chemistry through the lens of physics, or biology through chemistry, has generated great insights that have advanced human understanding. Each of these cases is reductionist, but they advance science and thus are intellectually useful. Indeed, the word "understand," literally "stand under," implies a connection to other levels of explanation. Similarly, there is no logical reason why human behavior cannot be studied through this lens as well, assuming of course that creationism is not true, that humans are not created as unique beings apart from the animal world.[147]

To capture the distinction between reductionism that advances science or that retards it, Dawkins introduces the concept of "hierarchical reductionism," which he contrasts with simple reductionism. Dawkins, always a skillful writer, is at his most compelling and colorful discussing this point: "To call oneself a reductionist will sound, in some circles, a bit like admitting to eating babies. But, just as nobody actually eats babies, so nobody is really a reductionist in any sense worth being against. The nonexistent reductionist," he submits, "tries to explain complicated things *directly* in terms of the *smallest* parts, even, in some extreme versions of the myth, as the *sum* of the parts!"[148]

In contrast, hierarchical reductionism recognizes that a complex issue, such as the causes of human behavior, may be usefully explained at a number of levels of analysis. Indeed, we may think of a hierarchy of explanation when we examine complicated phenomena. For human behavior, this hierarchy extends from concepts like economic class to subatomic particles. The explanations that are relevant near the top of the hierarchy, for example, macro explanations like class, may be poorly suited for understanding the lower-end, micro explanations like subatomic particles. The reverse obviously may be true as well. For instance, in order to understand the operation of a car it is more useful to comprehend its electronic ignition system than electrons.[149]

This hierarchy helps scholars understand immensely complicated issues, such as human behavior. As Daniel Dennett argues, "societies are composed of human beings, who, as mammals, must fall under the principles of biology that cover all mammals. Mammals, in turn, are composed of molecules which must obey the laws of chemistry, which in turn must answer to . . . physics."[150] For Dennett, to be a reductionist is to be a good scientist who has a "commitment to non-question-begging science without cheating by embracing mysteries or miracles at the outset."[151] I concur with Dennett's assessment, so long as we appreciate the limitations as well as the abilities of any level of analysis. For example, geology explains how tectonic plates move, while it must turn to physics to explain the interaction between the core and the mantle. Evolutionary theory is quite useful for explaining the origin of ethnic conflict, but is less so for explaining the origin of the struggle in Chechnya.

Second, the argument that evolutionary theory is reductionist is as off target as the genetic determinism argument. The same thinking that keeps modern evolutionary theorists from being determinists also prevents them from being reductionists: they understand that human behavior is caused by the interaction of genetic and environmental causes. As Dawkins explains, "it is perfectly possible to hold that genes exert a statistical influence on human behaviour while at the same time believing that this influence can be modified, overridden or reversed by other influences. Genes must exert a statistical influence on any behaviour pattern that evolves by natural selection."[152] For example, Dawkins argues in direct response to Lewontin, Rose, and Kamin, "presumably Rose and his colleagues agree that human sexual desire has evolved by natural selection, in the same sense as anything ever evolves by natural selection. They therefore must agree that there have been genes influencing sexual desire—in the same sense as genes ever influence anything."[153] Dawkins continues, "they presumably have no trouble with curbing their sexual desires when it is socially necessary to do so."[154]

The argument that applying evolutionary theory to human behavior constitutes reductionism is not valid because human behavior is, at least in part, determined by the genotype.[155] This alone is sufficient to legitimize such research, given the desire of social and natural scientists to understand human behavior. Moreover, Dawkins's concept of hierarchical reductionism helps scholars understand the conditions in which reductionism is useful for providing robust explanations for human behavior.

Attempting to exclude the use of evolutionary theory in social science by simply advancing the charge of reductionism will not serve the cause of advancing knowledge. To support such a strong claim, Lewontin, Rose, and Kamin must demonstrate why evolutionary theory does not provide an ultimate cause of human behavior; that is, why our social behavior is not affected by our genotype. This they have not done. Nonetheless, as with their charge of determinism, they have provided a valuable service for those who use evolutionary theory to inform explanations of human behavior: they have increased our sensitivity to the need to apply evolutionary theory accurately to these issues. However, properly understood, reductionism is not to be condemned. In the context of evolutionary theory, it contributes to understanding human behavior and, as such, is simply one form of a good scientific explanation.

Is Evolutionary Theory Tautological and Panglossian?

A third major critique of evolutionary theory is that it is tautological. Logicians use the term "tautology" to define a category of simple logical truths that may take the form *P* or *not-P.* The statement, "it is raining (*P*) or it is not raining (*not-P*)," is often used in logic textbooks as an example of a tautology. The statement is true due to the logical definitions of the terms "or" and "not." Philosophers also call it "analytic"; the truth of the statement emerges when one logically defines the terms it contains.

A second type of tautology is broader than the first. The statement: "for all *x*, if *x* is a bachelor, then *x* is married" is different because the logical terms of the statement do not guarantee that it is true. The knowledge of the nonlogical terms "married" and "bachelor" is also required. These statements are called "synthetic"; their truth or falsehood does not emerge simply by examining the logical terms.

Critics of evolutionary theory have argued that the concept of fitness in natural selection is a tautology. Biologist Bobbi Low shows how a superficial understanding of natural selection appears to be a tautology: "what works, works, so if you see it, it must be working."[156] But properly understood, it is not. Consider Sober's example of a serviceable definition of natural selection: "traits found in contemporary populations are present because those populations were descended from ancestral populations in which those traits were the fittest of the variants available."[157] He rightfully demonstrates that this definition is not tautological for two reasons: first, it is not a logical truth that present populations descended from ancestral ones; second, strictly speaking, the statement is false because a trait at fixation may have reached fixation through other evolutionary media, such as genetic mutation.[158] Thus, the theory of natural selection is not a tautology and we may move on to address a related argument, that it is Panglossian.

Gould and Lewontin advance the argument that evolutionary theory is Panglossian in an influential article, "The Spandrels of San Marco and the Panglossian Paradigm."[159] Their argument is not that evolutionary theory is tautological but that almost every outcome of the human phenotype, for example, need not be explained as the result of fitness or adaptationism.[160] They draw a comparison with the spandrels of San Marco, the triangular spaces formed by the intersections of arches supporting the dome of the cathedral; they submit that what appears to be a functional design element can in fact be the product of architectural constraints. To conclude that the spandrel was the intent, rather than the result, of the design would be a mistake. Reasoning by analogy, the conclusion that all biological forms evolved as adaptations to local environments is not correct because it fails to consider "architectural" constraints. The thrust of their argument is directed, first, at the adaptationist argument, as they define it, that atomizes organisms into parts and explains each as a direct, near-optimal path of adaptation to the environment; and second, at the adaptationist research strategy, which for Gould and Lewontin may be summarized as: "If one adaptive argument fails, try another one."

They agree that adaptationism plays an important role in biology, but they say it should be balanced with other approaches to evolution.[161] Gould and Lewontin argue that features like spandrels often arise as incidental consequences of evolutionary change, and thus are not selected by natural selection. One of their examples of a spandrel is the human chin, which did not evolve for any functional reason; rather, as Williams notes, it was a "prominence left behind when the dental arcades shrank from protohominid to a modern size."[162]

They use the term "Panglossian paradigm" for what they believe to be the excesses of adaptationism. "Dr. Pangloss" is a character in Voltaire's *Candide* who argued that any calamity—earthquake, flood, or disease—happened for the best. He is a parody of Leibniz's argument that, due to God's moral perfection, this is the best of possible worlds. Thus, for Gould and Lewontin, adaptationists who understand everything in terms of natural selection are arguing along Panglossian lines. They have also termed these adaptationist explanations "Just So Stories," after Rudyard Kipling's fanciful *Just So Stories,* which explained how the leopard got its spots, the camel its hump, the elephant its trunk, and so on.[163]

I propose four responses to these arguments by Gould and Lewontin.[164] First, it is important to keep their arguments in focus. They are not questioning that natural selection is a powerful force for evolution. However, evolutionary theorists do not support their alternatives. As Mayr writes, "Gould and Lewontin's proposals are not 'alternatives to the adaptationist program,' but simply legitimate forms of it," and like it "are ultimately based on natural selection."[165] Nonetheless, the criticism of Gould and Lewontin is valuable because it reminds adaptationists to be rigorous in their analysis.

Williams suggested the means to such rigorous analysis over thirty years ago. He argues—in his seminal work *Adaptation and Natural Selection*—that three considerations are relevant.[166] The first is that adaptation should not be invoked when other, lower levels of analytic explanation, such as those from physics, are sufficient. For example, a biologist does not question "the trajectory of a falling apple," because "the principles of mechanics" can explain it sufficiently.[167] But we would need natural selection if "we were asked how the apple acquired its various properties, and why it has these properties instead of others."[168] Next, we should not employ adaptation when a feature is the outcome of a general developmental requirement, such as the fact that limbs come in pairs or heads are attached to bodies.[169] Finally, we should not use adaptation to explain a feature when it is the by-product of another function.[170] For example, when we watch a bird use its beak to clean its feathers, "we don't need to give an adaptationist explanation" of this "since the features of the bird's beak are there for more pressing reasons."[171]

Second, my response to the charge of Gould and Lewontin, Rose, and Kamin that adaptationists are Panglossian or tell "Just So Stories," and thus imply that what exists is immutable and close to perfection, is merely to remind them that no modern evolutionary theorist suggests that evolution produces the "best" outcome for humans or other animals. Almost everyone wishes for some trait or characteristic that humans do not possess: a nervous airline passenger might wish for wings or a baseball player at bat may desire a second pair of eyes to better hit the ball.

Despite such desires, as I argued above, evolutionary theory makes no promise of perfection and no animal is perfectly adapted to an environment. Evolution causes animals to be relatively well adapted to their specific environment. As Darwin emphasized, "natural selection tends only to make each organic being as perfect as, or slightly more perfect than, the other inhabitants of the same country with which it has to struggle for existence."[172] Arguing along similar lines is evolutionary theorist George Simpson: "evolution is not invariably accompanied by progress, nor does it really seem to be characterized by progress as an essential feature. Progress has occurred within it but is not of its essence."[173] Instead, he says, "within the framework of the evolutionary history of life there have been not one but many different sorts of progress" depending on the environment.[174] Mayr submits that "evolution is opportunistic, and natural selection makes use of what variation it encounters"; moreover, natural selection is an "optimization process, but one of a very special kind."[175] Selection will optimize for a given environment, ceteris paribus, but is subject to many constraints in changing environmental conditions, and the environment is almost always changing. The frequency of extinction demonstrates this point. "More than 99.9 percent of all species that ever existed on earth have become extinct," including species that flourished spectacularly for long periods of time, such as "the trilobites, ammonites, and dinosaurs."[176] In the Permian extinction alone, some 250 million years ago, 96 percent of all species disappeared due to the impact of a comet, and the mass extinction at the end of the Cretaceous, which eliminated the dinosaurs and was probably caused by meteor impact near the Yucatan Peninsula, was similarly catastrophic for life on earth.[177]

What caused each mass extinction is not critically important, whether it was "competition, a pathogen, a climatic catastrophe, or an asteroid's impact"; what is critical is that in each case selection was unable to find an appropriate response.[178] Humans, like ammonites or dinosaurs, evolved to suit an environment, but this fact does not require or guarantee that humans are "better" now than in the past or will be "better" in the future. There is teleology in the sense that natural selection is operating on humans, but it is not necessarily a positive or directed selection. Biologist François Jacob captures this well: "Natural selection does not work as an engineer works. It works like a tinkerer—a tinkerer who does not know exactly what he is going to produce but uses whatever he finds around him . . . to produce some kind of workable object."[179] Therefore, natural selection does not guarantee that this is the best of all possible worlds for humans, only that we have sufficient ability to survive in the present one.

Third, if spandrels do arise, and Dennett's detailed discussion of the concept casts considerable doubt on the possibility, natural selection works on them as well.[180] To take the example of human chins, they may have originally been spandrels, but now that they exist "they can be acted on by natural selection" because "no matter how trivial the circumstances of its origin, the human chin has the potential for endless complexity by parameter optimization, given the requisite selection pressures."[181] A spandrel may become useful for a given environment as humans evolve and thus contribute to fitness, or it may simply continue as a spandrel. Thus the observation of Gould and Lewontin is useful but does not threaten the preeminence of natural selection for explaining evolutionary change.

Fourth, it may be an important intellectual exercise to ask the types of questions that Gould or Lewontin dismiss as "Just So Stories" because they may assist scientific inquiry and discovery. Williams makes this point most cogently. Many would agree with Gould when he writes, "our large brains may have originated 'for' some set of necessary skills in gathering food, socializing, or whatever: but these skills do not exhaust the limits of what such a complex machine can do. . . . Built for one thing, it can also do others."[182] Of course, Williams agrees with Gould and argues compellingly that the human brain no doubt arose for specific purposes, but that should not keep scientists from exploring adaptationist questions; it should not "dissuade people from asking what our special braininess was actually selected for," and from suggesting hypotheses, testing them, and so on to advance human understanding.[183]

In his book, *The Pony Fish's Glow,* Williams explores, appropriately enough, how the pony fish got its glow. In a telling reply to Gould, he explains that his story and a Kiplingesque "Just So Story" differ in significant ways: "mine was consistent with the known facts about the pony fish and its habitat, and I was careful to make it consistent with what might be called the Darwinian constraints."[184] He describes the constraints: "refer only to well-established material processes in formulating the story. This means avoiding any supernatural factors and always including a way in which natural selection . . . could maintain the proposed adaptation."[185]

Williams also points out that detective stories provide a closer analogy to adaptationist explanations than do Kipling's "Just So Stories."[186] A detective story entertains more than one proposed explanation and compares each one with the empirical evidence. Competing explanations can be checked and can be cited as evidence for validity. In addition, any explanation would have to be consistent with what is known about the particular trait. For example, any explanation for the pony fish's glow would at least have to be able to explain "when the light would be turned on" and "what kind of light it would emit."[187] Finally, Williams notes that the value of adaptationist explanations is not just that they explain why the pony fish has a light, or photophore, "but rather *why* the pony fish *keeps* its photophore."[188] As Williams concludes, "Gould's criticism notwithstanding, adaptationist story-telling continues to be a powerful method for the discovery of important facts about living organisms."[189]

Mayr also criticizes the way Gould and Lewontin characterize and dismiss the adaptationist research strategy as "If one adaptive argument fails, try another one." To the contrary, Mayr argues, doing precisely this is a sound research strategy that has advanced scientific knowledge: "the strategy to try another hypothesis when the first fails is a traditional methodology in all branches of science. It is the standard in physics, chemistry, physiology, and archeology."[190] It has generated important findings in biology as well: for example, in "the field of avian orientation in which sun compass, sun map, star navigation, Coriolis force, magnetism, olfactory clues, and several other factors were investigated sequentially in order to explain as yet unexplained aspects of orientation and homing."[191] In addition, Mayr points out we can test the adaptationist "Just So Stories." For example, "the hypothesis that valves in veins regulate blood flow can be tested very easily" and thus "to merely ask whether a structure has any possible function can never lead to an answer, unless one asks some more specific questions first. And that is what the adaptationist program does."[192]

The major criticisms advanced against evolutionary theory are useful because they help to illuminate dangers for natural or social scientists and challenge them to be aware of these concerns. However, although these criticisms are useful for illuminating dangers, they are not sufficient to impede, and should not impede, the use of evolutionary theory to explain human behavior.

Conclusion

In this chapter I have explained evolutionary theory, its application to social science, and the major criticisms of it. As I explained at its beginning, this chapter is essential because evolutionary theory is the intellectual foundation for the arguments of this study. In the chapters that follow, I will apply natural selection and it subdivisions, particularly inclusive fitness, to specific social scientific questions.

Of course, once individuals understand natural selection they can evaluate the common criticisms of evolutionary theory that I examined here. After reviewing the arguments that evolutionary theory is deterministic, reductionist, tautological, and Panglossian, I concluded that these criticisms are not sufficiently powerful to keep social scientists from using evolutionary theory to inform the questions or issues of interest to them.

Nonetheless, while social scientists should not be deterred from applying evolutionary theory to particular issues, they must use the theory appropriately. I suggested a metric of three factors. First, social scientists considering its use should consider whether an ultimate causal explanation is appropriate for the social science issue they address. Second, the issue or topic addressed is also important: is a particular issue or topic studied usefully through evolutionary theory and is any explanation also sensitive as to the impact of human culture on the issue or event studied? Third, choosing the proper level of analysis for any investigation is both obvious and important. Is the scholar interested in an ultimate causal explanation of behavior or proximate? Evolutionary theory helps us understand the origins of much of human behavior (ultimate), but it is less useful for explaining many major specific questions in international relations, such as why the Cold War ended or whether hegemony is necessary for stability in the international trade regime (proximate).

Finally, evolutionary theory provides a double-edged sword to social science. It is a valuable tool that reveals or permits investigation of human behavior, but social reality also becomes more complicated as a result. Conceivably, almost any social behavior may be explored at an evolutionary level of analysis. While doing so may improve our comprehension of social issues, social scientists must also labor to understand evolutionary theory. This makes their job more difficult. Nevertheless, the intellectual benefit—and the ability to produce knowledge that facilitates better decision making and the crafting of more efficacious policies—outweighs the costs.

2 Evolutionary Theory, Realism, and Rational Choice

In this chapter I explain some of the theoretical benefits that evolutionary theory may provide social science. To this end, I apply evolutionary theory to two important theories commonly used in studying international relations as well as American politics, economics, and comparative politics: realism and rational choice.[1] Evolutionary theory can contribute to each because it explains the ultimate cause of the egoism upon which each depends. In discussing realism I explain the ultimate causes of the traditional realist argument and then provide an evolutionary explanation for the origins of egoism and domination. Next, I explain why evolutionary theory provides a superior foundation for the realist argument, and why it also contributes to a modification of realist theory called "offensive realism." Finally, my discussion of rational choice also builds on the evolutionary explanation of egoism. Like realism, rational choice theory assumes that individuals are egoistic; thus it also benefits from an evolutionary explanation of egoism and of the evolution of the rational mind. In addition, evolutionary theory can help rational choice theorists better predict individual behavior because it allows a more nuanced explanation of human preferences.

While I explain how evolutionary theory can strengthen these theories, I do not want to imply that this is the limit to such support. Evolutionary theory explains the origin of social behavior in humans and other animals and can help social scientists to develop or critique theories that depend on egoism or on conceptions of what is termed "human nature." For example, an epistemological assumption of the political philosophy of liberalism is that individuals are born, in the philosopher John Locke's felicitous phrase, as a tabula rasa, a blank slate or neonate mind. Evolutionary theory reveals that this is not the case, as both cognitive neuroscientists like Steven Pinker and linguists like Noam Chomsky have shown. The human animal's neural network is born with much that is innate, or more precisely cognitive adaptations, including the abilities to learn language, to respond to patterns, and to reason in a logical and sophisticated manner.

Furthermore, I do not wish to imply that the intellectual intercourse between evolutionary theory and social science is a one-way street. As I discuss below in my analysis of rational choice, evolutionary theorists and ethologists can benefit from understanding the laws of economics and the sophisticated models of behavior developed by economists. Indeed, one of the major recent insights of evolutionary theory is Robert Trivers's development of reciprocal altruism: the idea that genetically unrelated individuals provide benefits for one another in the expectation of reciprocal behavior. This process may be understood in economic terms as "gains from trade"; that is, what each party receives, now or in the future, is worth more than the cost of reciprocating.[2] In this and other matters, an intellectual cross-fertilization between economics and evolutionary theory, and more broadly, social science and evolutionary theory, is rewarding and will become more so as increasing numbers of social scientists become familiar with evolutionary theory and as evolutionary theorists and ethologists gain more understanding of social science.

The Traditional Ultimate Causes of Realism and Neorealism

Realism is a theory of international relations associated in contemporary times with George Kennan, Henry Kissinger, Hans Morgenthau, and Reinhold Niebuhr. It is sometimes called power politics, *Machtpolitik, raison d'etat,* or *Realpolitik.* The essence of the theory is that states seek economic and military power to compete with others in the international system; they do so because they are composed of individuals who are egoistic and strive to dominate others.[3] These are the proximate causes of the realist argument. In chapter 1, I distinguished between ultimate and proximate causation: ultimate causes are universal statements that explain proximate causes. Proximate causes are deductively derivable from ultimate causes and focus on explanations of immediate occurrences. Realists widely use egoism and a drive to dominate as proximate causes of behavior, but few have explained their ultimate causation. Two theorists who have offered explanations are Niebuhr and Morgenthau.

The first explanation or ultimate cause of this behavior is expressed by Niebuhr and grounded in theology: Humans are evil. Human evil is the primary cause of human behavior, especially of the desire to dominate others. Niebuhr argues humans possess "unlimited and demonic potencies of which animal life is innocent."[4] Evil manifests itself in sin, or human refusal to accept inherent limitations: "Man is ignorant and involved in the limitations of a finite mind; but he pretends that he is not limited."[5] Furthermore, all human activity is tainted with a narcissistic self-love that Niebuhr sees as the essence of evil.[6] Self-love, or pride, causes humans to seek power because "the ego does not feel secure and therefore grasps for more power in order to make itself secure. It does not regard itself as sufficiently significant or respected or feared and therefore seeks to enhance its position in nature and society."[7]

The recognition that humans are finite creatures causes them to seek power: "man is the only finite creature who knows that he is finite and he is therefore tempted to protest against his fate."[8] Niebuhr laments that "one form which this protest takes is his imperialistic ambition, his effort to overcome his insignificance by subordinating other lives to his individual or collective will."[9]

The recognition of human sinfulness manifests itself in Niebuhr's consideration of international relations. Pride and a desire for power exist not only among individuals, but also among states. And because national pride is capable of causing greater evil it is especially dangerous.[10] Niebuhr argues that the traditional realist mechanism of stability, the balance of power, is the only force capable of causing justice in the world. The balance of power is necessary because the "natural weakness of democracy as a form of government when dealing with foreign policy is aggravated by liberalism as the culture which has informed the life of democratic nations."[11] As Niebuhr explains, "In this liberalism there is little understanding of the depth to which human malevolence may sink and the heights to which malignant power may rise."[12]

The second ultimate cause of egoistic and dominating behavior is given by Morgenthau: Humans behave as they do because they possess an *animus dominandi.*[13] That is, they seek power because human nature is fundamentally egoistic and malignant. Thus conflict and war occur because human nature is bad.[14] Thomas Hobbes provided the foundation for this second, secular pillar of realist thought: humans are ruled by their insatiable desire for power.[15] As he describes in *Leviathan:* "I put for a generall inclination of all mankind, a perpetuall and restlesse desire for Power after power, that ceaseth onely in Death."[16] This lust for power has created a state of war, in which humans live in reciprocal and permanent fear of violent death, and in which peace is always precarious.[17]

Like Hobbes, Morgenthau believes that humans possess an inherent *ani-*

mus dominandi, "the desire for power," which manifests itself as the desire to dominate others.[18] An individual's "desire for power . . . concerns itself not with the individual's survival but with his position among his fellows once his survival has been secured"; he continues, "his lust for power would be satisfied only if the last became an object of his domination, there being nobody above or beside him, that is, if he became like God."[19] So encompassing is this desire for power that the tendency to dominate "is an element of all human associations, from the family through fraternal and professional associations and local political organizations, to the state."[20] This sentiment was perhaps most clearly expressed by the British imperialist Cecil Rhodes when he stated: "These stars that you see overhead at night, these vast worlds which we can never reach. I would annex the planets if I could. I often think of that. It makes me sad to see them so clear and so far away."[21]

Two types of behavior are the proximate causes of the realist argument: egoism and domination.[22] Egoism refers to the individual's tendency to place his interests before those of others, his own and his family's interests before those of more distant relatives, and those of relatives before those of his community, state, and so on like ever-expanding concentric circles.[23] The desire to dominate, realists believe, is inherent in people, and this drive often turns into latent or physical aggression against those who oppose one's objectives.

Leaders of states are expected to mirror this ordering, to place the interests of their state before the interests of others or of the world community, and to strive to dominate other states. Realists argue that only by possessing power can individuals attack and conquer others, in addition to deterring attacks and defending themselves. The principal result of this process is that balances of power will form and re-form cyclically, producing both periods of stability and intense security competition in international relations. Indeed, as Morgenthau insists, the desire to attain a maximum of power is universal among states and is one of the "objective laws that have their roots in human nature."[24]

Despite its long history as a theory of international relations and its widespread use by scholars and policymakers, such as E.H. Carr, George Kennan, and Henry Kissinger, the traditional realist argument rests on weak foundations. This is so because the ultimate causations that Niebuhr and Morgenthau offer are noumenal, outside the realm of what science can investigate and demonstrate. It is not possible to test the argument that humans are evil, nor can scientists judge whether an *animus dominandi* motivates humans. Neither Niebuhr nor Morgenthau explains how the proximate causes of egoism and domination may be derived logically from the ultimate causes they offer or how they can be tested scientifically. Rather, they assert that individuals are evil or possess a drive to dominate and thus must be egoistic or strive to dominate others. The result is that neither theory has a scientific ultimate cause.

Kenneth Waltz placed realism on a more scientific foundation by introducing a new realist theory: "neorealism," or "structural realism." Neorealism points to international anarchy, a phenomenon we can evaluate, as the ultimate cause of state behavior. This more scientific foundation permits us to reach realist conclusions about international relations, such as the importance of power in interstate relations, without having to believe in the theological or metaphysical concepts that Niebuhr and Morgenthau suggest.[25]

Waltz argued in *Theory of International Politics* that the international political system is anarchic. That is, there is no ultimate authority in international relations comparable to a domestic government that can adjudicate disputes and provide protection for citizens.[26] Without governmental authority, Waltz argues, the international system is anarchic, making international relations a dangerous environment, and a self-help system, where states must provide for their own protection through arms and alliances. Anarchy allows Waltz to argue that international relations is dangerous and states must rely on the balance of power and behave much the way Morgenthau or Niebuhr expected. And he advances these arguments without arguing that individuals or individual states are evil or possess an *animus dominandi.*

While anarchy provides the ultimate cause of state behavior, Waltz also uses a structuralist analysis in his argument. Structuralism is a method of study that focuses on the interaction of the parts, or units, of a system. The interaction of units in a system is what is critical to study rather than the individual units themselves.[27] Waltz uses structuralism to demonstrate how the distribution of power in international relations is vital for understanding whether war is more or less likely. He argues that a world where power is largely distributed to two poles, bipolarity, such as during the Cold War, is more stable than multipolarity, where power is about equally distributed to three or more great powers, such as in Europe before World War II.[28] This is because superpowers achieve security more easily in bipolarity. To maintain its security, each superpower will balance against the other. It can do so more effectively in bipolarity than multipolarity because it faces only one other major threat: the other superpower. By wedding anarchy as an ultimate cause and structuralism as a method of analysis, Waltz created a new realist theory—neorealism—that improves upon the realism of Morgenthau or Niebuhr in two ways. First, it does not rely on noumenal ultimate causation; and second, it can explain and predict better than realism the likelihood of great power war in international relations.

John Mearsheimer's more recent contribution to neorealism is also significant. In *The Tragedy of Great Power Politics,* he argues, like Waltz, that the anarchic international system is responsible for much trouble—wars, suspicion, fear, and security competition—in international relations.[29] Also like Waltz, Mearsheimer argues that bipolarity is more stable than multipolarity for three reasons: first, it provides more opportunities for war between or among great powers; second, it allows larger imbalances of power between great powers; and third, there is more potential for great power miscalculation.[30]

However, unlike Waltz, who feared that too much power for a state would lead to balancing against it and thus actually threaten its security, Mearsheimer argues that the international system requires states to maximize their offensive power in order to be secure and to keep rivals from gaining power at their expense.[31] In fact, this systemic incentive is so powerful that states would be hegemons if they could: "A state's ultimate goal is to be the hegemon in the system."[32] Only by being the hegemon can the state be absolutely sure of its security. For Mearsheimer, states should behave this way not because they are aggressive but because the system requires it: this behavior is the best way to maximize security in an anarchic world.

Mearsheimer's argument is a major contribution to a growing body of literature within realist thought called offensive realism.[33] In general, offensive realists argue that states are compelled to maximize their relative power because of competition in the international system.[34] They will be secure only by acting in this way. As Eric Labs argues, "a strategy that seeks to maximize security through a maximum of relative power is the rational response to anarchy."[35] As the theory of offensive realism is now formulated by Mearsheimer, it is a type of neorealism because the principal causes of state behavior are rooted in the anarchic international system; however, as I show below, offensive realists need not depend on the anarchy of the state system to advance their argument.[36]

Evolutionary Theory's Contributions to Realism

Evolutionary theory makes two contributions to the realist theory of international relations.[37] First, it places the theory on a scientific foundation for the first time. No longer must realists ground their theory on noumenal ultimate causes, because evolutionary theory can explain why egoism and domination evolved as human traits; it therefore strengthens realist theory. Second, it allows realists to advance a theory of offensive realism without depending on international anarchy to do so. This allows realists to recognize that offensive realism explains more than state behavior and is thus a more powerful theory. It also may be applied to other cases to explain the behavior of non-state actors, both today and before the creation of the state system in 1648. The theory might be used to explain the behavior and actions of many entities: empires like Rome, indigenous tribes in Papua New Guinea or North America, institutions like the Catholic Church in the Middle Ages, or commercial organizations from the East India Company to Coca Cola. In sum, I argue that evolutionary theory explains why humans seek gain and domination, whether they are acting as an individual like Pizarro, as an agent of an institution like the Pope or Bill Gates, or as the head of a state like a U.S. president.

Evolutionary Theory as an Ultimate Cause of Realism

In my discussion of Niebuhr and Morgenthau I demonstrated that realism lacks a scientific ultimate cause for the proximate causes of behavior that those authors identify. Evolutionary theory can provide this foundation for realism and moreover supply ultimate and proximate causes that are both logically coherent and testable. Philosophers of science agree that a theory is better, that is, more scientific, if its ultimate and proximate causes are logically coherent and testable. A realism anchored on evolutionary theory meets these criteria for the first time.

In this section I show that evolutionary theory is logically coherent according to two common metrics used in philosophy of science. I then explain how the traits of egoism and the need to dominate may be derived from evolutionary theory. As I explained in chapter 1, evolutionary theory is the result of specific processes, and evolution by natural selection is particularly important for the development of human traits, including egoism and the need to dominate. This process is testable, as Darwin proved in his famous example of the species of mocking-thrushes—or finches—of the Galápagos. It does account for observed differences in animal behavior and is rightly accepted as a scientific fact.[38]

This strongly suggests that human evolution is a better ultimate cause of the behavior expected by realists than the ultimate causes presented by Niebuhr

Table 2.1. Realism and Evolutionary Theory			
	Realism	*Neorealism*	*Evolutionary Theory's Contribution*
Ultimate Cause of State Behavior	Evil *Animus dominandi*	Anarchic international system (produces a dangerous environment for states)	Behavior explained by human evolution in the anarchic environments of the Pliocene, Pleistocene, and Holocene (environments of scarce resources and many threats)
Proximate Cause of State Behavior	Egoism Domination	Desire for security Self-help system	Egoism Domination In-group/out-group distinctions produce fear, threats, and desire for power
Solution for State	Maximize relative power (Morgenthau)	Develop sufficient power for defensive realism (Waltz) Maximize relative power for offensive realism (Mearsheimer)	Maximize relative power

or Morgenthau. But this can also be demonstrated with two common standards that philosophers of science use to evaluate theories: The deductive-nomological (D-N) model developed by Carl Hempel and Karl Popper's conception of theory falsification, which is known more formally as critical rationalism.[39]

Using Hempel's criteria for the D-N model, evolution is a better ultimate cause because it meets all the criteria of the D-N explanation, unlike Niebuhr's evil or Morgenthau's *animus dominandi.* For Hempel, and indeed almost all philosophers of science, good theories can be logically deduced from their antecedent conditions.[40] Formally, the D-N model requires that the explanation, or explanans, comprises statements of the antecedent conditions, *C,* and general laws such as laws of nature *L*. The explanandum, *E*, is the description of whatever is being explained, predicted or postdicted. *E* must follow deductively from *C* and *L*.

Evolution provides a better ultimate causal foundation according to the D-N model because it tightly fits this model on two levels. First, it explains how life evolves through the evolutionary processes (natural selection, gene mutation, etc.) described in chapter 1 that provide the general laws of evolution and specific antecedent conditions affecting these laws. This theory of how nature evolves may be applied and tested against specific evidence, for example, about how early primates and humans lived and continue to do so, which may confirm evolutionary processes. Second, proximate causes of human (or other animal) behavior may be deduced from it. That is, if the evolutionary process is valid, then much of human behavior must have evolved because the behavior contributed to fitness in past environments. Accordingly, evolutionary theory provides an adequate causal explanation for realism because if the antecedent conditions are provided the ultimate cause logically produces the proximate causes (egoism and domination) of realism.

Measured by Popper's method of falsification, evolutionary theory is also superior to the ultimate causes of Niebuhr and Morgenthau because it is falsifiable.[41] That is, scholars know what evidence would not verify the theory. Popper argued that if a theory is scientific, then we may conceive of observations that would show the theory to be false. His intent was to make precise the idea that scientific theories should be subject to empirical test. In contrast to good scientific theories that can be falsified, Popper suggested that no pattern of human behavior could falsify Marxism or Freudian psychoanalytic theory.

More formally, Popper's criterion of falsifiability requires that a theory contain "observation sentences," that is, "proposition *P* is falsifiable if and only if *P* deductively implies at least one observation sentence *O*."[42] Falsifiable theories contain predictions that may be checked against empirical evidence. So according to Popper, scientists should accept a theory only if it is falsifiable and no observation sentence has falsified it.

Evolutionary theory is falsifiable. That is, the conditions under which the theory would be disproved can be derived from the fundamental theory, along with the empirical evidence that would show it to be false. However, Popper himself once charged that evolutionary theory was "not a testable scientific theory."[43] Popper's argument with respect to evolutionary theory is incorrect and seems to have stemmed from confusion about its complete contents.[44] He later reversed himself and declared it to be falsifiable.[45] Indeed, evolutionary theory is a testable scientific theory that possesses many falsifiable claims. For example, the key components of evolution, natural selection and genetic variation, have been shown to be falsifiable by Michael Ruse, Elliott Sober, and Mary Williams, among others.[46] Natural selection has been tested against alternative theories of evolution, such as Lamarckism, saltationism, creationism, and orthogenesis and found to possess more logical coherence and to account better for empirical evidence.[47] Although scholars may find this hard to appreciate today, an intense struggle occurred among these competing theories a century ago.[48] However, genetic variation within populations and between and among species has been demonstrated beyond doubt. Thus, in the marketplace of ideas, natural selection has properly won its predominant place.

In addition to the ability to test the key or macro concepts of evolution, a specific trait may be tested as well, such as sex ratio—the mix of males and females found in a population. Evolutionary theorist R.A. Fisher predicted that several factors, including the higher mortality rates of males, will cause slightly more males than females to be conceived and born, and the sex ratio will become 1:1 at the age when the need for parental care ends.[49] Clearly, this explanation can be tested. A rather more alarming example of natural selection is the relatively rapid rise of new drug-resistant strains of diseases that were once defeated or effectively suppressed by antibiotics. These diseases, including cholera, influenza, malaria, methicillin-resistant *Staphylococcus aureus,* and tuberculosis, pose an increasingly significant danger for the world population.[50] Even relatively common diseases such as campylobacter, *E. coli,* and salmonella are becoming more common in the United States as they become resistant to the anti-bacterial agents found in many popular hand soaps and the too liberal use of antibiotics.[51] As Ernst Mayr argues, there are "hundreds, if not thousands, of well-established proofs, including such well-known instances as insecticide resistance of agricultural pests" and "the attenuation of the myxomatosis virus in Australia."[52] This virus was introduced to kill rabbits and at first caused enormous mortality, but soon the rabbits became immune to the virus as a result of natural selection.

Measured by the two metrics of the D-N model and falsification, evolutionary theory offers a widely accepted scientific explanation of human evolution that is both logically coherent and testable. Thus, it provides a solid foundation for proximate causes of human behavior, such as egoism and domination, and in turn for the realist theory of international relations.

The Origins of Egoism

Evolutionary theory offers two sufficient explanations for the trait of egoism. The first is a classic Darwinian argument: Darwin argued that an individual organism is concerned for its own survival in an environment where resources are scarce. It has to ensure that its physiological needs—for food, shelter, and so on—are satisfied so that it can continue to survive. The concern for survival in a hostile environment also requires that in a time of danger or great stress an individual organism usually places its life, its survival, above that of other members of the social group, the pack, herd, or tribe.[53] For these reasons, egoistic behavior contributes to fitness.

The selfish gene theory of evolutionary theorist Richard Dawkins provides the second sufficient explanation for egoism. As I discussed in chapter 1, Dawkins focuses his analysis on the gene, not the organism. Beginning with chemicals in a primordial "soup," different types of molecules started forming, and in time efficient copy makers emerged.[54] They made mistakes, however, and these contributed to fitness, such as the formation of a thin membrane that held the contents of the molecule together to become a primitive cell. Over time, these "survival machines" became more sophisticated due to evolution. Some cells became specialized, creating organs and ultimately animal bodies. But again, as I stressed in the previous chapter, there is no intentionality in this process. Genes did not want to create or inhabit people, but the process continued nonetheless. The fundamental point here is that "selfishness" of the gene increases its fitness, and so the behavior spreads.

The gene creates an instinctual or genetic basis for egoism because it is concerned only with satisfying its wants, principally reproduction and food consumption. The organism evolved largely to satisfy the wants of the gene, and in a similar manner egoism evolves through a population. Egoism thus becomes a trait or adaptation in animals, such as humans, that aids survival.

Evolutionary theorists now recognize, as a result of William Hamilton's idea of inclusive fitness, that egoism is more complex than Darwin envisioned. Hamilton recognized that individuals are egoistic, but less so in their behavior toward genetic relatives, in parent-offspring and sibling relationships. This is because close relatives share at least fifty percent of their genotype—one-half for siblings and parents, one-quarter for aunts, uncles, and grandparents, and one-eighth for cousins. As the great evolutionary theorist J.B.S. Haldane wrote in 1955, the gene that inclines a man to jump into a river to save a drowning child, and thus to take a one-in-ten chance of dying, could flourish as long as the child were his offspring or sibling.[55] The gene could also spread, albeit more slowly, if the child were a first cousin, since the cousin shares an average of one-eighth of his genes. Indeed, Haldane captured this point well when he wrote that he would give his life to save two of his brothers (each sharing half of his genotype) or eight of his cousins (each sharing one-eighth of his genotype).

As a result of the ideas of Darwin, Dawkins, and Hamilton, evolutionary theory provides an explanation for what is commonly known, that individuals favor those who are close genetic relatives. Consequently, complex social behavior among unrelated individuals can be seen as the interaction of selfish individuals, and most evolutionary theorists expect no tendency toward solidarity, cooperation, or altruism beyond what is in the interests of the animals. Similarly, realists and, as we will see below, rational choice theorists also do not expect individuals or states to show this type of behavior beyond their own self-interest. Thus, evolutionary theory can explain egoism and suggests why cooperation between unrelated individuals is very often difficult and remarkably unlike the behavior one encounters within the family.

The Origins of Domination

Evolutionary theory can also explain domination. Like egoism, the desire to dominate is a trait. In the context of evolutionary theory, domination usually means that particular individuals in social groups have regular priority of access to resources in competitive situations. For most social mammals, a form of social organization called a dominance hierarchy operates most of the time. The creation of the dominance hierarchy may be violent and is almost always competitive. A single leader, almost always male (the alpha male), leads the group. The ubiquity of this social ordering strongly suggests that such a pattern of organization contributes to fitness.

Ethologists categorize two principal types of behavior among social mammals in a dominance hierarchy: dominant and submissive.[56] Dominant mammals have enhanced access to mates, food, and territory, increasing their chances of reproductive success.[57] Acquiring dominant status usually requires aggression.[58] Dominance, however, is an unstable condition; to maintain it, dominant individuals must be willing to defend their privileged access to available resources as long as they are able. Evolutionary anthropologist Richard Wrangham and ethologist Dale Peterson explain why an individual animal is motivated to vie for dominant status: "The motivation of a male chimpanzee who challenges another's rank is not that he foresees more matings or better food or a longer life."[59] Rather, "those rewards explain why . . . selection has favored the desire for power, but the immediate reason he vies for status . . . is simply to dominate his peers."[60]

Dominant animals often assume behavior reflecting their status. For example, dominant wolves and rhesus monkeys hold their tails higher than other members of their group in an effort to communicate dominance. A dominant animal that engages in such displays is better off if it can gain priority of access to resources without having to fight for it continuously.[61] Such signaling behavior is found among submissive social mammals as well. They often try to be as inconspicuous as possible and recognize what is permitted and forbidden given their place in the hierarchy. This behavior shows that the subordinate accepts its place in the dominance hierarchy and—at least temporarily—will make no effort to challenge the dominant animal.

Ethologists argue that dominance hierarchies evolve because they help defend against predators, promote the harvesting of resources, and reduce intragroup conflict.[62] A species that lives communally has two choices. It can either accept organization with some centralization of power, or engage in perpetual conflict over scarce resources, which may result in serious injury and deprive the group of the benefits of a communal existence, such as more efficient resource harvesting.[63] Ethological studies have confirmed that a hierarchical dominance system within a primate band minimizes overt aggression and that group aggression often increases when the alpha male is challenged.[64]

For primates and especially humans, the dominance hierarchy may have produced a fortuitous result: great intelligence.[65] As cognitive psychologist Denise Dellarosa Cummins argues, it has had a profound effect on human evolution: "The fundamental components of our reasoning architecture evolved in response to pressures to reason about dominance hierarchies, the social organization that characterizes most social mammals."[66] Her study and others have found that dominance hierarchies have contributed to the evolution of the mind, which in turn has contributed to fitness.

According to Cummins, submissive individuals can detect, exploit, and circumvent the constraints of domination. If an animal can take what it wants by force, it is sure to dominate the available resources—unless its subordinates are smart enough to outwit it. To survive, a subordinate must use other strategies: deception, guile, appeasement, bartering, alliance formation, or friendship. Thus intelligence is particularly important to the survival of subordinates. "The evolution of mind emerges," Cummins writes, "as a strategic arms race in which the weaponry is ever-increasing mental capacity to represent and manipulate internal representations of the minds of others."[67]

From their studies of chimpanzee societies, ethologists have learned that the struggle for survival is best characterized as a struggle between those who are dominant and those seeking to outwit them, i.e., between recognizing an opponent's intentions and hiding one's own. The following example illustrates how a subordinate chimpanzee, Belle, who knows the location of hidden food, attempts to deceive Rock, who is dominant:

> Belle accordingly stopped uncovering the food if Rock was close. She sat on it until Rock left. Rock, however, soon learned this, and when she sat in one place for more than a few seconds, he came over, shoved her aside, searched her sitting place, and got the food. Belle next stopped going all the way [to the food]. Rock, however, countered by steadily expanding the area of his search through the grass near where Belle had sat. Eventually, Belle sat farther and farther away, waiting until Rock looked in the opposite direction before she moved toward the food at all, and Rock in turn seemed to look away until Belle started to move somewhere. On some occasions Rock started to wander off, only to wheel around suddenly precisely as Belle was about to uncover some food. . . . On a few trials, she actually started off a trail by leading the group in the opposite direction from the food, and then, while Rock was engaged in his search, she doubled back rapidly and got some food.[68]

Despite the "arms race" described by Cummins to outwit dominance, the subordinate members of the group continue to participate in the dominance hierarchy because doing so increases the chances of survival. As ethologist David Barash explains, if subordinates "are more fit by accepting" subordinate "ranking than by refusing to participate, then some form of social dominance hierarchy will result."[69]

Humans and other primates evolved a mental architecture to address the difficulties they encountered in dominance hierarchies. These problems, which "directly impact survival rates and reproductive success," include two crucial needs: "the necessity to recognize and respond appropriately to permissions, obligations, and prohibition," and the necessity "to circumvent the constraints of hierarchy by dint of guile, particularly through successfully forecasting others' behavior."[70] Because human mental architecture was created through evolution, it remains part of human behavior today, as cognitive psychology studies show.[71]

One result of the evolution of our mental architecture is the ability to indoctrinate humans. As E.O. Wilson writes: "human beings are absurdly easy to indoctrinate—they *seek* it."[72] Three factors cause this ease of indoctrination. First, survival in an anarchic and dangerous world dictates membership in a group and produces a fear of ostracism from it. Second, an acceptance of or conformity to a particular status quo lowers the risk of conflict in a dominance hierarchy. Third, conformity helps keep groups together.[73] If group conformity becomes too weak, the group could fall apart and then die out because of predation from its or another species.[74] Thus, for most primates, belonging to the group is better—it increases chances of survival—than existing alone, even if belonging requires subordination.

These understandings have great consequences for the study of politics. Irenäus Eibl-Eibesfeldt, Albert Somit and Steven Peterson, E.O. Wilson, and psychologist Donald Campbell, among others, suggest that humans readily give allegiance or submit to the state, or to ideologies like liberalism or communism, or to religion, because evolution has produced a need to belong to a dominance hierarchy.[75] An overview of human history provides context: much of it is a record of threats of force or wars to gain territory and resources.[76] Political institutions, whether monarchies or aristocracies, and leaders such as Julius Caesar, Louis XIV, or Somali warlord Mohamed Farrah Aidid, typify dominance hierarchies—as do the modern state and its many institutions, such as government bureaucracies and the military.[77]

These political examples are readily evident, but dominance hierarchies also have more subtle effects even among the young and between the sexes. They help explain why people obey authority and intensify the significance of birth order. Research on children's social interactions has shown that children as young as three organize themselves into dominance hierarchies. The dominance of males over females in nearly all primate species, the Bonobo perhaps being an exception, is what has been termed sexism in our own species.[78] Stanley Milgram's famous psychological experiments show that ordinary citizens will obey those they recognize as dominant even when they are using their power for clearly malevolent ends.[79] Frank Sulloway's analysis of birth order shows that dominance structures within the family influence personality, with first-born dominant siblings seeking to maintain the status quo through conformity, and later-borns, as subordinates, seeking to rebel against those constraints.[80]

In this section of the chapter I have shown that evolutionary theory provides a more useful explanation of realist behavior than do the arguments advanced by Niebuhr or Morgenthau. As discussed above, evolutionary theory provides a scientific foundation for realism because it yields both logical and testable ultimate and proximate causes. As a result, it is superior as judged by commonly used metrics in the philosophy of science, such as those developed by Hempel and Popper, and it is desirable from a social science perspective to provide realism with the scientific foundation it has been lacking.

Evolutionary Theory as an Ultimate Cause of Offensive Realism

Evolutionary theory allows realists to advance offensive realist arguments without seeking an ultimate cause in either the anarchic international state system or in theological or metaphysical ideas. Realism based on evolutionary theory reaches the same conclusions, but the ultimate causal mechanism is different: human evolution in the anarchic and perilous conditions of the late-Pliocene, Pleistocene, and most of the Holocene epochs. Specially, evolutionary theory explains why humans are egoistic, strive to dominate others, and make in-group/out-group distinctions. These adaptations in turn serve as a foundation for offensive realism.

The central issue here is what causes states to behave as offensive realists predict. Mearsheimer advances a powerful argument that anarchy is the fundamental cause of such behavior. The fact that there is no world government compels the leaders of states to take steps to ensure their security, such as striving to have a powerful military, aggressing when forced to do so, and forging and maintaining alliances. This is what neorealists call a self-help system: leaders of states are forced to take these steps because nothing else can guarantee their security in the anarchic world of international relations.

I argue that evolutionary theory also offers a fundamental cause for offensive realist behavior. Evolutionary theory explains why individuals are motivated to act as offensive realism expects, whether an individual is a captain of industry or a conquistador. My argument is that anarchy is even more important than most scholars of international relations recognize. The human environment of evolutionary adaptation was anarchic; our ancestors lived in a state of nature in which resources were poor and dangers from other humans and the environment were great—so great that it is truly remarkable that a mammal standing three feet high—without claws or strong teeth, not particularly strong or swift—survived and evolved to become what we consider human. Humans endured because natural selection gave them the right behaviors to last in those conditions. This environment produced the behaviors examined here: egoism, domination, and the in-group/out-group distinction. These specific traits are sufficient to explain why leaders will behave, in the proper circumstances, as offensive realists expect them to behave. That is, even if they must hurt other humans or risk injury to themselves, they will strive to maximize their power, defined as either control over others (for example, through wealth or leadership) or control over ecological circumstances (such as meeting their own and their family's or tribe's need for food, shelter, or other resources).

Evolutionary theory explains why people seek control over environmental circumstances—humans are egoistic and concerned about food—and why some, particularly males, will seek to dominate others by maintaining a privileged position in a dominance hierarchy. Clearly, as the leaders of states are human, they too will be influenced by evolutionary theory as they respond to the actions of other states and as they make their own decisions.

In this chapter, I have already discussed these two elements of my argument: that evolutionary theory allows realist scholars to explain why state decision makers are egoistic and why those state leaders strive to dominate others when circumstances permit.[81] These adaptations were critically important in the course of human evolution and remain a significant ultimate cause of human behavior. Recalling that biology is good probability, not destiny, we should expect that leaders of states and major decision makers will possess these traits and are not likely to suppress them the way a saintly individual, like Mother Teresa, might. Indeed, if leaders suppress them, they are likely doing so only for tactical reasons or as required by specific circumstances.

In fact, a state's elites—the captains of industry and media, and military and political leaders—are more likely than average to show these traits in abundance since most leaders rise to the top of their respective hierarchies through a very competitive process. This is almost always the case for political leaders. The rise to the top is the result of an often arduous, and perhaps physically dangerous, process; those who triumph, whether Mao or Clinton, are almost certain to be egoistic individuals who are used to dominating the individuals or institutions around them.

Even those who inherit the throne have discovered time and again that keeping it involves almost endless struggle and sacrifice. Shakespeare has King Henry remind us in *Henry IV,* Part 2 (Act III, scene i), that "uneasy lies the head that wears a crown." Moreover, the king struggles against internal as well as external enemies, as the Bard captures so well in *MacBeth* (Act II, scene ii) with MacBeth's murder of Duncan, and with his suspicion of what the murder would cost him:

Methought I heard a voice cry, "Sleep no more!
MacBeth does murther sleep"—the innocent sleep,
Sleep that knits up the ravell'd sleave of care

A third cause of offensive realist behavior in humans is the in-group/out-group distinction commonly made in psychology, anthropology, and sociology, which I will discuss in more detail in chapter 5 when I explain the origins of ethnic conflict. The basic points here are that humans divide their worldview into an "Us," the in-group, versus "Them," the out-group, in order to simplify the human mind's information processing. Psychologists refer to the in-group as one's own group, toward which one is positively biased, perhaps a family, tribe, organization, or state. They argue that in-groups develop from a need for self-definition, both positive and negative. The in-group identity provides people with meaning, purpose, and a sense of community, but also with knowledge of what they are not—the out-group. The out-group may also be another family, tribe, organization, or state.

The in-group/out-group distinction is important to evolutionary theory's explanation of offensive realism because it explains why humans created these ideas and why we often fear out-groups. When the out-group is another state, we can see it as a threat to our state's resources or territory, or to our elite's political or economic interests. By explaining the origins of the distinction between in-groups and out-groups, evolutionary theory shows why humans make the distinction readily, whether it is a sports fan's trivial in-group/out-group distinction, or a nontrivial, life-or-death distinction held by a Hutu or a Tutsi, a kulak or a commissar, a friend or foe.

Humans make in-group/out-group distinctions for three reasons. First, humans seek resources—food, water, and shelter—to care for themselves and relatives, and they seek mates to reproduce their genotype; in sum, they are egoistic for the reasons advanced by Darwin, William Hamilton, and other evolutionary theorists, as I described in chapter 1 and in the discussion above. They are unlikely to assist those who are not related, but may do so occasionally, expecting reciprocal behavior. Humans behave in these ways because resources were scarce in the late-Pliocene, Pleistocene, and Holocene environments in which we evolved. In that environment, it is easy to understand why humans would prefer more resources to fewer: more strength is preferable to less strength, more wealth to less wealth, domination to being dominated. Most people do indeed prefer more resources to fewer; the rich want even more wealth, and seldom say they are too wealthy. Rather, they seem to worry about protecting their wealth from those who may take it from them, such as revolutionaries or the government. In essence, in prehistoric times when there was too little to go around, humans discriminated between self and others, family and others, tribe and others, in-groups and out-groups. This behavior remains today. We humans are likely to perceive out-groups as threats to our resources, the resources we need to maintain ourselves and our families and extended in-groups such as the tribe or state.

Second, living and evolving in dangerous environments, humans, like other animals, need the ability to assess threats rapidly and react quickly. The in-group/out-group distinction may be thought of as the human mind's immediate threat assessment. It is a mechanism for determining whether or not nonrelated conspecifics presented a threat. In sum, our mind rapidly debates: no threat/threat. Is the outsider a threat to oneself or to one's family? As a result, over the course of human evolution, strangers were first likely to fear one another, at least until they became familiar.

Third, in addition to the immediate threat posed by conspecifics, humans also learned to assess fairly quickly whether the outsider was a threat in the long run to the in-group, or to one's position in the dominance hierarchy. Members of the in-group might ask whether the presence of an individual or group would be a threat to their future resources—the scarce and precious resources needed to survive.

The evolutionary origins of this distinction suggest that elites, as people, will also fear other states because they are the out-group and may have the power to take control if the elites cannot deter an attack. Thus, evolutionary theory explains why people will be egoistic, strive to dominate, and create in-groups and out-groups, and this explains why people, especially leaders of states, are likely to behave as offensive realism expects.

Having advanced the argument for an evolutionary explanation of offensive realism, it is important to stress what I am not arguing. First, offensive realists agree on the proper mechanism to achieve domination: maximizing state power. Second, I am not arguing that the anarchy of the international system is a weak force in international relations. No question: it is a powerful force affecting state behavior. Nor am I arguing that it is an insufficient or flawed foundation for offensive realism. It certainly is not. My argument is only that human evolution greatly informs the behavior of all humans, even leaders of states, and thus provides an excellent, scientific foundation for offensive realism. My explanation of offensive realism should not be viewed as an alternative to Mearsheimer's argument, but rather as a complement to it. Human evolution produces the type of behavior expected by offensive realists. The existence of the anarchic system of states only makes this situation worse.

This contribution to offensive realism is significant for two reasons. First, offensive realists no longer need to depend entirely upon the anarchy of the international state system to advance their argument. As I have explained, egoism, domination, and in-group/out-group distinctions are sufficient to explain offensive realist behavior. As a consequence, this makes the theory more powerful. Offensive realists can now explain more than state or great power behavior. When anchored on evolutionary theory, the theory of offensive realism will become more powerful than they have realized thus far. It will allow them to elucidate why sub-state groups—individuals, tribes, or organizations—will also often behave as their theory predicts, and to explain this behavior before the creation of the modern state system in 1648. When it depends on anarchy as its ultimate cause, offensive realism needs a more specific condition, the state system, to obtain. In fact, offensive realists do not need a state system. They only need humans. Wherever humans form groups, be they tribes, organizations, or states, the argument will apply. When used in those circumstances, it might be useful to apply the term "evolutionary realism" in order to distinguish a broader and more robust realism from the theoretical label "offensive realism," as it is traditionally defined and applied in the discipline.

Second, Mearsheimer's offensive realist explanation argues that people gain power to yield security. That certainly may be. However, an overtone of this argument is that power or domination is distasteful for leaders—that they tolerate it only for the sake of their state's security. They are forced to maximize power when perhaps they would rather cooperate or share power with others.

Certainly, this may be so. And it seems that most countries have a leader who appears to disdain power—a figure like Calvin Coolidge—at some point in their histories. But I think it is more likely that leaders value power because it allows them to work their will, to dominate. Rarely do retired presidents, prime ministers, or dictators complain that they had too much power while in office. Indeed, I would submit, leaders generally want more power to further their aims, which they always cast in a benevolent light: for a political leader, more power means he is better able to work his will, which results in a better future for the populace. If we reflect upon history, a recurring theme is that leaders strive to gain more power, and more resources, perhaps through aggression, and this is not distasteful to leaders of states.

That said, we should expect that political leaders will without doubt carefully consider the risk of aggression for gain. After all, any competent politician recognizes that a great loss could cause him to lose face (or his life), and thus increase challenges to his power and perhaps generate unrest in the state. However, the prospect of gain balances the risks.

It is also important to address whether anchoring offensive realist behavior on evolutionary theory or on the anarchic state system is significant for the study of international relations. This book makes an important theoretical contribution to this issue, but we will need focused, empirical testing to determine which insights arise from an offensive realism based on evolutionary theory.[82] In particular, offensive realism based on evolutionary theory is likely to inform explanations, for example, of the conditions in which state leaders choose to aggress or expand their power. The reasons they choose to expand are often complicated, but they seek more power for their own egoism or vainglory as well as for the security of their state. Cortez sought glory and riches for himself, as well as for his king; a CEO seeks to expand his corporation's market share to increase the shareholder value but also, one suspects, to prove that he can personally conquer and expand markets.

In addition, this book should inform analyses of why leaders can often generate popular support for expansion with relative ease; or why external or internal threats have been such powerful motivators to develop national solidarity and to mobilize the resources of a society or a group. It can also explain social imperialism: a state's elites may use in-group/out-group distinctions to maintain their power by inflating the threat posed by other states to generate fear among the masses and thus preserve a status quo in domestic politics. Charismatic politicians, like charismatic military leaders or business executives, are often expert at manipulating in-groups and out-groups.

Finally, while the neorealism of Waltz and Mearsheimer uses anarchy as the cause of much state behavior, my use of evolutionary theory anchors the theory on the firm intellectual foundation of evolutionary theory. However, the application of evolutionary theory to realism should not be seen as an effort to supplant neorealism. The individual scholar should use either approach depending on the particular usefulness of the theory for a given research question.

The Contribution of Evolutionary Theory to Rational Choice

Rational choice theory has developed from its origins in the 1940s and 1950s into a well-developed paradigm centered on several core ideas.[83] The theory is widely used in political science, sociology, and especially economics to explain and predict human behavior. In this section of the chapter I will describe the core assumptions of rational choice theory, briefly discuss some major criticisms of the theory, and then, drawing in part from my discussion of egoism above, explain how evolutionary theory can assist rational choice. My intent in this book is not to debate the usefulness of rational choice for the study of social issues—I want neither to bury nor to praise it. It is only to suggest how evolutionary theory can improve rational choice. It can do so because it explains egoism and rationality, and so increases the logical coherence and testability of rational choice.

The Core Components of Rational Choice Theory

Rational choice theory explains individual expectations, goals, and preferences, how people attempt to realize them by their own actions, and how they are helped or hindered by the actions of others and the structure of the situation.[84] I will describe the core components and assumptions of rational choice theory necessary for my argument. I will not present or judge the totality of the assumptions, the standards of explanation, or the appropriateness of the methodology; nor will I evaluate the many debates within rational choice theory.

Rational choice theory contains three closely related and crucial assumptions. First, individuals are rational. Rationality is usually defined as follows: "when faced with several courses of action, people usually do what they believe is likely to have the best overall outcome" in advancing their goals.[85] In sum, the theory posits that individuals choose actions they believe will advance, or allow them to achieve, their goals. As rational choice theorist Jon Elster writes, "there are strong *a priori* grounds for assuming that people, by and large, behave rationally."[86] Evolutionary theory will allow those "*a priori* grounds" to be anchored in science.

Second, individuals are egoistic. The best overall outcome is often defined as the self-interest of the actor: "to act rationally is to do as well for oneself as one can."[87] This is often described as maximizing utility; it assumes that individuals seek to maximize their utility, which is often assumed to be wealth, although it need not be. Utility may be defined in many ways, including happiness or the pursuit of wealth.[88] How it is defined is not critical here. What is significant is the core assumption made by rational choice theorists that individuals seek to maximize their self-interest.

Third, rational choice theorists assume that their models apply equally to all people. As economists George Stigler and Gary Becker argue, decisions, rules, and tastes are stable over time and similar among people.[89] Rational choice theorists clearly do not argue that all people have the same preferences, or that people place the same value on an item or idea across cultures. The claim is that individuals can maximize their utility in a predictable manner, once utility in a particular context for an individual is understood. This assumption is rooted in the first and second assumptions, since it requires that individuals are rational and egoistic irrespective of culture.

In addition to these assumptions, rational choice theory also presumes that actors have choices, even if these choices are constrained by the structure of the situation, and that preferences are consistent. That is, people can rank their preferences and the ordering is transitive: If they prefer A to B, and B to C, they must prefer A to C. Finally, the methodology of rational choice theory is often game theory or formal modeling, which are useful for analyzing situations where one player's payoff is affected by the actions of one or more other players.

The core belief that individuals are rational and pursue their self-interest is also the root of the collective action problem defined by Mancur Olson.[90] This problem is also called the problem of free riding or of voluntary provision of public goods. Olson recognized that "rational, self-interested individuals will not act to achieve their common or group interest" in the absence of coercion or "some other special device to make individuals act in their common interest."[91] Building on Olson's work, Russell Hardin reinforces this point when he writes that the logic of a collective action problem "is based on the strong assumption that individual actions are motivated by self-interest."[92] Becker says this assumption, along with market equilibrium and stable preferences, is the "heart of the economic approach."[93]

As with the realist theory of international relations, the scientific grounding of rational choice is important because it is desirable from the perspective of science to have a theory that is logical from its core assumptions through its secondary or mid-level assumptions. Clearly, a theory whose core assumptions are noumenal or simply postulated will be weaker scientifically than a theory that is scientifically grounded, no matter how logically consistent its argument or how powerful the empirical insights it generates. Rational choice is an important theory in many disciplines of social science. It has produced a fertile research agenda. Among the issues usefully studied through its lens are voting behavior, deterrence, political negotiation, crisis bargaining, the spread of ethnic conflict, institutional behavior, and the significance of relative gains considerations in international relations.[94] It also has helped explain and solve collective action problems such as group behavior and conflict, revolution, and state formation.[95] However, as I will discuss below, the use of evolutionary theory will improve rational choice theory by assisting with some criticisms now directed at it.

Criticisms of Rational Choice Theory

Rational choice theory is criticized from several different perspectives, including its reliance on egoism, its empirical usefulness, and its ability to predict an individual's preferences. Economist Amartya Sen argues the dependence of rational choice upon assumptions of egoism is problematic because it presumes too much: "because choice may reflect a compromise among a variety of considerations of which personal welfare may be just one."[96] This includes, he argues, the fact that much behavior is based on commitment that necessitates a sense of moral obligation that often belies egoistic behavior. In many respects, this is a modern formulation of a frequent criticism of the way classical economists understood the motives behind human behavior. However, Sen says that were rational choice theorists to recognize these points, the theory would possess a richer structure for evaluating rational behavior.

Political scientists Donald Green and Ian Shapiro submit that rational choice has little value when it is actually applied to political issues and tested: "few theoretical insights derived from rational choice theory have been subjected to serious empirical scrutiny and survived."[97] They re-examine many of the issues famously studied by rational choice theorists, such as individual and legislative voting behavior and free-riding problems, to reveal the complexities of these issues that the rational choice theorists simply did not capture when they examined these issues.

Stephen Walt, a major theorist of international relations, argues that while formal theory may improve the logical consistency of an argument, it is neither necessary nor sufficient, and it does not provide original or strong empirical insights when it is applied to the study of international relations.[98] Critiquing the insights provided by rational choice theorists who examined such major international political issues as the origins of war, he shows they are often modest at best.

Psychologists Daniel Kahneman and Amos Tversky argue that in certain circumstances rational choice does not provide an adequate foundation for decision making. This is because cognitive processes often make it difficult for individuals to reach rational decisions, particularly when the same choice is presented or framed differently. They have developed "prospect theory" to explain decision making in certain conditions. Prospect theory is a two-stage theory. In the first stage, an individual frames an issue. In the second, the evaluation phase, the person often makes choices on the basis of comparative value. Their research has revealed that people often misevaluate probabilities, particularly because they rely on psychological heuristic rules concerning probabilities that can lead to errors.[99] In particular, they have found people often place too much psychological weight on low probabilities, while they underweigh moderate and high probabilities, and the latter effect is psychologically more distinct than the first.

Finally, economist Geoffrey Brennan argues that rational choice requires knowledge of the individual's ends or purposes and not only the assumption that the agent is rational. Without knowledge of ends or purposes, rational choice does not rule out any action, "for virtually any action there exists some purpose for which that action is best. Even nontransitive choices can be rational if the agent's ends are sufficiently finely individuated."[100] So for rational choice to be able to predict an individual action, the user needs more knowledge than just that the agent is maximizing utility. Brennan says that knowledge about the actual content of the agent's utility function is necessary as well.

The Contribution of Evolutionary Theory for Rational Choice

Evolutionary theory can make three major contributions to rational choice.[101] At root, rational choice theory assumes a specific view of human nature: individuals are rational, egoistic maximizers of utility. As rational choice theory is presently defined, these are assumptions of human motivation. As assumptions, they are not anchored on an ultimate cause. Evolutionary theory explains why individuals are, first, egoistic and, second, rational. Applying an evolutionary explanation allows us to ground these assumptions in scientific explanation and thus improve rational choice. Third, evolutionary theory working with rational choice theory can allow scholars to predict individual behavior more accurately.

Of course, economists have been aware of evolutionary theory for years; they have often used concepts like natural selection as metaphors for the major questions in neoclassical economics, for example, why some firms survive in a competitive environment.[102] Nonetheless, evolutionary theory has great potential to support neoclassical economics and rational choice theory, and too few scholars have used its scientific foundation of human behavior for the assumptions that these fields share. Notable exceptions include Peter Corning, Richard Nelson, and Sidney Winter, as well as Robert Frank and Jack Hirshleifer, whose work I discuss below.[103]

However, while the intellectual benefits for rational choice are great, I do not want my discussion of evolutionary theory and rational choice to imply that the intellectual benefit need only flow in one direction. Rational choice theorists have developed sophisticated models of equilibria, multiple-level games, and decision making; they may be useful as analytic tools for evolutionary biologists. In addition, while ethologists may study social organization in multiple forms of life, economists have created advanced models that ethologists may usefully adopt as analogies or substantive models to describe animal behavior. Moreover, economists have addressed the problem of scarce resources in social organization, and the limited nature of materials and energy in the face of almost limitless wants; they have also identified many types of costs in exchanges, such as opportunity and transaction. Ethologists may also use these concepts to further their comprehension of animal behavior.

Explaining Egoism

Evolutionary theory can explain why egoism evolved as I described it above in my analysis of realism. Individuals are expected to behave egoistically given their own preference, which includes anticipating the preferences of others and acting in a manner that lets them realize their own goals. As was true for realism, an evolutionary explanation of egoism is also useful to economists and rational choice theorists because these theorists have assumed that individuals are egoistic, but they have not had an ultimate explanation of why individuals should be so. Evolutionary theory explains why individual animals who must compete in a dangerous world to satisfy their physiological needs will be egoists. Without evolutionary theory to serve as a foundation for his arguments, Adam Smith time and again postulated the importance of egoism for understanding human behavior in *The Theory of Moral Sentiments:* "the loss or gain of a very small interest of our own appears to be of vastly more importance, excites a much more passionate joy or sorrow, a much more ardent desire or aversion, than the greatest concern of another with whom we have no particular connection."[104] More famously, in *The Wealth of Nations* he wrote: "It is not from the benevolence of the butcher, the brewer, or the baker, that we expect our dinner, but from their regard to their own interest."[105] Economists Robert Boyd and Peter Richardson summarize nicely how evolutionary theory can assist economists and rational choice theorists, as well as those using other theories that assume egoism, when they state that evolutionary theory "predicts that natural selection should produce *Homo economicus.*"[106]

Moreover, by relying on inclusive fitness, rational choice theorists may extend their argument to explain scientifically that individuals are egoistic in their behavior toward unrelated individuals. As I discuss below, one result of this recognition is that it allows rational choice theorists to include familial relationships in their explanations of egoistical behavior. In addition, they will be able to make more precise predictions about individual behavior in familial relationships.

The conception of egoism provided by evolutionary theory assists rational choice in a response to Sen's criticism. It does not permit a direct rebuttal, of course, because his argument is correct: one may have other motivations for a particular preference, including a commitment based on moral obligation. Evolutionary theorists recognize that there are often multiple proximate explanations for a particular animal's behavior. Thus evolutionary theory cannot be used to respond directly to Sen's critique. However, the theory can explain why egoism is an adaptation and so why, in complex social interactions involving unrelated individuals, one can reasonably expect to see few tendencies toward solidarity, cooperation, or altruism.

Sen's key question is: Should we modify the assumption of egoism to include other types of behavior, at the cost of explanatory power or ability to generalize? This is a subjective choice that some may be willing to make and indeed should make if the issue studied requires it. However, evolutionary theory explains why egoism is a human trait and thus possibly the most powerful motivator of human behavior. As a result, explanations of human behavior based on egoism will, a priori, have greater explanatory power than explanations based on what most evolutionary theorists would regretfully agree are less frequent motivators, such as altruism. Thus, egoism is well grounded as a central assumption of rational choice theory. An important issue for scholars of rational choice remains beyond the scope of this book: should rational choice theory incorporate a moral commitment and thereby produce a theory with greater explanatory power in some circumstances at the cost of its parsimony and considerable general application?

The Rational Mind

Evolutionary theory also provides an ultimate causation explanation for the rational mind upon which rational choice depends. Lest this argument be misunderstood, it is important to recall the conception of rationality used by most rational choice theorists. The argument is not that individuals are perfectly rational, that they analyze all their goals and actions with the intellectual acuity of Immanuel Kant. The conception is rather less ambitious: an individual is considered rational if the actions he chooses are directed toward achieving his goals.

Any explanation of how the rational mind evolved must note Cummins's argument that the human mind has evolved out of an arms race between dominant and submissive individuals. This is certainly part of the explanation that must begin with the obvious but profound recognition that animals move. Psychologist Patricia Churchland succinctly explains how this affects the origin of the mind in the animal kingdom: "If you root yourself to the ground, you can afford to be stupid" but "if you move, you must have mechanisms for moving, and mechanisms to ensure that the movement is not utterly arbitrary and independent of what's going on outside," and so "neurons" and their aggregation in brains, "are evolution's solution to the problem of adaptive movement."[107]

Also to aid movement, primates evolved sophisticated sight. Pinker argues that vision contributes greatly to the evolution of mind. Visual information in primates provides depth perception that "defines a three-dimensional space filled with movable solid objects" and color that "makes objects pop out from their backgrounds."[108] According to him, these factors have caused the primate brain to divide the flow of information into "two streams: a 'what' system, for objects and their shapes and compositions, and a 'where' system, for their locations and motions."[109] Pinker continues, "It can't be a coincidence that the human mind grasps the world . . . as a space filled with movable things"; for example, "we say that John *went from* being sick *to* being well, even if he didn't move an inch. Mary can *give* him *many pieces* of advice, even if they merely talked on the phone and nothing changed hands."[110]

In addition to the crucial role of vision in processing information in a three-dimensional world, Pinker also notes the importance of groups. Living in groups presents unique cognitive challenges, such as dominance and the outwitting of dominance and how to send and receive signals that help in defense, resource gathering, and sexual access. Pinker notes that the hand is a third cause of intelligence: "Hands are levers of influence on the world that make intelligence worth having. Precision hands and precision intelligence co-evolved in the human lineage, and the fossil record shows that hands led the way."[111] Finally, hunting is also critically important for the development of an advanced intelligence because it provides concentrated nutrients: "meat is complete protein containing all twenty amino acids, and provides energy-rich fat and indispensable fatty acids."[112] In fact, "across the mammals, carnivores have larger brains" compared to "their body size than herbivores, partly because of the greater skill it takes to subdue a rabbit than to subdue grass, and partly because meat can better feed ravenous brain tissue."[113]

Evolutionary theory explains intelligence and a rationality inextricably associated with it in the following, narrow sense: rationality is understood as the mental logic necessary for understanding cause and effect, symbols, spatial relationships, contiguity in time or place, for recognizing faces and emotions, and detecting cheating, as well as for language acquisition and speech comprehension.[114] These might be thought of as the "components of rationality," or the tools necessary for primates and other advanced mammals such as dolphins to navigate successfully, that is, long enough to reproduce and raise young, in a competitive and dangerous world.[115]

Many examples of such components are present in the animal kingdom. Wedge-capped capuchin monkeys in central Venezuela use millipedes (*Orthoporus dorsovittatus*) as a mosquito repellent, rubbing them around their faces and into their fur.[116] Millipedes offer effective relief from mosquitoes and other insects because they possess benzoquinones—powerful defensive chemicals. Imo, a Japanese macaque, became famous among ethologists for her unique behavior: she washed the sweet potatoes they provided before she ate them, and this behavior was emulated by other macaques in her group. Later, when researchers provided wheat to her group she devised a novel solution to a problem they confronted: how to separate the wheat from sand. When Imo tossed her sand and wheat mixture into water, the sand sank, leaving only the wheat on the surface. Imo's group of macaques copied this behavior, indicating to some ethologists that animals transmit cultural behaviors.[117] Ethologists and cognitive psychologists have demonstrated that chimpanzees and human babies can understand basic reasoning, symbols, spatial relationships, and contiguity in time and space.[118] These findings benefit rational choice theory because they place on the bedrock of evolution what heretofore have been assumptions: first, individuals can think in a goal-directed manner, and second, preferences are transitive.

In addition to these components of rationality, evolutionary psychologists Leda Cosmides and John Tooby argue that the human mind evolved "reasoning instincts" or many reasoning, learning, and preference circuits in the mind. These circuits, commonly called "instincts," make "certain kinds of inferences just as easy, effortless, and 'natural' to humans as spinning a web is to a spider or building a dam is to a beaver."[119] They are present and reliably developed in all human beings without conscious effort or formal instruction, and are specialized for solving the problems that our hominid ancestors encountered, including mate choice, threat recognition, alliance relations, parenting, object permanence, predator avoidance, disease avoidance, social exchange, and bluff and double-cross recognition.[120] "Each cognitive specialization is expected to contain design features targeted to mesh with the recurrent structure of its characteristic problem type, as encountered under Pleistocene conditions"; as a result "one expects cognitive adaptation specialized for reasoning about social exchange to have some design features that are particular and appropriate to social exchange, but that are not activated by or applied to other content domains."[121]

The research of Cosmides and Tooby has two important implications. First, these complex, cognitive adaptations are ancient; humans have likely engaged in social exchange for most of their existence. Social exchange behavior is "both universal and highly elaborated across all human cultures—including hunter gatherer cultures—as would be expected if it were an ancient and central part of human life," according to Cosmides and Tooby.[122] Were it otherwise, they say, "like writing or rice cultivation, one would expect to find evidence of its having one or several points of origin, of its having spread by contact, and of its being extremely elaborated in some cultures and absent in others."[123]

In addition, ethological studies also offer important evidence. Ethologists know that macaques and baboons engage in reciprocal exchange, and chimpanzees seem to be masters of it.[124] Thus the cognitive adaptations necessary for social exchange were probably present in "the hominid lineage as far back as the common ancestors that we share with the chimpanzees, five to ten million years ago."[125]

Second, and of particular interest for economists, Cosmides and Tooby argue that humans are relatively good at detecting cheating in social exchanges because of a specialized reasoning procedure. They submit that: "human reasoning is well designed for detecting violations of conditional rules."[126] Humans have a relatively acute mental ability to facilitate exchange: we can specify who in a given situation is an agent in the exchange, the costs and benefits and who benefits, who is entitled to what, and under what conditions the contract is broken.[127] The human ability to detect cheating in social exchange lends support to the rationality assumption of rational choice theory.

However, while evolutionary theory provides a scientific explanation for the rational mind upon which rational choice theorists depend, it also lends support for the human propensity to miscalculate in the circumstances studied by Kahneman and Tversky and to commit the multiple errors in reasoning that have been detected by psychologists.[128] On the whole, however, evolutionary theory provides support for the assumption of rationality as used in mainstream rational choice theory.

Better Understanding Preferences

A third contribution of evolutionary theory is that it helps rational choice theorists better understand preferences in given situations. Brennan's criticism is sound: greater knowledge of an individual's goals and preferences clearly helps us understand and predict behavior. Traditionally, economists and rational choice theorists have forsworn attempts to study the causes and content of individual tastes or preferences. They have usually treated them as given or inexplicable: *de gustibus non est disputandum.* However, as Hirshleifer argues, "Economists today would do well to go back to the master, Adam Smith, who did not regard the fundamental drives of men as arbitrary and inexplicable, who clearly understood that human desires are ultimately adaptive responses shaped by man's biological nature and situation on earth."[129]

As my discussion of egoism demonstrated, the prediction of egoistic behavior is supported by evolutionary theory. So is the prediction of rational behavior, properly understood. The Darwinian explanation of egoism and inclusive fitness's expectation of self-sacrificing behavior for genetically related kin may be used to make predictions about when individuals might be significantly influenced by emotions in conditions not expected by traditional rational choice theory.[130] These insights provided by evolutionary theory help in the following respects.

First, Hirshleifer notes that economists tend to resist applying biological ideas to human beings, but they should appreciate that physiological factors caused by natural selection do affect preferences in nontrivial ways: "hunger and sex, though not *all*-important, are scarcely *un*important motivators of human beings . . . [as] love and rage and family feeling and group loyalty [can be]."[131]

This is more significant than the simple recognition that hot drinks will sell better, per capita, in Anchorage than in Atlanta, or of the ubiquity and permanence of food markets and the world's oldest profession. It permits the recognition that humans evolved in a competitive realm where resources necessary to meet our physiological needs were scarce. The result of our evolution in the Pliocene, Pleistocene, and Holocene epochs has direct implications for understanding human decision making. This includes decisions about what we consume, want, and value. Studies conducted by economists and evolutionary psychologists illuminate how this is so.

Starting from the understanding that humans evolved in a dangerous environment, humans should value a present resource—even in a smaller amount—than a greater amount at a later date. Studies have found this true even for a delay of a few days. To help account for what psychologists call "time inconsistent" choices, Richard Herrnstein developed the concept of the "matching law."[132] One of its properties is that the attractiveness of the delay is inversely proportional to the delay; thus we heavily discount rewards in the distant future in favor of more immediate ones. This is not the case when we choose between two rewards to be received at the same time; then we prefer the larger of the two. "The time-discounting feature of the matching law," Frank argues, "is one of the most robust regularities in experimental psychology."[133] For example, when a pigeon is given a chance to peck "one of two buttons to choose between a morsel of food 30 seconds from now and a much larger morsel 40 seconds from now, it takes the latter. But when it chooses between the same morsel now and the larger morsel 10 seconds hence, it often picks the former."[134] Other animals behave the same way "and so, much of the time, do humans," which suggests to Frank: "this feature is apparently part of the hard-wiring of most animal nervous systems."[135]

Of course, multiple factors may explain why a particular individual values a lesser amount sooner than a greater amount later. But the strength of the evidence amassed by psychological studies suggests an evolutionary explanation. This seems intuitively plausible as well when one reflects upon conditions in the Pleistocene when, in fact, a bird in the hand was indeed worth two in the bush since one could never be certain about the source of the next meal, a condition all too frequent today in many underdeveloped states.

A second way that evolutionary theory helps economists and rational choice theorists think about preference formation begins with the recognition that individuals are egoists except in matters concerning their genetically related kin; then, evolutionary theory suggests, they will be motivated to perform significant acts of altruism toward them. This behavior, in turn, is motivated by the drive for reproductive success. When food is scarce, mothers feed their babies instead of feeding on them, as a strict rational choice model might predict; otherwise the infants would die and the mothers would be unable to replicate themselves.[136] Genes that caused mammalian mothers to care for their infants were replicated more successfully than those that caused mothers to eat their infants when hungry. As a result, in the conditions under which mammals evolved, a pure egoism will likely fail when it competes with nepotistic egoism.[137]

The powerful drives that affect so much of our life, such as the desire to see one's children do well in life, the desire to ensure the well-being of one's parents, nepotism, and the attraction to beauty and high status symbols or resources, are common preferences that flow from human evolution. My suggestion allows economists and rational choice theorists to explain behavior within the family as well as outside of it. This is a significant development because neoclassical economists had largely avoided individual behavior within the family, treating it as an economic unit and ascribing totally altruistic behavior within it and selfishness outside it. In fact, an evolutionary perspective permits economists to recognize that altruistic behavior within the family is not an aberration from their assumption of egoism but in fact is in accord with it.

As this brief discussion shows, evolutionary theory can inform a discussion of preferences. Contradicting the belief of traditional neoclassical economics, preferences are often not arbitrary but can stem from definite and permanent motivations explained by evolutionary theory. This understanding permits economists and rational choice theorists to produce more refined models of human behavior.

Finally, the use of evolutionary theory can strengthen the argument made by some economists and rational choice theorists that their explanation of behavior is universal because people strive to maximize their utility. Evolutionary theory demonstrates that some human actions are universal either because an action is cross-culturally valid like a scientific truth (what anthropologists term an etic universal), such as the fact that objects fall down instead of up or the perception of the color green, or because it is part of the conceptual system of all people (what they call an emic universal), such as incest avoidance and, perhaps, the maximization of utility.[138] Although I will not consider this issue here, evolutionary theory facilitates an understanding of human universals, so it will allow cross-cultural studies to research arguments for the universality of the behavior that economists and rational choice theorists assume. While evolutionary theorists are always careful to caution that human behavior is the product of both the evolutionary process and culture, we can expect important insights from research that arises to determine whether utility-maximization is truly universal behavior.

Conclusion

Evolutionary theory may make a similar contribution to both realism and rational choice: for the assumptions upon which they rely, it may provide scientific ground and do so without making changes to the body of either theory. Thus it will change the substructure rather than the superstructure of either theory. As a result, both will have a scientific foundation for the first time because the ultimate causation upon which both rely will be logically coherent and testable.

Realism benefits even more than rational choice from the use of evolutionary theory because scholars of international relations have moved away from the theory since the 1979 publication of Waltz's *Theory of International Politics*. A major reason why this occurred is that Waltz's more scientific argument improved upon the noumenal foundation of realism as defined by Morgenthau and Niebuhr. Evolutionary theory can change the perception that realism is not a scientific theory because of its noumenal foundations: its expectations of egoism and domination are explained by evolutionary theory. Consequently, scholars may increasingly find it relevant for their research questions, particularly scholars concerned with advancing offensive realism and exploring its applications for state behavior.

My analysis of evolutionary theory and realism is not intended to be the final word. As others apply evolutionary theory to realism and to the types of questions realists address, they will reveal important insights. For example, an interesting point for realists that I did not discuss is an evolutionary explanation for the difficulties of cooperation. Evolutionary theory explains why cooperation among unrelated individuals is difficult and remarkably unlike the behavior one encounters within the family. Realists and neorealists expect cooperation to be difficult in international relations because of the uncertainty and the often divergent security interests of states that raise concerns over the relative gains from trade and other forms of cooperation.

However, evolutionary theory provides another explanation for the difficulty of cooperation. Just as individuals are willing to share resources within the family rather than with outsiders because they are concerned with the health and well-being of kin, so too are individuals within states reluctant to trade scarce resources that the state itself needs for its own development or security. Naturally, states trade such resources all the time; I do not claim that trade is impossible. Rather, I believe, individual emotion (e.g. anger, envy, or hatred) can be a significant barrier to cooperation and one that is not well recognized by realism or other theories.

Ethnocentrism often leads individuals to perceive their country as something of an extended family; they will be jealous or resentful if they see necessary resources leaving the country, especially if the benefits of the trade are directed to only a few or are visible only in the long run, or become imperceptible because they are divided among many people. I will discuss the evolutionary origins of ethnocentrism in detail in chapter 5. The connection I posit here between ethnocentrism and an emotional barrier to cooperation is one that deserves to be researched. Reactions to trade are often framed in the media as workers concerned over the loss of their jobs. No doubt such concern is significant, but I suspect a deeper cause. Many individuals may also be concerned that resources are leaving the "family" and seem to be benefiting outsiders.

Evolutionary theory, and ethnocentrism in particular, may also explain why some resources are valued more highly than others. One may recall the resentment in the United States in the 1980s and early 1990s when Japanese corporations bought major landmarks in American cities. The fact that the purchasers were Japanese seemed to trouble Americans more than the larger investments by British and Dutch concerns. Similarly, in many cultures men perceive women as a valued resource, and women who date or marry the enemy or foreigners—especially across race—may cause significant friction. Think of the French public's humiliation of French women who had relationships with occupying German troops, or the tensions that arose in America's alliance relationships with Japan when U.S. servicemen raped schoolchildren in Okinawa or those that occasionally erupt in South Korea between South Korean men and American servicemen who are dating South Korean women.[139] These tensions in what are cooperative alliance relationships are caused in part by human emotions. This suggests that the difficulties in cooperation are multiple. Realists have done an excellent job in defining some of them, but I suspect that a better understanding of human evolution will permit more to be identified for research.

Economists have used the assumption of *Homo economicus* without major concern about their theory's scientific foundations. This is in accord with Milton Friedman's famous argument that the assumptions of a theory are less important than its ability to "yield sufficiently accurate predictions" of behavior.[140] However, their critics have expressed greater concerns; I have described some of these above. Using evolutionary theory, economists and other users of rational choice theory will have a scientific foundation for the assumptions of egoism and rationality and so will be able to respond directly to the critics on each of these points. This significantly improves the theory, but will not wholly mollify the critics. Even an improved rational choice theory that is anchored on an evolutionary foundation may still be criticized for placing too much emphasis on egoistic behavior while slighting forms of selfless behavior and for reducing complicated events or behavior to issues of egoism. Since evolutionary theory cannot assist rational choice theory in these criticisms, the debate between rational choice theorists and their critics will continue.

Evolutionary theory also enables us to understand human preferences in more detail. Economists and rational choice theorists may recognize that preferences are not necessarily arbitrary; rather, any explanation may be informed by an understanding of universal human desires and the agent's personal relationship to other actors. Clearly, the individual's behavior is likely to be different in a personal relationship than in interaction with a stranger.

Finally, I want to stress that rational choice theory and the discipline of economics can also contribute to evolutionary theory and ethology. There is no hierarchy of science in which biology defers to physics or the social sciences are subordinate to the natural. Intellectual exchange can go both ways, and it is important for the growth of knowledge that scholars understand as broadly as possible the key concepts of other disciplines. The problem of scarcity of resources in social organization, the limited nature of materials and energy, the conception of gains from trade, and the multiple types or forms of costs in exchanges may be useful for ethologists to adopt as analogies or to use to inform their substantive models as they describe animal behavior.

3 Evolutionary Theory and War

In the last chapter I showed how evolutionary theory assists realism and rational choice, but the life sciences can also contribute to comprehending major issues studied in the discipline such as the origins of war and ethnic conflict. In this chapter, I examine how evolutionary theory and ecology can contribute to the study of warfare. My objectives are, first, to demonstrate how evolutionary theory explains the origins of warfare, and second, to examine evidence from past and extant premodern societies to support this argument.

I do not intend to use evolutionary theory and ecology to explain the precise origins of a specific war in human history, such as the Punic Wars, the Russo-Japanese War, or World War II. Nor can evolutionary theory explain the conduct of those wars: why Hannibal acted as he did at Cannae, or why Napoleon succeeded brilliantly at Austerlitz but failed at Waterloo. As with the analyses of realism and rational choice, evolutionary theory provides an ultimate causal explanation for these issues. It explains how warfare contributed to fitness in the course of the evolution of *Homo sapiens*.

Scholars have offered other theoretical approaches to the study of the origin of war: Samuel Huntington described the future of international relations as divided along civilizational lines, with increasing clashes between and among civilizations; J.A. Hobson showed the association between imperialism and war; and Kenneth Waltz's theory of neorealism analyzes the relationship between the distribution of power in international relations and the likelihood of great power war.[1] These and other theoretical approaches to war are valuable because they identify and analyze an important causal mechanism that we need to explain the origins and even the likelihood of war in the specific circumstances the theorist defined. But evolutionary theory's contribution is greater because it answers the fundamental question: Why would *Homo sapiens* fight wars? It explains the purpose of war in our species' distant past.

An Evolutionary Explanation for Warfare

Among topics in international relations, warfare has one of the most extensive literatures. Many have studied the theoretical origins of war; the cause of a particular war, such as World War I; or the conduct and consequences of warfare or specific wars. The literature alone on the Allied campaign against Germany in World War II can fill a small university library. This is appropriate: to prevent future wars we must understand war's causes and the conditions that make it more or less likely. In addition, because warfare has been so crucial throughout history, understanding how wars were conducted and why they ended at specific times are often critical for truly comprehending a nation's past and present policies.

While the origins of war have been usefully analyzed from multiple and often interdisciplinary perspectives, the contribution that evolutionary biology can make toward understanding warfare is not yet adequately recognized. Warfare has been studied from the perspectives of anthropology, psychology, regime type, and the international system, and each generates important insights for understanding this phenomenon.[2] For example, the anthropological scholarship of Harry Holbert Turney-High yields important insights into the motives for warfare in tribal, or "primitive," cultures.[3] He analyzes tribal warfare through the lens of contemporary military principles, including mass, surprise, simplicity of plans, utilization of the terrain, and the command structure of premodern groups.[4] Lawrence Keeley thoroughly analyzes the strategies, tactics, and weapons of prehistoric warfare, revealing as well the acute violence of such conflict.[5]

The psychological study of war begins with Sigmund Freud. Freud interpreted aggressive human behavior as the outcome of a drive that constantly seeks release. The desires to dominate and to aggress are a result of two diametrically opposed instincts: the life instinct (or life force, *Eros*) and the death instinct (*Thanatos*).[6] Freud argued that war results from many motives, "some of which are openly declared and others which are never mentioned," but "a lust for aggression and destruction is certainly among them."[7] Building on Freud's work, Erich Fromm argued that humans are subject to a unique death instinct that leads to pathological forms of aggression beyond those found in animals.[8]

Many scholars working in international relations find the psychological approach useful. Robert Jervis has revealed the importance of perceptions in international relations by explaining how states and other international actors draw inferences about the behavior of others and why this can cause war.[9] Yuen Foong Khong demonstrates the importance of historical analogies, such as the appeasement of Hitler at Munich and the American and Chinese involvement in Korea, for influencing the thought of decision makers and how they frame international political issues.[10]

The type of regimes involved—whether authoritarian or democratic, capitalist or socialist, status quo or revisionist—also have a significant impact on the causes of war. Jack Levy, Jack Snyder, and Stephen Walt have written particularly valuable studies that reveal how specific attributes of states make them more likely to go to war. Levy examines the effect of a state's economic structure and nationalism on the origins of war.[11] Snyder submits that a state's regime type, specifically if it is authoritarian rather than democratic, makes it more susceptible to the belief that it can gain security through expansion.[12] Stephen Walt's comprehensive study of the impact of mass revolutions on war demonstrates how revolutionary ideology contributed to interstate war.[13] In the French, Russian, and Iranian Revolutions, he shows how each state's revolutionary ideology made the revolutionary state seem more threatening to its neighbors, contributing to the outbreak of wars and interventions.

From the systemic perspective, Waltz argues in *Theory of International Politics* that the distribution of power in international relations determines whether great power war is more or less likely.[14] Power was about evenly divided between two poles, or superpowers, during the Cold War, making war less likely in part because it reduced the risk of miscalculation. John Mearsheimer's masterful study of the conduct of international relations explains how the international system forces states to maximize their military power, contributing to intense security competition that can escalate to war.[15] Christopher Layne provides an exceptional analysis of the relationship between the international system and the grand strategy of the United States.[16] How the United States conceives of its interests and the threats to those interests is informed by the international system as well as by the ideology of liberalism, says Layne, and in the future the United States' grand strategy of primacy will intensify the security competition with great power rivals, such as China. Finally, Stephen Van Evera's impressive study of the causes of war examines the impact of the offense-defense balance: is it easier for states to gain territory or defend it? He demonstrates the balance's impact on international relations and the incentives for war when the offense is dominant.[17]

Though the causes of modern war are often intricate, its prevalence throughout human history suggests that it is not caused principally by modern developments, such as imperialism or ideology, although these no doubt contribute to the scope, if not necessarily the intensity, of conflict. As I discuss below, studies of war in the ancient world reveal that it was absolute for those populations involved, and often included massacres of surviving males. So warfare in the past could be no less total than it is today, although clearly modern technology makes it easier to kill more people more efficiently. Thomas Schelling captures this point well with his typically insightful, but unsettling, comment: "Against defenseless people there is not much that nuclear weapons can do that cannot be done with an ice pick."[18]

Evolution provides a perspective not provided by any of those above. It acknowledges that warfare is an ancient human activity, predating historical records. As biologist Paul Ehrlich argues, "war" seems to have been "a major and sanguinary human activity long before our ancestors got civilized."[19] It improves upon the insights of Freud and Fromm, because it explains the origin of war in human history without depending on "lusts" for destruction as a causal mechanism. It also improves upon ethological studies of aggression, such as those of Konrad Lorenz or Robert Ardrey, because it focuses directly on the question of evolution and warfare—demonstrating how warfare contributes to fitness—rather than on instincts for aggression.[20] In response to the war-as-instinct argument, as E.O. Wilson argues, evolutionary theory permits genuine insights into the problem of human aggression, and it "cannot be explained as either a dark-angelic flaw or a bestial instinct."[21]

The evolutionary study of warfare is the only approach that may be described as underdeveloped, despite the contributions of some excellent scholars, such as Robert Bigelow, Peter Corning, Vincent Falger, Azar Gat, Paul Shaw and Yuwa Wong, John Strate, and, most importantly, Johan van der Dennen.[22] While this scholarship is excellent, significant theoretical and empirical work must still be accomplished to demonstrate evolution's contribution to this subject.[23]

Evolutionary theory's major implication for the study of the origin of warfare is that it provides an ultimate causal explanation for warfare. Warfare contributes to fitness in certain evolutionary conditions. To make this argument in this chapter, I explain how warfare could contribute to fitness in the proper circumstances. The essence of my argument is that people wage war to gain and defend resources. Thus evolutionary theory provides a sufficient cause of warfare. Of course other factors cause war: resource-abundant societies start wars, and resource-poor societies may do so for other reasons. However, my analysis focuses solely on the evolutionary origins of warfare. The many other causes of warfare familiar to anthropologists, economists, historians, and political scientists are beyond the scope of my project.

I then test my argument in the next section of the chapter by examining the empirical evidence, testing it against the behavior of past and present tribes as documented by ethnographers, historians, and human ecologists. My argument is subject to verification by examining the actual behavior of tribal societies. The argument may also be disproved if the patterns of conflict in the specific context regularly contradict its expectations. Of course, evolutionary and social explanations may complement each other as we examine any particular war or the behavior of a particular clan, tribe, or state. Indeed, as anthropologists describe the cultural practices of warfare, they often build upon or reinforce evolutionary explanations.

In discussing the empirical evidence offered by primitive warfare, I also examine the three main forms of primitive warfare: the raid, the pitched battle, and the massacre. In this discussion I note the great similarities to modern military practices. The grand strategies and tactics of primitive warfare are familiar to modern militaries and states. At the level of grand strategy, as with modern states, tribes are acutely aware of their own interests and relative power, and those of their neighbors and adversaries. They recognize that their attempts to defeat the enemy quickly and decisively may fail, and so they often prepare for a protracted conflict. They are also fully aware that war aims may change during a conflict, especially if they fail to win a rapid victory, and must be carefully considered before the conflict begins; otherwise internal tensions or debates within the tribe may weaken its ability to prosecute the war or force them to end it before they meet their stated objective. These issues are carefully considered by leaders from Papua New Guinea to Washington, D.C.

Also like states in contemporary international relations, they seek allies, wholly cognizant that their allies may be disloyal and betray their plans or defect either before or during combat. A tribe may have less incentive for combat and may attempt to "buckpass" to its erstwhile allies in an effort to make them pay the costs of deterring the enemy or of warfare should deterrence fail. In addition, tribes recognize, as the United States did during the bipolar Cold War, that their reputations with their allies are important, and this may force them to act against their preferences, for example, by continuing to back a feeble ally.

Finally, in the chapter's conclusion I comment on the importance of resources as a cause of conflict today. Evolutionary theory explains why tribes fight over resources they need for survival and reproduction. Human ecology provides the specific resources over which humans fight. These resources may concern food, such as good fishing or hunting grounds, or good cropland; or the reproduction of the group, such as females or children from neighboring tribes; or better terrain for living that is relatively free of insects, disease, or dangerous animals.

In some respects, the motivation of warfare has changed little. States still wage war for resources, whether they are defined as critical natural resources (such as oil) or territory (which might provide resources or assist in defending the state) or vital geographic points (such as a bend in a river—think of the location of the Tower of London on the Thames—or of an important strait, such as the Strait of Gibraltar).

Warfare and Fitness: General Points

In offering an evolutionary explanation of warfare I encounter directly the impact of the divide between the natural and social sciences that I discussed in the introduction and that has needlessly hindered both. One effect of the divide is that most researchers into the origins of warfare are quite well versed in the cultural or political explanations for its origin, but less so in the evolutionary one. As a result, an evolutionary explanation for warfare may strike many students of warfare as a rather curious, and perhaps even dangerous, explanation for its origins.

Why might it seem a curious explanation? It is difficult to understand, at least prima facie, how war could be the product of evolution: War, after all, involves death and destruction, both of which rather obviously hinder humanity. So how could warfare be the result of a natural process? Warfare seems clearly dysgenic.

This opinion is understandable among social scientists who study war through the lens of culture or politics. Well attuned to the horrific effects of warfare, these scholars can certainly understand war arising from a religious or ideological dispute, a diplomatic blunder, an aggressive military, technological innovation, or tightly interconnected alliances; important scholarship explains how each of these factors, in specific circumstances, may result in warfare. For those not familiar with evolutionary theory, the concept itself seems to suggest life and its growth and spread, and warfare reasonably suggests the opposite. Thus, few in mainstream social science understand evolution's potential to explain warfare. One reason why is that few scholars have understood the causal connection between the two.

This argument might also be considered dangerous because it may seem, again prima facie, to support a social Darwinist argument. Some scholars might assume, incorrectly but understandably, that what is "natural" is "good"; thus an evolutionary explanation for warfare implies that warfare is good or makes individuals or states who engage in it "fitter," more robust, or more virile. This is understandable because social Darwinists made precisely these types of arguments.[24]

For example, Sir Walter Bagehot, writing in the late 1860s, explained how "conquest improved mankind by the intermixture of strengths; the armed truce, which was then called peace, improved them by the competition of training and the constant creation of new power," he continues, "since the long-headed men first drove the short-headed men out of the best land in Europe, all European history has been the history of the superposition of the more military races over the less military."[25] In 1880, German Chief of the General Staff Helmuth von Moltke (the Elder) wrote: "Eternal peace is a dream, and not even a pleasant one. War is a part of God's world order."[26] Shortly before World War I, General Friedrich von Bernhardi published works that continued this theme.[27] He saw war as Darwinian "natural selection" where the strong rule and the weak and "unwholesome" are eliminated.[28] In 1931, Sir Arthur Keith famously and approvingly described warfare as "nature's pruning hook" that eliminated the weak.[29]

While the social Darwinists perverted Charles Darwin's arguments, the very fact that these arguments were used to justify policies that led to the Holocaust and other gross human rights abuses will keep natural and social scientists on guard against social Darwinist arguments. Recognizing the concerns of many social scientists, I suggest we keep in mind three points. The first is the central issue of the chapter: what insights into the origin of war can evolutionary theory provide? That is, could warfare indeed contribute to fitness in humans and the other animals that practice it? To explore this question I formulate an explanation of warfare, determine if evidence exists to support it, and analyze some implications of this argument.

Second, recall from chapter 1 the naturalistic fallacy: one cannot derive an "ought" from an "is." I seek to understand how evolutionary theory might explain warfare. My argument does not concern the "naturalness" or "goodness" of warfare. There is no normative content in this explanation. As Darwin wrote in *The Descent of Man,* "we are not here concerned with hopes or fears, only with the truth as far as our reason permits us to discover it."[30] An evolutionary explanation for warfare can make a valuable contribution to science and to understanding how to prevent or mitigate it, but, again, it cannot be used to justify an "ought," such as a political opinion.

Third, the social Darwinists distorted evolutionary explanations because they misunderstood Darwin's ideas and were ignorant of or consciously chose to ignore the naturalistic fallacy. Those who use evolutionary theory to explain aspects of human behavior must recall the social Darwinists' errors. Doing so makes it possible not only to avoid repeating errors but also to advance scientific understanding.

Having addressed these important general points, I can now present the major arguments of the chapter. There are three sufficient evolutionary explanations for the origin of war in *Homo sapiens.* These are the three subdivisions of natural selection: classical Darwinian, inclusive fitness, and group selection. I will consider classical Darwinian natural selection and inclusive fitness together, as individual selection, because they overlap significantly. Then I will explain how group selection explains warfare. While both individual and group selection offer sufficient explanations for warfare in *Homo sapiens,* individual selection provides a stronger explanation because the empirical evidence better supports it.

Before this discussion, I will describe four specific precursors of an evolutionary explanation for warfare, without which warfare cannot evolve. All four conditions are common among animals. The first is a rudimentary intelligence that can react to certain environmental stimuli. This might be as simple as the chemical cues that motivate ants, or a much more sophisticated intelligence found in chimpanzees and humans. Second, group living is necessary since solitary animals cannot wage warfare. Group living is, of course, common in the animal kingdom because it aids survival of the individual.[31] Third, the species must have a conception of group identity, with the ability to discern who belongs within the group and who is an outsider; of course most animals possess this ability. Fourth, the species must be territorial or willing to defend its resources from outsiders within its own species as well as from other species.

The Evolutionary Explanation for Warfare at the Level of Individual Selection: Classical Darwinian Natural Selection and Inclusive Fitness

In this section of the chapter I will show how the ultimate causation for warfare is anchored in Darwinian natural selection and inclusive fitness. I will present the logic of these explanations and then explain why, in specific conditions, warfare can increase both the absolute and relative fitness of humans.

From the classical Darwinian perspective, warfare contributes to fitness because individuals who wage war successfully are better able to survive and reproduce. In addition, in his inclusive fitness theory, William Hamilton suggests that we measure reproductive success in terms of the individual animal but also of its genetic relatives; thus an individual's self-interest can be served by assisting genetically related individuals.[32] Hamilton's insight implies that groups may go to war to gain resources to help the survival and reproduction of relatives as well as the individual.[33]

Recall from chapter 1 that natural selection is constantly operating in living species. Variation may arise in a random fashion, perhaps through mutation, but the differential survival and reproduction of a beneficial variant will not be random and will enable those who possess it to survive or reproduce more effectively, perhaps at the expense of others who do not possess the variation.

From the perspective of individual selection, the origin of warfare is as an adaptation to a competitive environment. Because resources are almost always limited, competition occurs both within and among species.[34] For the vast majority of their history, humans have faced a shortage of resources. In order to understand this fully, we must recall the environment of the Pliocene, Pleistocene, and most of the Holocene. Two points are significant here.

First, multiple environments actually succeeded one another during this time as many environmental factors changed. As a result, our ancestors had to cope with an environment that was highly variable over time. Indeed, geologists have discovered that the full climatic cycle in the Pleistocene is composed of two major parts, glacial and interglacial, which affected the average surface temperature and sea levels.[35] From the start of the Pleistocene to about 900,000 years ago, a full cycle—from beginning of glacial to end of interglacial—occurred about every 40,000 years. From 900,000 to 450,000 years ago, the average cycle lengthened to about 70,000 years. The last four full cycles have averaged about 100,000 years. Tables 3.1 and 3.2 illustrate the last full climatic cycle of the late Pleistocene and early Holocene and the general climatic conditions found 250,000 to 30,000 years ago. Of course, climatic conditions continue to change in our present time, often with profound results for life, such as extinctions.[36] In addition, ecologists and geologists now recognize significant variations in micro-climates, and in the relative abundance of flora and fauna, over relatively short periods of time.[37] Even over the last thousand years there have been mini-Ice Ages, such as the one between A.D. 1300

Table 3.1. The Last Full Climatic Cycle of the Middle Pleistocene to the Holocene		
Climatic Conditions	*Estimated Years Ago*	*Percentage of Last Cycle*
1. Interglacial	130,000–115,000	12.5
2. Early glacialtemperate/cool	115,000–75,000	33
3. Early glacialcool/glacial	75,000–30,000	37.5
4. Full glacial	30,000–13,000	14
5. Late glacial	13,000–10,000	2.5
6. Postglacial (Beginning of a new interglacial period)	10,000–present	New cycle

Adapted from: Christopher Stringer and Clive Gamble, *In Search of the Neanderthals: Solving the Puzzle of Human Origins* (New York: Thames and Hudson, 1993), Table 19, p. 45. Numbers do not total due to rounding.

to 1800, and periods of warming, such as the Little Climatic Optimum or Medieval Warm Period from A.D. 500 to 1100; these affected human history by permitting or preventing exploration, settlement, and trade, and contributing to the survival of those in subsistence economies.[38]

The second significant point is that humans moved from Africa to other continents and thus encountered different—in places rather dramatically different—local environments. The variations in food availability and competitor species, both proto-hominid and others, in local environments in turn affected human evolution.[39] Potable water, of course, was always critical. Anthropologist Rosemary Newman observed that the early human may be described as a "thirsty, sweaty animal" that probably needed as much as two quarts of water per hour while in pursuit of prey on a hot day.[40] The number of predators also mattered considerably, a point all too easy to forget in our present conditions. Our human ancestors faced a formidable number of carnivores: bears, cheetahs, crocodiles, hyenas, lions, leopards, tigers, wild dogs, wolves, and even eagles and other birds of prey. Humans may be thought of as di-

Table 3.2. Climatic Conditions and Human Evolution 250,000–30,000 Years before Present

Climatic Conditions	*Estimated Years Ago*	*Hominid Evolution in Africa and Eurasia*	*Hominid Evolution in East Asia*
Mainly temperate or cool with some glacial intervals (Large oceans, small ice caps)	250,000–180,000	*Homo erectus, H. neanderthalensis,* and *H. sapiens.* Possibly *H. heidelbergensis* in Asia.	*H. erectus* in Java. *H. erectus* and *H. sapiens* in China.
Full glacial with some milder intervals (Small oceans, large ice caps)	180,000–130,000	All three present. *H. heidelbergensis* either becomes extinct or evolves into *H. sapiens* and/or *H. neanderthalensis.*	Both present. No confirmation of the presence of *H. neanderthalensis.*
Last interglacial, warm conditions (Large oceans, small ice caps)	130,000–115,000	*H. sapiens* present and *H. neanderthalensis* in Europe and perhaps Middle East.	*H. sapiens* in China. Late *H. erectus* in Java?
Temperate/cool	115,000–75,000	Rise of Anatomically Modern Human (A.M.H. or *Homo sapiens sapiens*) in Africa and Middle East. Neanderthals in Europe and possibly in Middle East.	*H. sapiens* in China, possibly Australia and New Guinea. Late *H. erectus* in Java?
Cool/glacial (Ice caps increasing)	75,000–30,000	*H. neanderthalensis* extinct (possibly as late as 40,000–28,000 years ago), A.M.H. in Europe.	A.M.H. in East Asia, Australia, and New Guinea. *H. erectus* extinct.

Adapted from: Paul R. Ehrlich, *Human Natures: Genes, Cultures, and the Human Prospect* (Washington, D.C.: Island Press, 2000), pp. 83, 96–100; and Christopher Stringer and Clive Gamble, *In Search of the Neanderthals: Solving the Puzzle of Human Origins* (New York: Thames and Hudson, 1993), Table 20, p. 47.

minutive apes in our evolutionary past (our earliest proto-hominid ancestors were less than three feet tall) and thus vulnerable to many animals we do not now perceive as threats. Venomous and large constricting snakes, of course, were omnipresent threats as well. As a result, hominid evolution should be regarded as a complicated process of "branches and stages" containing no one simple pathway through which humans evolved.[41]

The process of evolution is sometimes described metaphorically as a corridor: the ancient ancestor of the great apes and Man enters at one end and *Homo sapiens* emerges at the other. Rather than conceiving of it as a long corridor, it is better to think of it as the combination and interplay of chance, what was favored by natural selection, and how humans coevolved with other species. In addition—and significantly—evolution produced multiple forms of humanity, but these other humans, part of our "family" if you will, died out. In our species, only *Homo sapiens sapiens* successfully adapted to all its environments in the past; *Homo erectus* and *Homo neanderthalensis* became extinct.

Now we are in a position to address the question of why warfare would begin. To answer this question it is important to consider what lengths people will go to assist their relatives and what purpose war would have served in the environment of evolutionary adaptation. The answer is that natural selection caused humans to have a strong instinct to preserve themselves, related individuals, and members of the tribe. Although it may not be pleasant to think of warfare in this way, it is one mechanism that permits preservation in the right circumstances. Successful warfare would serve two purposes: it would increase the absolute fitness of a group by providing them more resources and it might also increase their relative fitness. To answer the question "why would warfare begin," evolutionary theory suggests that warfare is an egoist's solution to the problem of gaining resources.

Warfare to Increase Absolute Fitness

An ultimate causal explanation for warfare based in evolutionary theory begins with the recognition that warfare contributes to fitness in certain circumstances because successful warfare lets the winner acquire resources. For evolutionary biology, a resource is any material substance that has the potential to increase the individual's ability to survive or reproduce. As such it may be food, shelter, or territory, especially high-quality soil or wild foods; abundant firewood; or territory free of dangerous animals, such as lions, or insect infestations, or disease; and also status, coalition allies, and members of the opposite sex.[42] These resources are necessary to maintain a healthy life for oneself and one's family, and these desires are the strongest feelings one possesses. Of course, humans are not unique in this respect. All animals share the desire or instinct for self-preservation, and many have strong feelings for their offspring.

What an evolutionary perspective allows students of warfare to understand is that the origins of warfare and the functions of warfare are interconnected. The origins of warfare are rooted in the requirement to gain and defend resources necessary for survival and reproductive success in dangerous and competitive conditions. Competition for resources may result in situations where consumption by one individual or group diminishes the amount available for others, or where resources are controlled by one individual or group who determines distribution and thus can deny them to others.

The origins of warfare are grounded in one's egoism—the human desire to gain or defend the resources needed to feed and protect a family, other relatives, and then one's group. These desires or emotions (as traits) evolved because they contributed to human survival and reproductive success. My argument is not that they necessarily evolved specifically for warfare; however, if they did not, they and other emotions, such as hatred, certainly could have contributed to the origins of warfare in *Homo sapiens* and perhaps also in *Pan troglodytes,* the chimpanzee.

From the Pliocene until very recent history many resources were scarce for all humans.[43] Indeed, people in many parts of the world today face shortages of what may be considered basic resources, such as land and clean water. Scarcity may take two forms. Resources may be rare or hard to come by, or they may be plentiful enough but access to them is controlled, perhaps by a hostile tribe.

Any group facing a shortage of resources may adopt one or a combination of three basic strategies. The first strategy is the group eliminates or reduces consumption to make the resource last. This is clearly less acceptable for food and water than for some other resources. Second, the group can seek an alternative for the resource, perhaps through technological innovation. Again, this may be more of an option with some resources than others. Third, they can acquire more of the resource from outside of their territory through migration to uninhabited areas, trade, theft, or warfare.

Although warfare is certainly costly in terms of those members of a tribe killed and wounded as well as the resources and time expended, it might become the sole choice if migration is impossible due to unsuitable terrain or hostile neighboring groups, and where trade is also not possible, perhaps due to the nature of the resource. For example, a group seeking a fixed source of water may be unable to trade for it, due either to an unacceptable price from those in control or a lack of any resource to offer in exchange. Warfare might be necessary then for offensive purposes, to plunder resources from others.

In these circumstances, an individual becomes fitter if he can successfully attack to take the resources of others.[44] But warfare also may be waged for defensive reasons, such as to defend resources from the advances of others.[45] E.O. Wilson captures these points succinctly, saying that humans would fight wars "when they and their closest relatives stand to gain long-term reproductive success," and he continues, "despite appearances to the contrary, warfare may be just one example of the rule that cultural practices are generally adaptive in a Darwinian sense."[46] Evolutionary theory, then, expects that humans would possess specific behavioral traits that contributed to fitness in the past, such as a desire to fight to gain the resources necessary so that the individual, his family, and extended family group would continue to survive, or to defend those resources from others.

Of course, success in attack is crucial. Bobbi Low captures this point well: "potentially lethal conflict is only likely when the possible reproductive rewards—mates, status, resources for mates—are high."[47] If a tribe did not think that a successful attack were likely, it would refrain from so doing unless dire conditions forced it to defend itself against even an obviously superior force. Living in groups consisting of extended families obviously increases the chance of a successful attack. It also reduces the risk of injury to any specific animal, and provides a greater chance of success in defense against an attack. These benefits are important enough to overcome the costs of group living, such as the greater likelihood that disease or parasites will spread.

Indeed, a major figure in evolutionary theory, Richard Alexander, argues that humans came to live in groups for protection from external attack, to create what he terms a "balance of power" against attack. This process occurred in three stages: first, humans lived in small, polygynous bands for protection against large predators; second, as time passed they continued to live in similar bands to defend against predators and to bring down large game; in the third stage, these bands stayed together to protect themselves against other groups of humans located nearby.[48] In his insightful study of warfare, Peter Corning makes a related argument.[49] Both agree that the origins of human sociability might be anchored in our need for protection.[50]

As is often the case, we see that Darwin anticipated this argument. In *The Descent of Man,* Darwin describes how more resources would accumulate to more powerful individuals: "The strongest and most vigorous men,—those who could best defend and hunt for their families, and during later times the chiefs or head-men,—those who were provided with the best weapons and possessed the most property, such as a larger number of dogs or other animals, would have succeeded in rearing a greater average number of offspring, than would the weaker, poorer and lower members of the same tribes."[51]

In addition, Darwin explained how one group might conquer others through a technological advance that was then emulated: "Now, if some one man in a tribe, more sagacious than the others, invented a new snare or weapon, or other means of attack or defence, the plainest self-interest, without the assistance of much reasoning power, would prompt the other members to imitate him"; he continues, "if the invention were an important one, the tribe would increase in number, spread, and supplant other tribes."[52]

Evolutionary theorists William Durham and Wilson build on Darwin's thinking, arguing that aggression is anchored in the need for resources.[53] Durham fully develops this argument, and submits that intergroup aggression develops as a behavioral adaptation to conditions of scarcity and the competition for resources.[54] War is one means by which individuals "may improve the material conditions of their lives and thereby increase their ability to survive and reproduce."[55]

Of course, an individual or group must be able to wage defensive war when competitors threaten his or the group's resources. But Durham argues that offensive war is a key solution to the problem of scarce resources. His research confirms that a group can expand its resource base by aggressively seizing resources from other groups.[56] Pressure might be particularly acute if population size is increasing faster than resources. A group may wage war to gain access to fixed resources, such as a spring or other watering hole, or an area with fewer predators. Certainly, as sociologist Peter Meyer argues, such resources may have been the only types of territory especially valuable before the advent of food production, which greatly increased the value of arable land.[57] Second, people may wage war to gain tribute or human labor power from another group. Third, a tribe may wage war to gain revenge or to enhance the position of decision makers within the tribe. Fourth, a war may occur to preempt an attack by another tribe; that is, warfare may be waged for defensive as well as offensive purposes.

Finally, the scholarship of ethnographer Andrew Vayda is particularly important for understanding how population ecology affects the availability of resources, which in turn affects the causes of warfare. In particular, Vayda argues that knowing whether a population is waxing or waning is important for understanding why it would wage war because the motivations for warfare are different in either condition.[58] An expanding population living in premodern conditions and based on a swidden, or slash-and-burn, agricultural economy faces serious constraints on its ability to produce food.[59] Thus successful warfare would help the tribe gain resources, and for a swidden agricultural economy land is critically important. But warfare might provide another benefit for tribes in these conditions: the war itself may kill enough people to reduce the population pressure on scare resources. In addition, warfare may also serve decreasing or stable populations by allowing them to adjust their male-female ratios by capturing women or children from other tribes.[60]

Once begun, it is likely that warfare would spread because behavior that increased military efficiency would be made uniform in a region for two reasons. First, the less efficient tribes would be eliminated through attrition, the dispersal of survivors, or absorption into the victorious tribe. The second reason is familiar to students of international relations and was advanced forcefully by Waltz.[61] Just as states have an incentive to socialize or emulate successful states in contemporary international relations lest they fall behind and jeopardize their security, so too primitive tribes would have had great incentives to emulate the successful tribe if they could do so. If they could not, then migration to escape a more efficient tribe might be their only hope for survival.

Warfare to Increase Relative Fitness

In addition to increasing fitness in absolute terms, fitness might be improved if a group could increase its relative access to resources. An individual may gain by reproducing more successfully than others. This may simply involve reproducing in larger numbers than any rival. But it may also result from denying resources to rivals in order to hinder their reproductive success. This in turn would allow one group to reproduce more successfully than its rivals without necessarily producing many offspring, provided the costs of such behavior were sufficiently low. Clearly a group that could hinder successful reproduction by its rivals while maximizing its own reproduction could become dominant relatively quickly. Thus warfare might be waged to improve one's

Table 3.3. Type and Motivation of Warfare		
	TYPE *Offensive Warfare*	*TYPE* *Defensive Warfare*
MOTIVATION *Warfare for Absolute Gain of Resources*	Warfare conducted to gain resources for group.	Warfare may be necessary for protection of resources.
MOTIVATION *Warfare for Relative Gain of Resources*	Warfare conducted to gain resources or hinder enemy's ability to reproduce by denying resources.	Warfare may be necessary for the protection of present or future resources.

own resources but also to deny them to other groups of humans or animals. This is the concept of warfare for relative gain.

Table 3.3 illustrates how individual selection provides ultimate causation for warfare and how warfare among humans occurs in types, offensive and defensive warfare, and is waged to bring about an absolute or relative advantage over foes. It is important to stress that warfare might also be waged for defensive purposes. One example is a preemptive attack to thwart the enemy's plan, like the Israeli attack against its Arab neighbors in 1967.

This type of behavior is found in the animal kingdom and among humans, as I discuss in detail below. An individual animal may successfully disrupt the efforts of others to breed or raise young through nest destruction, infanticide, or direct attacks on adult rivals in order to reduce the number of competitors. In warfare, tribes kill enemy children and steal females to support their own reproduction and to hinder the recovery of the enemy. They also destroy the enemy's crops, animals, and food stores, and strive to displace the enemy from particularly valuable territory.

While individual selection suggests the circumstances in which warfare would assist fitness, it also suggests conditions in which more peaceful relations between groups may be expected. First, if resources are abundant or alternatives can be acquired at an acceptable cost, then groups need not compete for them. Second, if the demand for resources is held in check by other factors, such as a high mortality rate due to disease, parasites, or predators, and if a resource is widely distributed spatially or temporally, it may require either the reciprocal sharing of resources, or migration, in which case competing groups are unlikely to come into contact.[62] For example, the absence of war reported among the Netsilik Inuit in the Arctic may be due to the low population densities and great regional variation in food supply that requires migration, as well as the sharing of food resources, which is common to this culture, and the relative ease of food storage in arctic conditions.[63] In his insightful analyses of peaceful societies, van der Dennen finds that peaceful societies are "essentially small, local, face-to-face communities" and "do not maintain an exclusive monopoly over an area of land."[64]

Unfortunately, these circumstances are rare. Only infrequently are resources so abundant or adequate alternatives so available that tensions do not arise, population densities so low for generation after generation, or geographic conditions so propitious that groups can cooperate over resources consistently. Alexander writes of those who "interpret as relatively nonaggressive behavior on the parts of the hunting and gathering societies that remain today in a few places like the Australian . . . [desert] and the Arctic"; they argue that for "99 percent of their history our ancestors lived as these people do."[65] But Alexander argues that their argument fails to recognize that "such people survive today only in marginal impoverished habitats that support only the lowest of all densities of human population and also represent physical extremes that by themselves require cooperation among families for mere survival."[66] In addition, he argues, "hunter-gatherers survive today only because even the most advanced technological societies have found no way to use their homelands that would make it profitable to overrun or seize them by force."[67]

Group Selection's Explanation for Warfare

Just as individual selection provides an ultimate causal explanation for the origins of warfare, so too does the third subdivision of natural selection, group selection. Group selection helps us understand that war evolved through a process that favors the self-sacrificing tendencies of some warriors.[69] This may seem odd, but according to some evolutionary theorists altruistic or self-sacrificing attributes can evolve through natural selection even though they may appear to cause more individual costs than benefits.[70]

From the perspective of individual selection, individuals may be willing to fight against individuals of other tribes for the sake of themselves or their relatives. Fighting for the tribe, either aggressing or defending, would be fighting to preserve the family and the warrior's genes. But individuals, in general, would not fight absent these circumstances.

Evolutionary theorist David Sloan Wilson makes an argument for group selection that suggests why they would. According to Wilson, group selection is the component of natural selection that operates on the differential productivity of local populations within a global population.[71] That is, individual selection favors traits that maximize fitness within single groups, but group selection favors traits that maximize the relative fitness of groups.[72]

As human groups grew to include ever more distant relatives and unrelated individuals, wars to aggress and to defend the population were still necessary—so warriors were still needed. Some of these individuals were more willing to sacrifice for the group. But for this to happen, recall from chapter 1 that four conditions are necessary for group selection to obtain: (1) more than one group has to be involved; (2) because of variance among groups for a particular trait, some groups must have more altruists than others; (3) there has to be a direct relationship between the proportion of altruists in the group and the offspring of the group; and (4) the altruists have to reproduce and mix with other groups to create new groups.[73]

In war, the group that prevailed could conquer others because it had more warriors. It had more warriors because the population had more altruists. Because the warriors reproduced, their genotype continued, though it may have declined relative to the other members of the group during episodes of warfare. Thus group selection explains how a willingness to sacrifice contributes to warfare.[74]

Darwin also anticipated a group selection argument for warfare. As he explains in *Descent of Man*, "When two tribes of primeval man, living in the same country, come into competition, if the one tribe included (other circumstances being equal) a greater number of courageous, sympathetic, and faithful members, who were always ready to warn each other of danger, to aid and defend each other," as a result, "this tribe would without doubt succeed best and conquer the other."[75] Darwin concludes, "A tribe possessing the above qualities in a high degree would spread and be victorious over other tribes; but in the course of time would, judging from all past history, be overcome by some other and still more highly endowed tribe."[76]

E.O. Wilson further develops this insight in explaining the evolution of warfare. It might evolve as a result of "a process of group selection that favored the self-sacrificing tendencies of some warriors"; that is, the warriors of the tribe would fight battles "for the good of the group and . . . not therefore expect net benefits for themselves and their immediate kin."[77] Wilson agrees with Darwin's explanation of how this behavior might spread: a tribe might prevail if it were able "to expand by increasing the absolute number of its altruistic warriors, even though this genetic type declined relative to the other members of the tribe during episodes of warfare."[78]

According to Wilson, warfare may have begun when one group of early humans considered "the significance of adjacent social groups and [how] to deal with them in an intelligent, organized fashion. A band might then dispose of a neighboring band, appropriate its territory, and increase its own genetic representation in the metapopulation."[79] Furthermore, this band would retain the memory of the event and by repeating it would increase its control of resources.[80] The victories of the original band "might propel the spread of the genes through the genetic constitution of the metapopulation. Once begun, such a mutual reinforcement could be irreversible."[81]

In this chapter, I will not test group selection against the empirical evidence, although it is important for any study that seeks to bridge the gap between human evolution and social science to make explicit the logic of group selection. I have made this decision for two reasons. First, as the above discussion has shown, theoretically group selection can offer an explanation of the origins of warfare but individual selection offers a more logical explanation of the empirical evidence about why tribes fight. Second, individual selection also more parsimoniously explains the origins of warfare. This is because group selection depends on a mechanism of inheritance that first accepts individual selection but then requires an additional mechanism of inheritance not yet identified that would consistently, rather than sporadically, account for selection above the individual level.[82] Supporters of group selection acknowledge that group mutations do not occur spontaneously, and thus all mutations must spread through a population by way of individual selection and so must be advantageous to the individual, at least initially. Again, I do not suggest that group selection could not happen, but even were it to occur, it is not a consistent evolutionary force and its explanation is not sufficiently supported by the evidence I examine here.

Evidence from Human History and Tribal Societies

In this section of the chapter, I provide the empirical evidence on the causes of tribal wars, also known as pre-state or primitive wars.[83] I am examining only one set of the origins of warfare, not the universe of its causes. Warfare in tribal societies has multiple causes, as it does in industrial societies. These include accusations of sorcery, desire for prestige, and other cultural causes that are beyond the scope of my study.

The methodology I use in this analysis is straightforward. Individual selection suggests that tribes should fight to gain resources in conditions of scarcity. So I examine data from tribal warfare to determine if this is true. In order to separate ultimate material causes of warfare from the cultural or personal motives involved it is important, first, to use quantitative studies of primitive warfare across cultures and qualitative analyses of the causes of warfare for specific clans or tribes to document the argument. These qualitative studies have been conducted by anthropologists who possess the most detailed knowledge of the specific tribal society examined. They can expertly analyze the motivations for warfare within a particular clan or tribe and judge whether or not the group does indeed benefit from a successful campaign.

Second, in the analysis of evidence I also use a broad regional approach and study warfare in Melanesia, particularly New Guinea, as well as East Africa, and North and South America in order to determine if tribes on different continents wage war for similar reasons. Here, I am following the approach used by anthropologists Aubrey Cannon and Brian Ferguson, who stress that the advantage of studying a regional war pattern for a specific time is that it reveals who initiates the attack, who is attacked, and the intensity of the conflict, in order to ascertain if it matches expectations of material need.[84] As Ferguson argues, "if material considerations do account for the observed patterns . . . the most parsimonious explanation would be that the participants themselves are acting on these considerations," while other explanations would be also considered as causes if they produce fighting in the absence of material need.[85]

Using this methodology, I will be able to reach the conclusions of this study. Moreover, in this section of the chapter, I will briefly discuss as well the forms of primitive war, its common strategies and tactics, and the often terrific casualty rates of primitive warfare.

Introductory Points

As I begin to analyze the evidence I want to stress four points. First, we have no evidence that people in primitive societies engage in warfare because they do not appreciate its costs. Indeed, as I show below, in many tribes each generation knows its costs all too well. These societies often lament a war's beginning and mourn those killed. In sum, they know well the truth attributed to

Arthur Wellesley, later the Duke of Wellington, after the Battle of Waterloo: "next to a battle lost, the greatest misery is a battle gained."[86] Moreover, they appreciate another cost: Even victorious warfare can generate generations of revenge and feuding with those defeated. They know this as well as the French and Germans or the Israelis and Palestinians.

Second, we have no evidence to support an argument that primitive warriors are innately more aggressive or savage than individuals living in industrialized states. Indeed, apparently it would be hard to be more bellicose than European states, as Pitirim Sorokin revealed in his classic and monumental analysis of the histories of eleven European countries over periods that ranged from 275 to 1,025 years. He found that they were engaged in military action 47 percent of the time, or about one year out of two.[87] The United States, both during and since the Cold War, has been engaged in the use of force each year. So such an opinion about primitive societies is simply not correct.

Third, to test appropriately the argument advanced in this chapter, more scholarship by ethnographers and historians would be helpful, although some excellent cross-cultural studies and detailed ethnographic works of specific tribes already exist. These provide information about the resources available to these groups, their demographic, economic, and social conditions, and their explicit motivations for going to war. As this area of research becomes better delineated, and scholars recognize how useful evolutionary biology can be for social science, more researchers in anthropology, history, and related disciplines may begin to analyze the actions and relations between indigenous groups through this perspective.

Fourth, I do not want to imply that the causes of warfare identified by evolutionary theory are the sole reasons these tribes would go to war. Other causes of war, familiar to theorists of international relations, also affect these tribes. For example, the Yanomamö tribe (also called Yanomami, Yanomama, and Yanoama) of Brazil and Venezuela fully recognizes the horrors of warfare, and some of its members have stated to ethnographers: "We are tired of fighting. We don't want to kill anymore. But the others are treacherous and cannot be trusted."[88] Theorists of international relations recognize this fear among the Yanomamö as the security dilemma, which prevails under conditions of anarchy, conditions in which this tribe and many others often live. Briefly stated, the security dilemma is a condition in international relations in which one tribe or state arms to protect itself from threats in a dangerous world; neighbors (again either tribes or states) observe this and arm in turn, again for defensive reasons, thus reconfirming in the mind of the original party the hostile intentions of its neighbors.[89] Consequently, one party arms to become secure and those efforts defeat themselves. As a result of arming, everyone is better armed but no more secure.

In premodern societies, warfare is often motivated by the need to gain and protect the resources necessary to support life and reproduction, and at times to deny these resources to adversaries. Indeed, as I will show in the discussion of animal warfare in chapter 4, primitive human warfare is remarkable in its shared motivation with warfare found elsewhere in the animal kingdom. For *Homo sapiens,* however, we do not know when warfare began: was it known to the common ancestor we share with the chimpanzees, making it always part of our behavioral repertoire? Think of the scene in Stanley Kubrick's classic film adaptation of Arthur C. Clarke's *2001: A Space Odyssey,* in which a chimpanzee-like human ancestor turns an animal bone into a deadly weapon during a territorial display at a watering hole. Was it present in the other members of our family *Homo erectus* and *Homo neanderthalensis*? At this point, there is insufficient efficient evidence to answer these important questions.

Warfare in the Pliocene or Pleistocene epochs cannot be documented, but it can be in the Holocene in different parts of the world. From 10,000 B.C. to 7500 B.C. there is suspicion of warfare, and some evidence of it is present at the Shanidar Cave in Iraq and in the Levant at Erq el-Ahmar, Eynan, Mugharet El-Wad, Mugharet Kebara, Nahel Oren, and Shukba.[90] Archeologists generally accept that the earliest evidence for warfare is found in the massive fortified walls of Jericho in approximately 7500 B.C. at a site called Pre-Pottery Neolithic A Jericho.[91] Although this position is not without dissent, the evidence for warfare is based on the excavation of a wall and tower at the site, which were "purely military in planning."[92] Archaeologist James Mellaart says the defenses of Jericho are "thought to have been built against the people who finally took over the site Pre-Pottery Neolithic B, and there is evidence that they were already in Palestine at this period." He continues, "evidently their appearance was a threat to this town of Jericho, for no one builds walls this size against marauding bedouin."[93]

In the seventh millennium, evidence of warfare is found at more sites. Marilyn Keyes Roper documents evidence of warfare at Ras Shamra in modern Syria, and Çatal Hüyük East on the Konya Plain in modern Turkey.[94] In the sixth millennium, we have evidence of warfare at Tell es-Sawwan, in contemporary Iraq, and Hacilar in Turkey, as well as additional evidence of warfare at Ras Shamra.[95]

The earliest indications of warfare in China date from about 5000 B.C. This is during the early Banpo period that marks the beginning of an organized political system in that country.[96] Evidence from the Longshan period (3000 B.C. to 2000 B.C.) includes abundant evidence of warfare, including sophisticated structures, weapons, and forensic evidence.[97]

In the New World, the earliest evidence of war dates from the Andean region, followed much later by evidence from the Gulf Coast and Oaxaca Valley in Mesoamerica, if the evidence from the Kennewick Man (discussed below) is not considered. Warfare in the Andes begins by 3500 B.C. or 3000 B.C., long before the rise of the Incan Empire (ca. A.D. 1400 to 1533).[98] Along the Gulf Coast, warfare probably originated between 1150 B.C. to 900 B.C. and at approximately the same time among the tribes of the Oaxaca Valley, more than two thousand years before the Spanish conquest.[99]

Evidence in Support of Individual Selection

While these archaeological sites document the fact that warfare was present at the beginnings of civilization, we must examine the behavior of primitive societies in order to determine how evolutionary theory contributes to the origins of warfare. These tribes offer particularly valuable evidence for three reasons. First, many solid historical accounts of their behavior survive, as documented by those who were often the first anthropologists: early explorers, travelers, missionaries, and colonial or state officials. In addition, excellent ethnographic studies were often conducted of them, offering considerable insight into their behavior. Second, many of these tribes, such as the Mae Enga, Mendi, Maring, and other tribes of Papua New Guinea and Irian Jaya, had little or no contact with colonial administrations before ethnographers were able to study them, so modern societies probably had no chance to contaminate or significantly change their behavior. Finally, all are pre-state, rudimentary societies and thus are characteristic of the majority of human history. The way they behave must largely coincide with the way *Homo sapiens sapiens* lived in late-Pleistocene and Holocene conditions before the rise of modern civilizations.

Data from statistical studies of tribal societies support three arguments. First, almost all societies have a record of warfare. In a review of primitive societies, Keeley found that 90 to 95 percent had some past or present record of warfare.[100] Augmenting Keeley's overview of the history of conflict are four independent cross-cultural surveys of conflict in tribal societies and one study specifically of Indian tribes in western North America. These studies yield a consistent result: almost all tribal societies for whom data are available have a record of warfare. Quincy Wright demonstrated that of 216 hunting societies, 198 (92 percent) had warfare.[101] Anthropologist Keith Otterbein studied warfare among randomly chosen societies and found that only four out of fifty had no record of warfare, and of nine tribal societies seven (78 percent) fought wars.[102] In a study of thirty-one hunter-gatherer societies, Carol Ember found that twenty, or 65 percent, engaged in warfare at least once every two years, and eight, or 26 percent, did so less frequently, while only three, or about 10 percent, went to war rarely or had no record of it.[103] In a cross-cultural study of ninety tribal societies, Marc Ross found that only twelve, or 13 percent, engaged in warfare rarely or never.[104] Finally, Joseph Jorgensen comprehensively surveyed Indian tribes of western North America and concluded that only twenty-one, or 13 percent, of the 157 groups surveyed "never or rarely" raided more than once a year.[105] Of these twenty-one tribes, fourteen gave other evidence of having conducted or resisted occasional raids once every few years. Based on Jorgensen's study, only seven societies, or 4.5 percent, had no record of conflict.

Second, tribal societies go to war frequently.[106] According to Otterbein's study, 66 percent of the pre-state societies were at war every year and of nine such societies, seven (78 percent) fought wars, and Ross estimates that about 62 percent of pre-state societies fought that frequently.[107] These data are confirmed by Ember and Melvin Ember's 1992 study that found that 73 percent of forty-nine preindustrial societies went to war at least once every two years.[108] These societies fight more frequently than modern nation-states. From 1800 to 1945, the major European states—France, Great Britain, Germany, and Russia/the Soviet Union—were at war, according to Keeley's data, about one year in every five.[109]

Third, strong evidence shows that premodern tribes go to war to gain resources such as land, food, or mates. Cross-cultural anthropological studies confirm that warfare for resources lies at the root of warfare in most traditional societies.[110] Ember and Ember examine 186 mostly preindustrial societies and find that they go to war to take resources from adversaries, especially when the society fears a natural disaster.[111] They also conclude that the victors almost always—in 90 percent of sixty-two cases and 85 percent of thirty-nine cases—take resources from the defeated.[112]

Otterbein found that four of fifty societies did not fight war and three fought only for defense; nineteen fought for "plunder," to gain resources, as well as defense; fifteen fought for prestige (cultural honors such as trophies), plunder, and defense; and nine for political control, prestige, plunder, and defense.[113] The four societies that did not engage in war were driven from other areas and forced to seek refuge, so they probably should be considered refugees rather than pacific societies.[114] Raoul Narroll studied warfare among forty-three pre-state societies and found results similar to Otterbein's: nine societies fought for defense; six for plunder and defense; fifteen for prestige, plunder, and defense; and thirteen for control, prestige, plunder, and defense.[115] A study of forty-two similarly geographically diverse forager societies by anthropologists Joseph Manson and Richard Wrangham revealed that all the societies for which data were available (twenty-five of forty-two) fought to gain women or resources.[116]

In addition to statistical, cross-cultural studies, it is important to analyze the causes of warfare from historical and ethnographic accounts—because these accounts can avoid the sampling and coding biases of the statistical studies. This accuracy, in turn, is a result of the ethnographer's often exceptionally detailed study of the particular tribe's motivations for warfare, as well as the ethnographer's familiarity with neighboring tribes. For almost all of these accounts, the ethnographer lived with the tribe for many years, and thus possesses the most detailed knowledge possible about the tribe.

Inevitably, however, this record is not complete because the published material provides uneven coverage; for example, some wars are simply mentioned without elaboration, and for many tribes no detailed ethnographic studies exist that would explain the specific circumstances underlying their reasons for going to war. Despite these limitations, the detailed ethnographic studies cited below provide excellent documentation of what motivates humans living in primitive conditions to engage in warfare. To continue this analysis, in the next sections I provide data that explain the causes of warfare in tribal societies for the following world regions: Melanesia and Polynesia, North America, South America, and Africa.

Melanesia and Polynesia

Studying the causes of war among the Mae Enga of the western highlands of New Guinea, anthropologist Mervyn Meggitt found that warfare, especially for women or for resources such as land or food, is common for the Mae Enga and typical for New Guinean tribes.[117] He concludes that there were at least eighty-four conflicts among the Mae Enga between 1900 and 1950; of these, he could discern the cause of seventy-three, and almost all were conflicts over resources.[118] In order of frequency, the conflicts arose over land (forty-one, or 56.2 percent of reported causes of conflicts), theft of pigs (twelve, or 16.4 percent), homicides (eight, or 11 percent), failure to pay homicide compensation (three, or 4.1 percent), stealing pandanus nuts (three, or 4.1 percent), and once, or 1.4 percent, for each of the following: felling the trees of another group, theft of garden produce, rape, and the jilting of a suitor.[119] In addition, Meggitt's Mae informants listed other reasons for conflict: the destruction of crops by another clan's pigs, the theft of dogs or fowl, failure to meet bride price obligations, the removal or destruction of magical plants, and accusations of sorcery.[120]

Among the Mae Enga, land is a particularly valuable resource; not surprisingly it was the immediate source of well over half of these conflicts. Meggitt summarizes how the Mae Enga often believe warfare will help them acquire territory: "For 34 of these conflicts I have reasonable knowledge of the outcome. In only nine cases (26 percent) was the result a standoff in which the aggressors achieved none of their aims," while in "19 (55.9 percent) of these wars the invaders managed to secure at least a portion of their opponents' territory."[121] Finally, "the remaining six (17.6 percent) collisions led to the complete eviction of the losers and the seizure of their entire territory."[122] Indeed, the "Mae groups that feel themselves to be in need of more land to maintain their members have good reason to believe that violence is a fairly effective means of acquiring territory," because "the initiation of warfare generally . . . pays off for the aggressors," so "it is not surprising that by and large the Mae count warfare as well worth the cost in human casualties."[123]

When he conducted additional research with the Mae between 1961–1973, Meggitt found that land was an even greater source of warfare. During the latter period, land was responsible for forty-four (73.3 percent) of sixty conflicts. Other property issues caused nine (or 15 percent), homicide yielded four (6.6 percent), and other causes resulted in the remaining three (or 5.5 percent).[124]

Of course, like other warriors the Mae attempt to keep war casualties as low as possible. Meggitt emphasizes that the decision to go to war "is not taken lightly, for people are well aware of the discomfort and domestic dislocation that accompany mobilization, of the penalties that follow defeat, and the risk, win or lose, of significant casualties."[125] One mechanism to reduce casualties is to aggress only against the weakest clan: "by covertly shifting their boundary markers a few yards each time they bring a new garden into cultivation, they continually gnaw away at the fallow of the weakest agnatically unrelated clan that adjoins their own," and "because they are likely to have singled out for this treatment a clan considerably smaller than their own, there is a good chance that the victims will hesitate to push the matter to the point of violence for fear of provoking a full-scale invasion."[126] This state of affairs may last for years until "either the victims can no longer tolerate the losses and are forced to fight, or the usurpers, believing the odds favor them heavily, seize on any incident as a *casus belli* and attack."[127]

Another mechanism to gain land is to enter into an alliance with another clan to divide between them the territory of an adjoining, and often related, clan. *Realpolitik* is stronger than blood in these cases. When a clan is attacked on two fronts it is more likely to be routed before it has a chance to call upon allies to help stop the attack. Of course, as is often true for allied states at a war's end, the victors often have difficulty dividing the spoils of conflict, and this can easily lead to further tensions between them.[128]

Reflecting upon the frequency of warfare after contact, Meggitt states that the central Enga "are still conducting warfare on much the same scale as they did in the past," but "the main difference between past and present warfare concerns the duration and outcome of engagements" which is "directly attributable to the intervention of Administration forces."[129] In the past, before Western contact, "war between clans had its own 'natural' course of development, leading either to a stalemate, or, more commonly, to a partial or complete victory for the aggressors. . . . Nowadays, depending on the number of combatants and of officers and police, outbreaks of fighting are usually contained within one to four days."[130]

Turning now to another part of New Guinea, Vayda studied the Maring who live in the Bismarck Mountains of eastern Papua New Guinea. He emphasizes the importance of resources as a cause of war for this tribe. Of thirty-nine cases of warfare identified by his informants, murder or attempted murder was the most common cause, followed by "poaching, theft of crops, and territorial encroachment; sorcery or sorcery accusations; abducting women or receiving women who had eloped; rape; and insult."[131] Vayda also notes the importance of population size and ecological motivations for warfare among the Maring, including the value of cleared, crop-bearing land.[132]

Vayda's classic studies of the Iban of Borneo (Sarawak) and the Maori of New Zealand are particularly important for demonstrating the close relationship between population size and warfare in premodern societies. Specifically, he explains how striving for resources, population growth, migration, or the aggression of groups might set off a chain reaction in which "Group A" might expand into "the contiguous territory of Group B," which in turn "might lead Group B to expand into the contiguous territory of group C—and so forth until finally there would be a displacement of a group having territory contiguous to virgin land."[133] This might be dangerous for displaced tribes because they might be deprived of aid from contiguous and related tribes.

For a tribe to expand this way, waging war over the relatively scarce zones of secondary forests, even though primary or virgin forests are available, is actually beneficial: they see the costs of such warfare as lower than the costs of clearing land.[134] Vayda reports that clearing virgin land in the humid tropics requires great effort; given how hard it is even today with steel or iron tools, it must have been extremely hard with wooden or stone tools.[135]

The example of the Iban that Vayda provides "shows the warlike extension of territory as a means whereby a group can *avoid* experiencing any very great privations due to the pressure of population upon available resources."[136] For the Maori too, conquering neighboring groups was easier than expanding into new areas to cultivate resources. "If the time and effort required for clearing new virgin land were considerably more than were necessary for . . . conquest and the preparation of previously used land for cultivation, it follows that territorial conquests, such as some of those recorded in Maori traditional history, would have added more efficiently to the prosperity of particular groups than would peaceful dispersion."[137]

Ethnographer George Morren argues that among highland tribes in Papua New Guinea and the Irian Jaya province of Indonesia, specifically the Mountain Ok tribes, a great source of warfare is the desire for revenge because of "the murder or putative murder (by sorcery) of a group member by members of a neighboring group."[138] Other significant motivators are "pig stealing, interfering with women (unsanctioned fraternization, elopement, rape), or land claims based on putative ancestral tenure."[139]

Morren also argues that warfare in New Guinea is often at its worst because of the specific terrain, the fringe between the highlands and lowlands. In general, these tribes face major problems, such as "maintaining extensive territories for hunting by aggressively staving off expansive neighbors, facilitating access to new territories through extensive (often unilateral) raiding and genocide, and meeting more specific needs through plunder, abduction, and cannibalism."[140] These fringe tribes, specifically the Biami, Jalé, Kukukuku, and Miyanmin, are probably the most aggressive in New Guinea. Morren sees these tribes as "locked in a continuous cycle of defensive warfare with their higher altitude neighbors, while at the same time engaging in more sporadic, diffused, and smaller scale predatory warfare against weaker neighbors."[141] Finally, he recounts how tribes will cooperate with larger tribal groups in order to gain resources. The Miyanmin was one such tribal grouping in which men from southern groups would join northern ones for several months at a time in order to raid other groups: "This participation was seasonal . . . and resulted in manloads of fresh and smoked meat, including human flesh, along with male and female children, and other kinds of booty being brought back to the home populations."[142]

In his analysis of the Dani of the New Guinea Highlands, Karl Heider argues that warfare is fought principally over women, pigs, and land.[143] Paul Shankman's research echoes Heider's findings. Shankman studied the Dani years after Heider, and finds that increasing population pressure and the associated strain on resources for a given territory remain sources of warfare for this tribe.[144]

In her study of the origins of warfare in Melanesia, Camilla Wedgwood argues that conflicts between communities often begin over women or resources: "the stealing of women from a neighboring district, committing adultery with the wife of a man of that district, and disputes over fishing or garden rights . . . prove a fruitful source of strife."[145] In his ethological study of the Asmat, Gerard Zegwaard notes that when the Asmat raid neighboring villages, a woman or girl will be taken to the Asmat's village "to start or expand a family. When a village has a shortage of women, their abduction may be the sole objective of a raid."[146] He also explains how prestige is inextricably linked to warfare in this society. "In Asmat society all prestige, and therefore all authority, is ultimately derived from achievements in war. It is impossible to be a man of social standing without having captured a few heads."[147] According to Zegwaard, an Asmat can marry without successfully headhunting but will be often reminded of his low rank in the society: "His opinion will not be asked . . . [by boys and young men]; his own wife will pay little attention to him. When his wife wants to hurt him, she calls him *nas minu,* piece of meat; she declares that he is only meat, that he has no soul, no courage . . . [and] he is not considered a real man; he belongs to the category of women and children. He is not entitled to wear ornaments, has no share in the festive meals, stands no chance with other women [and remains] always the odd man out."[148]

Many tribes, however, recognize another motivation. The Kapauku Papuans in Irian Jaya understand that war is bad but that they must fight for the sake of relatives: "War is bad and nobody likes it. Sweet potatoes disappear, pigs disappear, fields deteriorate, and many relatives and friends get killed. . . . But one cannot help it. A man starts a fight and no matter how much one despises him, one has to go and help because he is one's relative and one feels sorry for him."[149]

Morren explains how among the Telefolmin tribes of New Guinea the type of warfare within the tribe differs from that waged against other tribes. "The Telefolmin appear to have observed a set of conventions when fighting other Telefolmin groups: objectives were limited, no captives were taken, and truces could be arranged," but warfare against non-Telefolmin tribes "lacked such limitations, with the historical record showing many cases of virtual annihilation."[150] Vayda also reveals that cooperation in offensive or defensive warfare among the Maori is determined to large degree by relatedness. He submits that close kin live near one another and cooperate in defense and other major tasks, while more distant kin or nonkin do not.[151]

North America

Among North American tribes the importance of resources was equally great. Ferguson notes the importance tribes placed on controlling the resources of rivers in his study of Indian tribes on the Northwest Pacific coast of Canada and the United States.[152] "Warfare . . . centered on control of resources associated with the rivers. Those controlling river mouths [the Tlingit, Tsimshian, and Bella Coola tribes] enjoyed a favored subsistence position. They usually had rich supplies of salmon and other river-spawning fish, but also could use ocean or seashore resources."[153] In contrast, upriver tribes did not have access to beached whales and other bounty from the sea and had to depend on the salmon runs, which Ferguson reports were not entirely dependable sources of food; as a result, the tribes "from exposed coasts or upstream locations regularly tried to displace those in more favorable, estuarine locations."[154] Indeed, anthropologist Franz Boas's informants described an eighteenth-century series of wars of extermination between the Tlingit and Tsimshian over control of the Nass and Skeena estuaries.[155]

These tribes were typical of primitive and indeed modern societies in their social stratification into two classes: elite and common tribe members. The status of leaders was marked as different from the rest of society by the number of wives leaders could possess, food consumption, burial arrangements, and other forms of deference behavior, as well as the number of slaves owned—slavery was a common feature among these tribes. Chiefs also used marriage to solidify alliance relationships; as Ferguson notes, "powerful chiefs took brides from their military inferiors, and gave as brides their close female kin."[156]

Another way to gain resources was to raid European ships. Many of the Pacific Northwest tribes raided vessels that arrived to trade for the pelts of sea otters and other animals in the late eighteenth and early nineteenth centuries. The Haida tribe attempted to raid five ships between 1791 and 1795, and the Nootka captured the first ship that visited them, "but they soon saw and clearly stated as much, that the wealth derived from a constant Western presence far outweighed the temporary abundance from one plundered ship."[157]

The fur trade made competition for resources more intense, given in particular the ecological consequences of eliminating sea otters. Their depletion led to the increased growth of herbivorous invertebrates, such as sea urchins, which reduced the algae near the shore, the basis of the food chain, and this in turn lowered the number of small fish and other animals that fed on the algae.[158] The final result for many tribes was a depletion of food sources that required them to become more dependent on trade with whites.

The tribes whose territory bordered inland waterways, such as the Coast Salish and Kwakiutl, were poorer than the tribes with extensive coastlines. They had no access to the valued sea otter pelts and so did not gain the wealth from trade that the Haida, Nootka, and Tlingit did. To gain resources, the Salish in particular had to raid, because they possessed no river trade routes and so had little access to salmon and furs. For them, "plunder and slaves would mean more than just living better. . . . Because wealth was needed to buy the weapons, and allies . . . needed to fight off . . . raiders, war booty probably meant life itself."[159] Slavery was a significant aspect of the Salish economy. It is estimated that slaves were about 10 percent of the total Salish population.[160]

Similarly, the Kwakiutl attacked their neighbors to gain resources, strategic territory, and slaves. One particular group, the Lekwiltok Kwakiutl, were especially feared because they raided the traders—other Kwakiutl and the Salish—who passed in the narrow channel between Vancouver Island and the mainland leading from the north to Victoria.[161] While the wars among the Pacific Northwest tribes that Ferguson studied were frequent and intense, they mostly ended with the growth of white settlement in the region; significant depopulation of the tribes due to disease, especially the smallpox epidemic of 1863; and the establishment of a permanent gunboat presence in the 1860s.[162]

Animals are also important resources and competition for them can result in war. In North America, the migrations and depletion of the buffalo were one cause of warfare among the Indian tribes of the Great Plains in the United States and Canada, since this was a great source of food for many of them. Neighboring or more distant tribes would often poach buffalo on an indigenous tribe's territory, presenting the indigenous tribe with a significant reduction in resources, but this condition became worse as the buffalo population declined from the 1820s through the 1860s due to the American demand for meat, buffalo floor coverings, overcoats, and lap robes. As one observer described the situation around 1851:

> Now each nation finds themselves in possession of a portion of these lands, necessary for their preservation. They are therefore determined to keep them from aggression by every means within their power. Should the game fail, they have a right to hunt it in any of their enemy's country, in which they are able to protect themselves.
>
> It is not land or territory they seek in this, but the means of subsistence, which every Indian deems himself entitled to, even should he be compelled to destroy his enemies or risk his own life to obtain it. Moreover, they are well aware that the surrounding nations would do the same and sweep them off entirely if they could with impunity, and each claims the same right.[163]

The migrations of these tribes often resulted in an initial competition for resources that could escalate to warfare. For example, the oral tradition of the Cheyenne includes the tale of their first battle on the Plains, when they encountered the Assiniboin (also spelled Assiniboine) tribe on a buffalo hunt:

> On this first battle it is said that during their journey westward, the people [Cheyenne] went out to kill the buffalo, and that while engaged at this the Assiniboines came up and tried to surround the same herd. The encounter caused a dispute, and soon afterward, in the night, the Assiniboines attacked the Cheyenne camp. The Assiniboines already possessed guns, while the Cheyennes had never seen any, and did not know what they were. The noise and effect of these strange weapons frightened the Cheyennes, and they ran away. The Assiniboines killed some of them, and cut off their hair; that is, scalped them.[164]

An additional source of warfare among the Plains tribes was horse raiding.

The horse was a critically important economic instrument, as well as a source of wealth and prestige for these tribes. Tribes would willingly conduct horse raids against enemy tribes, but also against any tribe that had good horses.[165] However, tribes that faced long winters and short growing seasons also experienced greater general attrition in their horse populations that forced them to raid. Ethnographer Thomas Biolsi reveals that there is much "ethnohistoric evidence of frequent and intense horse raiding of the Pawnee by the Sioux."[166] He continues, "certainly the need for horses and the resulting horse-raiding pattern of the Great Plains served as a continuous and ineluctable" source of hostility between these major tribes.[167]

Whatever the cause, archaeologist Douglas Bamforth's scholarship shows just how brutal warfare could be in the North American plains. Bamforth explains how archaeologists can help us understand the scale of warfare by examining human skeletal remains, defensive works, and other aspects of a site.[168] He reports on the significant and consistent tension between the Lakota and farming tribes of the Missouri River, such as the Arikara, Mandan, and Hidatsa.

At many sites Bamforth documents there is evidence of massacres occurring during which the victims were mutilated; but he often found insufficient evidence to determine which tribe or tribes conducted the attack. For example, at one site in South Dakota called Crow Creek he found evidence that five hundred people, of both sexes and all ages, were killed and their village burned around A.D. 1325. The denizens of this site were mutilated before or after death; over 90 percent were scalped, bodies were decapitated, tongues removed, and extremities cut from limbs.[169] The cause of warfare, he concludes, was likely food and other resources: "the osteological data from the Crow Creek victims indicate that the human population of the region was highly stressed, but that this stress was episodic rather than continuous."[170] The fact that that town's inhabitants were malnourished at the time of the attack suggests to Bamforth that "violence and subsistence stress" are likely to be related, especially given the evidence of a substantial population movement into the region probably as a result of the cooler, drier climatic conditions which obtained in the region after A.D. 1250.[171]

Finally, the discoveries of two early humans—"Ötzi" and Kennewick Man—are significant for this discussion because both were shot. A 5,300-year-old Bronze Age hunter, nicknamed "Ötzi," discovered high in the Italian Alps in 1991, had an arrow wound in his back, leading scholars to speculate about the circumstances surrounding his death.[172] Konrad Spindler of the

University of Innsbruck, who has studied "Ötzi" since his discovery, suggests that he was trying to escape an enemy by fleeing into the mountains; the many arrows in his possession indicated that he was prepared to defend himself.[173] Archaeologist Annaluisa Pedrotti of the University of Trento points to the copper ax found beside him as evidence that he did not die in front of his enemy, who would have certainly taken such a valuable item.[174]

The 9,500-year-old Kennewick Man discovered in July 1996 on the banks of the Columbia River in Washington is notable for both his means of death and his physical characteristics. First, a stone spear point embedded in the right side of his pelvis, as well as other injuries, suggests he was involved in fighting. The archaeologist and paleoecologist James Chatters, who first examined Kennewick Man, argues that it almost certainly was inflicted by a rival in his own tribe or by an enemy.[175] Based on his analysis of the wounds, Chatters says, "Kennewick Man probably saw the spear coming and tried to dodge it, which indicates he was facing his assailant. The depth of penetration—more than 2 inches (51mm)—indicates that the weapon came at a high velocity," and "was almost certainly propelled by an atlatl, or throwing stick."[176] Second, the analysis of Kennewick Man's remains, and another skeleton from approximately the same time, 9,000 years ago, called "Stick Man," suggests that the ancient people of the Columbia Basin did not resemble Indians. Indeed, they differ from all modern humans, leading to speculation about possible waves of immigration to the New World from East Asia or, less likely, even Iberia.[177] If this did occur, then perhaps the ancestors of Indians came to the New World later and absorbed these ancient people through intermarriage or destroyed them in conflict.

South America

South America also holds broad evidence of the relationship between resources and warfare. Having analyzed pre-Columbian warfare near the Atlantic littoral of Brazil, anthropologist William Balée argues that invasions were launched from areas of poor soil into more fertile areas near or on the coasts.[178] Balée notes the considerable evidence from missionary and other reports from the sixteenth and seventeenth centuries documenting wars over resources along the coasts and major rivers. Warfare was particularly acute along the Amazon River due to its abundant terrestrial and aquatic fauna: "on the basis of the primary documents . . . warfare was primarily between hinterland (interfluvial) and river bank peoples."[179]

This is confirmed by Donald Lathrap, who notes in his study of the Amazon basin that the richest agricultural and animal resources are found along the coasts and the rivers. Population growth placed pressure on the tribes who possessed this valuable territory and resulted in warfare, with the defeated tribes moving away from the territory into the interior or further up the river.[180] As he argues, the "fight for the limited supply of productive farm land" along the Amazon River "has been the most important single force" in the cultural "history of the Amazon Basin."[181]

Anthropologist Kalervo Oberg studied the lowland tribes of Central and South America and found that tribes raided each other for women and children to increase their population.[182] Among the Yanomamö as well, war is usually fought to gain or defend resources, like territory or food, or as a result of conflicts over women. This point is well captured in a conversation Napoleon Chagnon had in 1967 with the headman, Säsäwä, of one of the more distant Yanomamö tribes. Säsäwä coveted Chagnon's military knife. "He wanted me to tell him all about the knife, its origin, history, and how often it had been exchanged in trades."[183] Chagnon told him people of his "*teri*," or extended family group, used the knife when they raided their enemies.

> "Who [*sic*] did you raid?" he asked.
> "Germany-teri."
> "Did you go on the raid?"
> "No, but my father did."
> "How many of the enemy did he kill?"
> "None."
> "Did any of your kinsmen get killed by the enemy?"
> "No."
> "You probably raided because of women theft, didn't you?"
> "No."
> At this answer he was puzzled. He chatted for a moment with the others, seeming to doubt my answer.
> "Was it because of witchcraft?" he then asked.
> "No," I replied again.
> "Ah! Someone stole cultivated food from the other!" he exclaimed, citing confidently the only other incident that is deemed serious enough to provoke man to wage war.[184]

Chagnon also documents the reproductive advantage men may gain from warfare. Men who participate in warfare raids and ambushes have more wives and more children than others, and men who avoid warfare suffer reproductively. Yanomamö men who have killed as part of a war party are accorded the title *unokai* and have special privileges. The *unokais* are "socially rewarded and have greater prestige than other men"; this often lets them "obtain extra wives by whom they have larger than average numbers of children."[185] In fact they had, on average, "more than two-and-a-half times as many wives as non-*unokais* and over three times as many children."[186] Chagnon says another reason that the Yanomamö wage war is to maximize the number of offspring they or their close relatives have.[187]

Ferguson's study of the Yanomamö confirms Chagnon's argument that the Yanomamö wage war for resources, to destroy competitors for valuable resources, and to gain Western goods.[188] The Yanomamö often covet metal tools, such as machetes and cooking pots, and the control over access to them through contact with businessmen or miners becomes an important source of power in relationships between clans who have access and the more isolated subgroups who do not: "Yanomami who monopolize sources of Western manufactures are able to use their position as middlemen to obtain local products, women, and labor from more isolated villages."[189]

He also reveals how a Western presence and manufactured goods can make war more likely. This occurs in three ways. First, if some villages suddenly acquire a disproportionate amount of Western goods they are attractive targets for raids; groups will attack them either for plunder or to drive them from their privileged trading position. Second, a limited presence of Westerners may destroy established trade and alliance patterns. Third, a period of Western retraction, as occurred after the collapse of the 1920s rubber boom, can leave middlemen unable to meet existing obligations due to a scarcity of Western goods.[190]

Drawing on acute personal experiences, Chagnon explains that the Yanomamö particularly desired manufactured goods as well as the food and other supplies he brought to camp. "The hardest thing to learn to live with was the incessant, passioned, and often aggressive demands they would make. It would become so unbearable at times I would have to lock myself in my hut periodically just to escape from it," he writes. "I did not want privacy for its own sake; rather I simply had to get away from the begging."[191] This was particularly intense, as he explains:

> Day and night for almost the entire time I lived with the Yanomamö I was plagued by such demands as: "Give me a knife, I am poor!"; "If you don't take me with you on your next trip to Widokaiya-teri, I'll chop a hole in your

> canoe!"; "Take us hunting up the Mavaca River with your shotgun or we won't help you!"; "Give me some matches so I can trade with the Reyaboböwei-teri, and be quick about it or I'll hit you!"; "Share your food with me, or I'll burn your hut!"; "Give me a flashlight so I can hunt at night!"; "Give me all your medicine, I itch all over!"; "Give me an ax or I'll break into your hut when you are away and steal all of them!" And so I was bombarded by such demands day after day, month after month, until I could not bear to see a Yanomamö at times.
>
> It was not as difficult to become calloused to the incessant begging as it was to ignore the sense of urgency, the impassioned tone of voice and whining, or the intimidation and aggression with which many of the demands were made. It was likewise difficult to adjust to the fact that the Yanomamö refused to accept "No" for an answer until or unless it seethed with passion and intimidation—which it did after a few months. . . . I had to become like the Yanomamö to be able to get along with them on their terms: somewhat sly, aggressive, intimidating, and pushy.[192]

To combat this behavior, Chagnon had to become an expert in what the nuclear strategist Schelling terms the "manipulation of risk." He had to learn to threaten credibly his tormentors with unacceptable consequences should they start provoking him.

> It became indelibly clear to me shortly after I arrived there that had I failed to adjust in this fashion I would have lost six months of supplies to them in a single day. . . . In short, I had to acquire a certain proficiency in their style of interpersonal politics and to learn how to imply subtly that certain potentially undesirable, but unspecified, consequences might follow if they did such and such to me. They do this to each other incessantly in order to establish precisely the point at which they cannot goad or intimidate an individual any further without precipitating some kind of retaliation. . . .
>
> Many of them, years later, reminisced about the early days of my fieldwork when I was timid and *mohode* ("stupid") and a little afraid of them, those golden days when it was easy to bully me into giving my goods away for almost nothing.[193]

Africa

In Africa as well, the same motivations for warfare are present. The Maasai (also spelled Masai) of Kenya and Tanzania are famous for their fearsome behavior and frequent raids to gain cattle. However, anthropologist Alan Jacobs finds little evidence to support the argument that the Maasai are more bellicose than their neighbors or the other tribes of East Africa.[194] While they did fight major wars in the nineteenth century against their neighbors, and these wars remain important to the culture of the tribe, in the twentieth century Jacobs found evidence of only small-scale raids to gain cattle or retaliate for raids conducted by their Bantu neighbors.[195] Thus, despite the popular legend that the Maasai stopped the "road to Uganda for Arabs and Europeans alike," there is more evidence that the pathway to the interior of Africa from its East Coast was stopped more by the semi-arid terrain and some tribal hostility than by the reputation of the Maasai.[196]

In his study of Maasai and Samburu warfare, ethnographer Elliot Fratkin notes that "the history of East Africa is rich in accounts of warfare between various pastoralist groups competing for access to" and control of "livestock, grazing, and water resources."[197] Resources are critically important for these tribes, especially to feed growing populations. As a result of changes in population and scarcity of resources in times of famine, drought, and disease, the tribes found in Kenya today wage war to gain resources: "the warfare undertaken in the first half of the 19th century was motivated by the desire for territory and for livestock to feed a growing population; that of the second half of the century resulted from a need to feed a starving one."[198]

The Dassanetch of southwest Ethiopia raid for "cattle, young girls, and the prestige and other rewards that come from killing an enemy."[199] The looted cattle are distributed to relatives or bartered, while a captured girl is usually given to the uncle of the individual who captured her. For those who have killed an enemy, "if the killer smears the victim's blood on his body it is believed it will bring him and his cattle fertility," but the "most important 'reward' is the scarification which the killer is entitled to wear on his chest . . . and which is the only visible sign that distinguishes a killer from a non-killer."[200]

But wars among East African tribes, as elsewhere, often begin over control of the resources they need to survive. For example, in January 1972, the Dassanetch attacked a neighboring tribe, the Nyangatom of southwestern Ethiopia, because they had refused the Dassanetch access to their dry-season watering holes, which caused a complete breakdown in their relations. In June they attacked and "succeeded in exterminating" the Nyangatom "settlements at Kibish. . . . They fell simultaneously on the different settlements while their occupants were still asleep," and massacred "men, women, children, and old people, with guns, spears and clubs," and set fire to the huts.[201] The Dassanetch attack killed many Nyangatom and their cattle, and forced them to abandon their sorghum fields. This attack left the Nyangatom so desperate that they moved near the Kara tribe, with which they quickly developed hostilities, probably due to mutual thefts of sorghum. In January 1973, the Kara crossed the Omo River to attack a group of the Nyangatom, and the next month, the Nyangatom "set off to 'exterminate' the Kara" and destroyed one of their villages.[202] This set off further attacks and counterattacks.

The Boran, one of the Oromo tribes of Ethiopia and Kenya, fight to defend water and grazing rights but also to seize these resources: they organize raids "either when they covet a well or grazing of their enemies and wish to drive them from it, or to deter enemies who, they think, wish to do the same to them."[203] Anthropologist P.T.W. Baxter explains that all the young men are eager to participate in raids as warriors, for to be fit for marriage, a man must acquire a trophy, such as the genitals of a man he killed.[204] Moreover, "a man who did kill was given gifts of stock . . . lavished with sexual favors by the wives of elders . . . and allowed to wear ear rings, special necklaces and heavy ivory armlets . . . and, crucially, a successful warrior was allowed to grow a male hair tuft . . . from the top of his head."[205]

Animals are often a significant resource for tribal societies, both as measures of wealth and as food, both meat and milk. Among the Meru tribe of Kenya, livestock are used for bridewealth; to marry, a man must accumulate enough livestock to purchase a wife. Thus, considering the need to acquire cattle or other livestock, animals are also a source of war. Anthropologist Morimichi Tomikawa notes that Maasai cattle raiding is a frequent source of warfare between the Maasai and neighboring tribes, such as the Datoga and Iraqw in Tanzania.[206]

Finally, the Bible records instances of rape and forced marriage after war. In his *Sociobiology,* E.O. Wilson recalls Moses' instructions in the Book of Numbers (31:7,17–18) on the occasion of the victory over the Midianites: "Now kill every male dependent, and kill every woman who has had intercourse with a man, but spare for yourselves every woman among them who has not had intercourse."[207] In Deuteronomy (21:10–14), a similar exhortation occurs:

> When thou goest forth to war against thine enemies, and the LORD thy GOD hath delivered them into thine hands, and thou hast taken them captive, and seest among the captives a beautiful woman, and hast a desire unto her, that thou wouldest have her to thy wife; then thou shalt bring her home to thine house; and she shall shave her head, and pare her nails; and she shall put the raiment of her captivity from off her, and shall remain in thine house, and bewail her father and her mother a full month: and after that thou shalt go in unto her, and be her husband, and she shall be thy wife. And it

> shall be, if thou have no delight in her, then thou shalt let her go whither she will; but thou shalt not sell her at all for money, thou shalt not make merchandise of her, because thou hast humbled her.

Thus, there is significant evidence from premodern societies that successful warfare does increase the inclusive fitness of the surviving victors. These individuals do indeed stand to benefit in terms of their ability to reproduce successfully because of their increased resources. It is in this precise sense that warfare may be said to be adaptive. Of course, environmental factors may augment this with additional social benefits, such as heightened prestige for successful warriors.

The Forms and Tactics of Primitive Warfare

Primitive warfare takes three principal forms: the raid, the pitched battle, and the massacre. Modern militaries fight in broadly similar forms (as do chimpanzees, as I will explain in the next chapter), although naturally there are important differences, such as the decreasing importance of massacre in modern, conventional warfare. At a fundamental level, the similarities of primitive and modern warfare are striking. For example, the grand strategy and tactics of primitive warfare are familiar to a modern army.

Tribes are aware of their interests and their power, and they seek allies even while they know that allies might fail them. In an episode that mirrors countless events in the history of international relations, such as U.S. decision making during the Vietnam War or the Athenian reasoning captured by Thucydides in the Melian Dialogue, Chagnon explains why the leader of one Yanomamö clan (the Hukoshikuwä) felt compelled to demonstrate resolve lest his enemies and erstwhile allies think his leadership or clan weak. He writes, "The Shamatari allies even managed to demand a number of women from Hukoshikuwä's group in payment for girls they had given earlier. . . . In short, if Hukoshikuwä failed to put on a show of military determination and vindictiveness," then the situation he and his clan faced would be untenable: "it would not be long before his friends in allied villages would be taking even greater liberties and demanding more women. Thus, the system worked against him and demanded that he be fierce, whether or not he wanted to be," since this was the only way his small group could protect its sovereignty or risk being "absorbed by a greedy" and stronger ally.[208]

Finally, these tribes are well aware that postwar arguments with allies are common and may lead to further hostilities.[209] They are quite aware of the ancient adage "today's friend may be tomorrow's enemy," and can be just as sensitive to relative power considerations as the Americans, British, French, and Italians were after World War I, or once again after World War II, when Anglo-American cooperation with the Soviet Union came to an abrupt end with the defeat of their common adversaries.[210]

With respect to tactics, as with any modern army, primitive tribes emphasize good intelligence, knowledge of the terrain, care and maintenance of weaponry, speed, secrecy, concentration of force at the point of attack, and the requirements of logistical support. Planning is centrally important. The attackers typically have the attack route or routes well planned as well as egress routes. The tactics of tribal warfare are sound ones and recognizable by officers of modern militaries. Whether a military is fighting in modern day Iraq, or in Papua New Guinea ten thousand years ago, the tactics used will have strong similarities.

The Raid

Keeley argues that raids are the most common form of primitive warfare, and of course are still used today.[211] Raids usually involve an ambush and are well calculated to take advantage of surprise and an imbalance of power. Individuals in tribal societies often work and travel in small groups but must still be aware of an attack by a larger group from a neighboring hostile tribe. In surprise attacks at dawn, small numbers of men from the attacking tribe stealthily move into enemy territory to ambush a few of the enemy tribe at some distance from their village, or they may attack the village itself. A small tribe may mobilize in a few hours for a quick raid, in contrast to the weeks needed to equip and mobilize a modern army. "One common raiding technique" favored by groups as diverse as the Bering Strait Eskimo and the Mae Enga of New Guinea "consists of quietly surrounding enemy houses just before dawn and killing the occupants," either as they sleep by thrusting spears through the walls or by setting fire to the dwellings and killing the victims as they emerge.[212] After studying warfare from aboriginal Australia to the Alaskan Eskimos, historian Azar Gat stresses the importance of the raid in primitive warfare. Specifically, simple human societies prefer to conduct surprise first strikes to maximize the chance of catching the enemy helpless with asymmetrical numbers so that the attacker has an overwhelming advantage in number of combatants.[213]

Vayda also notes this practice among the Maring. "At dawn the raiders made fast the doors of as many . . . [houses] as possible and then shot their arrows and poked their long spears through the leaf-thatched walls"; any victims who succeeded in breaking out "were picked off by raiders waiting behind the fences."[214] Vayda concludes, "With a numerical advantage on the side of the raiders, these tactics could annihilate the manpower of the enemy clan."[215] Wedgwood also documents that the surprise attack or ambush at dawn is a common tactic of warfare in Melanesia. This "treacherous business" usually involves "the ambushing of a village at early dawn or the cutting down of an unsuspecting group of travelers, gardeners or fishermen, in which many people are killed and even whole hamlets wiped out."[216] Turney-High also notes the general preference for surprise attack among many tribes in Melanesia, and notes specifically that the Kiwai of New Guinea relied "on the dawn attack, surprise, and exploitation by wiping out every enemy life."[217] They accomplished this by entering the enemy village just before sunrise. As Turney-High describes it, "the Kiwai attempted to enter each bushman's house quietly and tapped the individual inhabitant silently but adequately on the head, which was removed on the spot."[218] Moreover, they also face a problem that plagues modern militaries: "friendly fire" or "blue-on-blue" casualties. These are deaths or injuries to one's own forces caused by one's own or allied forces. Interestingly, to help identify friend from foe, they painted themselves "with white pipe clay, even though some enemy might use the white stripes as an arrow target."[219]

Morren argues that in the highlands of Papua New Guinea and Irian Jaya, "warfare consists of small-scale ambushes and raids, which probably produce a majority of the casualties."[220] He also documents a decades-long period of raids between two tribal groups, the Telefolmin and Miyanmin.[221] Anthropologist Leopold Pospisil notes that among the Kapauku a single individual, whom he calls a "sniper," may sneak into an enemy village to kill a person. Waiting "at night at the door of an enemy's house until the man has to leave the shelter to urinate" facilitates an ambush.[222] "There is no better target than an erect, sleepy man standing in the door against the shimmering light of the fire inside the house."[223] A half a world away, the Icelandic sagas report similar ambush tactics. In *Eyrbyggja saga* and *Laxdæla saga* enemies are ambushed in exactly this way.

Meggitt explains how the Mae Enga often open hostilities "with a series of nocturnal raids intended to pick off notable warriors, thus demoralizing the enemy and weakening them for the subsequent engagement."[224] The Mae also adopt elaborate defensive measures; because they fear such assassinations as well as being incinerated in their houses by their enemies, they take great pains to protect their houses from attack through an elaborate network of ditches and fences surrounded by sharpened bamboo slivers.[225] Those who live near a declivity may build a tunnel so they can escape and perhaps turn the tables on the aggressors by attacking them as the blazing house illuminates them.[226]

Among the Yanomamö, Chagnon explains that the object of the raid is almost always "to kill one or more of the enemy and flee without being discovered."[227] In fact, for this tribe it is singularly important not to incur any losses: "If . . . the victims of the raid discover their assailants and manage to kill one of them, the campaign is not considered a success, no matter how many people the raiders may have killed before sustaining their single loss."[228] To bring this about, Chagnon describes the raiding tactics, which differ little from the raiding or guerrilla tactics of elite U.S. units, such as the Army's Special Forces and Rangers, or the Navy's SEALS. "The raiders travel slowly their first day away from the village. They have heavy burdens of food and try to pace themselves so as to arrive in the enemy's territory just as their food runs out."[229] Once near the enemy village, they try to camp close enough so that they may "reach his village at dawn—far enough so that enemy hunters will not discover their presence but close enough to the village that they can reach it in an hour or so from their last camp."[230] The men take great efforts to conceal their presence in enemy territory; they will even do without fire at night though fire decreases the risk of attack from both jaguars and unfriendly spirits.

To assist and protect the boys or young men who are on their first raid, "the older men will conduct mock raids, showing them how they are to participate."[231] They will often use "a grass dummy or soft log," while especially young men "will be positioned in the marching party somewhere in the middle of the single file of raiders so that they will not be the first ones to be exposed to danger should the raiders themselves be ambushed."[232] To attack the enemy, the raiders typically split into groups of about six men apiece, agree to meet afterward at a predetermined location between the enemy village and their own, conduct the attack, and then retreat. Each group retreats in a specific, leapfrogging manner. Two men lie in ambush in order to attack any pursuer; those two then flee while two more comrades lie in ambush to cover their retreat.[233] Just like the Kapauku, these groups ambush the enemy, often by lying in wait for them. "A good many of the victims of raids are shot while fetching water," or while relieving themselves near their villages.[234]

In addition to our knowledge about this tactic from premodern societies extant today or in the recent past, Keeley argues that substantial evidence from the prehistoric record supports a long history of the raid. The archaeological evidence for this is often marked by a specific pattern that includes four characteristics: burials of small groups of homicide victims; burials away from more established centers of habitation; women as a generally high percentage of the victims; and evidence that most wounds, even on adult males, were inflicted from behind.[235] To support his argument, Keeley reports precisely such evidence from archaeological sites in North America and other continents. "Several isolated prehistoric burials in central Washington State fit this pattern precisely. . . . Projectile points found embedded in these skeletons indicate that in some cases the killers were 'foreigners.'"[236] At another pre-Columbian site, "16 percent of the 264 individuals buried there met violent deaths and also fit the patterns expected for raid victims."[237] Keeley is confident these deaths are not from other causes, such as murders, for example, because this would require a homicide rate among these villagers about 1,400 times that of contemporary Britain or 70 times that of the United States.[238]

Thus, despite the costs of colonial rule, often one positive result of it was pacification and the termination of some of the risks of raid or ambush associated with primitive warfare.[239] For example, Australian colonial rule brought increased stability and safety to Papua New Guinea. One Auyana man there summarized the benefits of the *Pax Australiensis:* "life was better since the government had come" because "a man could now eat without looking over his shoulder and could leave his house in the morning to urinate without fear of being shot."[240]

The Battle

Battles are the second form of primitive warfare, and are as much a feature of premodern warfare as they are its modern form. They are less frequent than raids, and often not secretive but ritualized. Many tribes, including the Maring or Dugum Dani of New Guinea, send heralds to warn the enemy to expect battle at a given site and often include "considerable taunting and exchanging of insults" before battle.[241] The tactics of battle in premodern warfare are well suited to the technology available to these warriors, which Meyer says, is more or less evenly distributed across tribes in any given area.[242] Often one line of warriors will engage the other with spears or arrows at a common range of fifty or a hundred yards and the battle continues, as in modern warfare, until one side is broke or a truce or armistice is declared.

In the Trobriands, Wedgwood reports, battles are as formal as they would be under the Duke of Marlborough or Frederick the Great in seventeenth- and eighteenth-century Europe. "The chief of each district summoned to his village all the fighting men under his dominion. Then at a place midway between the principal villages of the contending parties a circular arena was cleared for the battlefield [and at this point] the two parties ranged up opposite each other at a distance of some thirty to fifty yards, the men armed with spears and shields only; the women behind them bearing refreshments and aiding their menfolk with verbal encouragement."[243] Then the conflict began, continuing until "one party had broken the line" of the enemy and then "ensued an onrush of the victors, killing everyone they could, destroying villages and laying waste to gardens."[244] Meggitt notes similar behavior in Mae Enga battles for both combatants and noncombatants: "the women watch anxiously in order to judge the course of battle and to determine when to flee with their children and pigs," while the young boys and old men "hold themselves ready to bring up supplies of arrows when ordered, or to remove and undertake emergency surgery on seriously wounded warriors."[245]

While the conduct of the battle may be as ritualized as the warfare of European kings, it may also be as lethal, if not more so.[246] As Keeley documents, one clan of the Maring numbered six hundred but lost "2 percent of its population in the rout that followed its loss of 3 percent of its people in the preceding battle."[247] He evaluates the situation: "This total may not seem very severe, but to produce equivalent figures," the France of 1940 with a population of 42 million, "would have had to lose over 1.2 million in its 1940 defeat and 840,000 civilians in the immediate aftermath," amounting to five times the total number of war-related French deaths during the whole war.[248] The percentage of combat deaths among the Maring is far higher than the World War II casualty rate of the United States, which lost about 300,000 soldiers, or about 0.2 percent of its population in 1945. Based on Keeley's research, tables 3.4 and 3.5 provide a general comparison, first, of the number of males mobilized for warfare in selected tribal societies and contemporary states; and, second, of the casualty rates from warfare, on an annual basis, again for tribal societies and contemporary states.

According to Vayda, a Maring clan suffered even worse casualties following a surprise raid by another clan: of three hundred people, twenty-three

Table 3.4. Combat Unit Sizes and Percentage of Males Mobilized for Selected Tribal Societies and Contemporary States

Group	*Maximum Unit Size*	*Male Population*	*% of Males Mobilized*
Huron	600	9,000–11,000	5–7
Mohave	100	1,500	7
U.S. in WWII	11,490,000	66,000,000	17
Modoc	100	500	20
USSR in WWII	20,000,000	91,000,000	22
Maori	350+	1,250–3,750	9–28+
Germany in WWII	10,800,000	34,250,000	32
Mae Enga (1 clan)	70	175	40
Miyanmin	200	< 500	40+
France in WWI	8,410,000	19,500,000	43
Tahiti	7,760	17,683	44

Source: Lawrence H. Keeley, *War before Civilization: The Myth of the Peaceful Savage* (New York: Oxford University Press, 1996), Table 2.6, p. 189.
Notes: All figures are estimates. States are italicized. Maximum Unit Size is the total number who served during a war. Male population is estimated by dividing total population in half.

were killed (fourteen men, six women, and three children), or 7.6 percent of the population.[249] The Mae Enga also suffered considerably from warfare. Meggitt studied the combat deaths of this tribe from 1900 to 1950 and found that male deaths attributable to warfare account for about 25 percent of all male fatalities.[250] Warfare almost always has a profound effect on a particular clan since the mean size of the clans in the area Meggitt studied was about 350 to 400 people, and in that fifty-year period he has casualty data from thirty-four conflicts, in which on average 3.7 men of the clan were killed per conflict.[251]

In addition, a greater number of men (up to 39 percent of the male population) are usually wounded in combat, most often by arrows, which frequently result in infection, thus further weakening the clan.[252] Among the Grand Valley Dani in the central highlands of Irian Jaya, Heider found that 100 of 350 (about 28.6 percent) deceased males and 5 of 201 (about 2.5 percent) de-

Table 3.5. Annual Warfare Death Rates for Selected Tribal Societies and Contemporary States

Society	*Region*	*Annual % Rate*
Dani—South Grand Valley (NPP)	New Guinea	1.00
Dinka (1928)	N.E. Africa	0.97
Fiji (1860s)	Polynesia	0.87
Chippewa (1825–1832)	Minnesota	0.75
Telefolmin (1939–1950)	New Guinea	0.74
Mtetwa (1806–1814)	S. Africa	0.59
Dugum Dani (1961)	New Guinea	0.48
Manga (1949–1956)	New Guinea	0.46
Modoc (NPP)	California	0.45
Auyana (1924–1949)	New Guinea	0.42
Mae Enga (1900–1950)	New Guinea	0.32
Dani—C. Grand Valley (1900–1960)	New Guinea	0.31
Yanomamö (1938–1958)	South America	0.29
Mohave (1840s)	California–Arizona	0.23
Gebusi (1940–1982)	New Guinea	0.20
Tiwi (1893–1903)	Australia	0.16
Germany (1900–1990)	Europe	0.16
Russia/Soviet Union (1900–1990)	Europe-Asia	0.15
Boko Dani (1937–1962)	New Guinea	0.14
France (1800–1899)	Europe	0.07
Japan (1900–1990)	East Asia	0.03

Source: Lawrence H. Keeley, *War before Civilization: The Myth of the Peaceful Savage* (New York: Oxford University Press, 1996), Table 6.1, p. 195. Central Grand Valley Dani Numbers are from Karl G. Heider, *Grand Valley Dani: Peaceful Warriors* (New York: Holt, Rinehart and Winston, 1979), p. 106.
Notes: All figures are estimates. States are italicized. Males and females are included. NPP stands for No Period Provided.

ceased females were killed in warfare.[253] These numbers are based on his construction of genealogies for this tribe, although he witnessed only one five-month war when he first conducted his fieldwork with this tribe in 1961.

Given the small population, the deaths of even a few warriors weaken a clan's ability to defend its resources against the depredations of its neighbors; as Meggitt explains, the "traditional modes of Mae warfare, both in their nature and in their frequency, take a relatively high toll of men's lives, whether in a clan through time or in the region as a whole, and many of the victims are bachelors," who are killed before they reproduce.[254] These high rates of death have political consequences because the clan's fighting strength "can quickly and radically erode" and thus "leave it open to the incursions of more fortunate neighbors."[255] Pospisil recounts disastrous wars among the Kapauku, including one that lasted over a year and left one village "practically without adult males."[256]

Of course, such losses are not limited to Melanesia. In Africa, ethnographer Serge Tornay studied warfare in the Lower Omo Valley of Ethiopia and Sudan in the 1970s, and documents how the Nyangatom tribe lost between four hundred to five hundred people in five years, out of a total population of five thousand.[257] Chagnon reveals that among the Yanomamö "approximately 40% of adult males participated in the killing of another Yanomamö. . . . approximately 25% of all deaths among adult males was due to violence" and approximately "two-thirds of all people aged 40 or older had lost, through violence, at least *one* of the following kinds of very close *biological* relatives: a parent, a sibling, or a child. Most of them (57%) have lost two or more such close relatives."[258]

Anthropologist Sterling Robbins studied the casualties of the Auyana in warfare with other tribes during a twenty-five-year period; he found that the total mortality rate from warfare for the Auyana during this period was approximately 4.2 per thousand.[259] If we applied this percentage to the United States for a roughly equivalent period, from 1940 to 1975, it would mean 630,000 American battle deaths per year.[260] In fact, all U.S. battle deaths in this period totaled approximately 373,000.[261]

Even during the American Civil War, the war with the highest casualty rates in American history, the battle death losses do not approach the Auyana casualty rate, which many other tribes have exceeded—see table 3.5. In 1861, the United States had a population of 31 million. For American losses to equal the Auyana rate (4.2 per thousand) would have required an annual mortality rate from warfare of about 130,000 for both the Union and Confederacy. In fact, the average was 51,500 battle deaths over each of the four years for both, or 0.17 percent of the total population. Even when we consider all casualties for the American Civil War, the numbers do not approach the Auyana rate; the Confederacy and Union lost a total of about 624,000 men in four years, an average of 156,000 total battle and nonbattle deaths per year, or 0.5 percent of the total population. Approximately 206,000 soldiers were actually killed on the battlefield or mortally wounded, and about 418,000 died from disease, while in prison, or from other causes.[262]

Of all industrialized states in the twentieth century, only the losses of the Soviet Union in World War II approach the Auyana loss rate. The Soviet Union is estimated to have incurred 7,500,000 battle deaths during World War II, or about 1,875,000 battle deaths a year, about 1 percent of the population per year.[263]

The early nineteenth-century Prussian officer and scholar Carl von Clausewitz wrote of the horrible costs of warfare in his classic work, *On War.* He termed the costs the "blood price" of warfare and warned: "the character of battle, like its name, is slaughter [*Schlacht*], and its price is blood. As a human being the commander will recoil from it."[264] Tribal warfare certainly captures well Clausewitz's point about the violence of war and its "blood price." Thus, it is certainly less technologically sophisticated than modern warfare, and narrower in scope—often confined to a specific field or valley rather than occurring on multiple continents simultaneously. But it can be just as deadly for the participants and as catastrophic for the societies affected by it as the modern, industrialized warfare characterized by World Wars I or II. In the modern era, only nuclear or biological warfare could unambiguously exceed the casualties, on a per capita basis, that are often associated with warfare in tribal societies.

The Massacre

The third form of primitive warfare is the massacre. In the massacre, members of one tribe attack an enemy village with the intention of killing either all of its denizens, or the adult males and boys while capturing the women for wives or slaves. While difficult to execute successfully, an advantage of the massacre is that it reduces the chances of subsequent wars of revenge and feuding since the target group is effectively eliminated.

There are numerous accounts of massacres among tribal societies. Ethnologists Clifton Kroeber and Bernard Fontana document a fascinating eyewitness account provided by stagecoach workers of an 1857 massacre of Maricopas by Mohaves, Quechans, and Tonto Apaches near modern-day Phoenix, and the retaliation taken by the Maricopa and their Pima allies.[265]

Keeley reports that when Spanish explorers first reached the coastal Barbareño Chumash tribe in modern California, they had "just had two of their villages surprised, burned, and completely annihilated by raiders from the interior, representing a minimum loss of 10 percent of their tribal population."[266] The Kutchin Inuit of subarctic Canada would attack their major enemy, the Mackenzie Inuit, trying to kill everyone in the village, sparing only one male, who could spread word of the attack.[267] In the Pacific Northwest, Ferguson describes how a Moachat chief slaughtered forty of his enemies at a feast.[268]

In her study from the 1930s, Wedgwood reports that people on San Cristoval have two words for war, "*heremae*" and "*surumae.*" The former may be understood as formal battle, but the *surumae* was a secret expedition; warriors aimed to kill and eat as many of the enemy as possible.[269] Morren also reports massacres as a feature of warfare in the Mountain Ok region of New Guinea.[270] Similarly, Heider recounts a 1966 massacre that claimed 125 people out of 2000, or 6.25 percent of the population of one of the Dani clans.[271] Among the Maring, Vayda explains how a defeated group would flee "to their hamlets" to gather "their women and children, and then flee to seek refuge, while the routers wreaked as much death and destruction as they could," including the destruction of homes and gardens.[272] His informants said this occurred in nineteen of the twenty-nine wars of which they were aware, suggesting it is not uncommon in their warfare.[273] Pospisil also notes this feature among the Kapauku, who will kill "all males from the enemy's camp except a baby boy in the arms of his mother and a very old man."[274]

Chagnon describes what the Yanomamö call a *nomohori* (dastardly trick). He recounts one episode from 1967 when some missionaries explained to him why "a village in Brazil . . . had a critical shortage of women."[275] Years earlier a distant group had come to abduct women. "They went there and told these people that they had machetes and cooking pots from foreigners, who prayed to a spirit that gave such items in answer to the prayers," and they "then volunteered to teach these people how to pray. When the men knelt down and bowed their heads, the raiders attacked them with their machetes and killed them. They captured their women and fled."[276]

In human history, the massacre appears to have been relatively infrequent when compared with the other two forms. One reason might be the difficulty in planning and executing such an attack, which would require massing all available combatants from the attacking tribe, and catching the target village unaware or unprotected, for example when its warriors are away hunting. Despite the difficulties, this type of warfare clearly was practiced. For example, archaeologists George Milner, Eve Anderson, and Virginia Smith analyzed skeletal remains from the Oneota tribe in Illinois dating to A.D. 1300 and concluded that warfare in this time was marked by repeated ambushes: "the osteological and archaeological information shows that outbreaks of violence occurred regularly."[277] Also on the basis of skeletal remains, these archaeologists found that "people apparently were killed either while alone or in small groups, and the attack continued throughout the period of cemetery use. Single-sex groups, presumably work parties, were often attacked," and those who killed them "often lingered at the scene of the attack, as shown by many examples of decapitation and postcranial dismemberment."[278]

Interestingly, in many tribes while the men plan the raid or battle, they often exclude the women from decision making to maintain secrecy. This would be wise in general: the fewer people of either sex who know of an attack the better. This is obviously as true today as it would have been in ancient times. Equally important, those planning the raid might fear the women's divided loyalty, since the women may have come from the neighboring tribes against whom the attack is being planned, and may still have relatives there. Thus, a woman in a primitive tribe may face a difficult situation if her husband is planning to attack her father and brothers. Given the dangers posed by divided loyalty, we can see why prudence would require excluding women. In addition, fighting males may be excluded and even restrained if they are thought to possess a divided loyalty.[279]

While evolutionary theory provides insights into the origins of warfare, it is important to emphasize two points. First, it is critical to understand how important the ecology of a specific area or time period is to this explanation. The ecological conditions of the human evolutionary environment shaped human behavior significantly. Second, the culture of the group is also essential for understanding why it maintains war. Among the significant features are the history of the group, its rituals and rewards for its seasoned warriors or veterans, its music and folklore, and its sense of uniqueness, passed on in stories about the group's struggles and victories. Evolutionary and cultural motivations for warfare are intertwined, just as smaller cables often form bigger ones, or discrete threads make up a rope. Recognizing this provides an understanding of how deeply anchored in human behavior the problem of war is and why efforts to solve this most pernicious problem must acknowledge this but continue undaunted.

Conclusion

Evolutionary theory and human ecology allow us to understand why tribes fight over the resources they need to survive and reproduce. In some fundamental respects, the motivation for warfare has changed little. States still wage war for resources, whether they are defined as critical natural resources, territory, or vital geographic points, such as straits. Demonstrating an acute sensitivity to the importance of vital geographic points, British admiral Jackie Fisher once described the five strategically important straits (Singapore, the Cape of Good Hope, Alexandria, Gibraltar, and Dover) as the "keys" that "lock up the world"; as long as Britain held them, he said, it would be secure.[280] However, not only military officers have had such sensitivity. Describing his "Heartland Theory," the geographer Sir Halford Mackinder argued that the state that dominated eastern Europe could command the "Heartland," central Asia, could rule the "World-Island," Eurasia and Africa, and thus the world: "Who rules eastern Europe commands the Heartland. Who rules the Heartland commands the World-Island. Who rules the World-Island commands the world."[281] The Panama and Suez Canals also remain strategically important. Moreover, countries that possess key natural resources, such as the oil-producing Persian Gulf states, are strategically important today for the United States, just as the grain-producing Black Sea steppes were for Athens, or Egypt's Nile Delta for Rome.

In addition, states still wage war over land, which a nation may need to survive economically. Having analyzed the 1969 war between El Salvador and Honduras, the "Soccer War," Durham determined that its fundamental cause was land scarcity in both states.[282] This problem will become more acute as the world's population grows following improved agricultural productivity, and as modern sanitary and medical technologies continue to reduce death rates, especially in nonindustrialized countries. At the time of Christ, the world's population was between 200 million and 300 million people; in 1650 it was about 500 million. In the first half of the nineteenth century, world population reached 1 billion, by 1930 it was 2 billion, and in 1975 4 billion. As the world's population grows from 6.1 billion in 2000 to 7.2 billion by the year 2015, land will become a more precious resource in many countries due to its scarcity and demographic pressures, and thus will remain a cause of war.[283] This is especially true in the thirty countries, mostly in Africa and the Middle East, which will experience a disproportionate growth in youth population, or "youth bulges," by 2015.[284]

Furthermore, land is often inextricably linked to ideology, nationalism, or the history of a state. Hitler's desire for *Lebensraum* was a major reason for attacking the Soviet Union in 1941. Egypt and Syria waged war against Israel in 1973 over their attachment to territory. Conflicts or terrorism in Chechnya, the Democratic Republic of Congo, Kashmir, Northern Ireland, Spain, and Sri Lanka, are anchored to a great degree in disputes over land. The Serbs remained committed to Kosovo because of the cultural significance of the land, and departed from it only after NATO's intense bombing campaign during Operation Allied Force in 1999.

Population growth has meant that some of the motivations for warfare that affect primitive tribes are of little concern to the modern world, for example raiding for women. On the other hand, environmental degradation may make war more important than it has been for modern societies in the recent past. Thomas Homer-Dixon and Michael Klare demonstrate how current environmental crises—global warming, deforestation, population growth, the decreased quality and quantity of renewable resources, and unequal access to resources—can lower economic productivity, create population expulsions, and weaken the power of states.[285] These factors, in turn, can raise the risk of ethnic conflict, coups d'état, or warfare.[286] Of course, particularly valuable resources may create conflict between and within countries. For example, the Kono diamond fields are the primary source of wealth and political power in Sierra Leone, leading to warfare between competing groups. The copper mines of Bougainville create wealth for the government of Papua New Guinea, and this is a major reason why it and a rebel movement, the Bougainville Revolutionary Army, are fighting over the independence of Bougainville.

In addition to land, water will become a more precious resource, particularly in those countries, mostly in Africa, Asia, and the Middle East, that are "water-stressed," or have less than 1,700 cubic meters of water per capita per year. China and India are expected to be in this situation by 2015. Israel remains concerned over water and has repeatedly warned Lebanon and Syria about the value it places on this resource, even threatening military action against these states. Lack of water will also affect agriculture. In the Third World, agriculture consumes about 80 percent of available water, and this is an unsustainable situation.[287] Water shortages will threaten agriculture, creating tension particularly where water resources are shared among states, such as major river basins like the Nile and Indus, as well as in the region of the Jordan, Tigris, and Euphrates Rivers.[288]

The demand for energy also will quickly increase in the Third World, driven by economic growth and increasing population. By 2015, the demand for energy is expected to increase by 50 percent, and fossil fuels will remain the dominant source of energy despite concern over global warming. The total demand for oil will increase from 75 million barrels of oil per day in 2000 to over 100 million barrels in 2015. By that time, Asia is predicted to surpass North America as the leading energy consumer, using an estimated 75 percent of all Persian Gulf oil as Western markets use only 10 percent.[289] Thus, ever-growing demand will make the control of fossil fuel resources even more important than at present, and a continuing source of internal tension as well as a potential subject of strategic interest for the United States and other great powers. Three regions are of particular concern, given their proven or suspected oil reserves and political instability: the Caspian Sea, the Persian Gulf, and the South China Sea.[290]

Though the Caspian Sea basin produces relatively little petroleum at present, it has great reserves of oil and natural gas. It also has great reserves of political tension: conflicts between Armenians and Azeris, repercussions from the Kurdish conflict with both Turkey and Iraq, fighting in Uzbekistan and in Kyrgyzstan, and Chechnya and Dagestan, as well as the conflicting interests of China, Iran, Russia, Turkey, and the United States.

The Persian Gulf region contains nearly two-thirds of world petroleum supplies and is fragmented by religious divisions, territorial disputes, and regional power rivalries. The significant U.S. military presence in the region suppresses some of the conflicts and thus far has ensured the free flow of oil and natural gas from the region. Still, the region remains volatile.

In the South China Sea, undersea oil and gas reserves are subject to the contested claims of Brunei, China, Malaysia, the Philippines, Taiwan, and Vietnam. The reserves are significant, perhaps as much as 130 billion barrels of oil, more than the combined reserves of Europe and Latin America. These claims could result in conflict, especially since Beijing's 1992 claim to the entire Spratly and Paracel Islands and to all resources adjacent to them in the South China Sea. Oil for Japan and South Korea also passes from the Strait of Malacca through the South China Sea, making the outcome of these claims a matter of concern for Japan, South Korea, and the United States. Moreover, U.S. interests are affected, not only because of its allies. With increasing tension in the Sino-American relationship, some in the United States use the dispute over the South China Sea as a barometer of Chinese intentions. If China resolves these claims through negotiation, China looks like a power oriented toward the status quo; if not, China looks like an aggressive, rising power that should be contained by U.S. power.

Thus, resources remain a significant source of conflict in international relations, just as they were in humans' evolutionary past. As I have shown in this chapter, while the origin of warfare is informed by evolutionary theory, warfare has multiple forms and is greatly influenced by culture and the international system. To understand why, we must move beyond the bounds of this chapter, from ultimate causes of warfare to proximate ones. As E.O. Wilson describes it, "the particular forms of organized violence are not inherited. No genes differentiate . . . headhunting from cannibalism, the duel of champions from genocide."[291] Each culture gives a specific form to its warfare, from the Greek phalanx to the precision weaponry of information warfare.[292]

Moreover, almost every society creates cultural emoluments for its warriors. These are as varied as the honorific military uniforms worn by modern soldiers that often testify to the individual's skill and expertise, or the tattoos worn by Maori warriors, or even the often grisly war honors, or *coups,* such as scalps, among the Crow, Sioux, and other Native American tribes. As Milner, Anderson, and Smith conclude based on their archaeological study of an Oneota cemetery, "corpse mutilation that involved the removal of body parts as trophies heralding the attacker's success was common."[293] Moreover, for warriors in many tribes, success in warfare often translates into greater opportunities to mate or win a bride. As Turney-High explains, "it was very easy for a young Crow who had just counted *coup* not only to have an affair with one of the young girls . . . [and] with almost any married woman," and among the Flathead, "a youth who had counted *coup*" would seldom "be refused by any girl's parents, even though they might have objected before."[294]

Finally, of course, as I have stressed at several points within this and earlier chapters, it is always important to recall the significant influence of culture—or more broadly the social environment—on warfare. This effect is indispensable to any explanation of warfare. Cultural explanations usefully augment evolutionary explanations for warfare and help explain the particular form that warfare takes for a tribe or state. In addition, and much more significantly for efforts to stop warfare, culture allows warfare to be either suppressed or exacerbated.

Therefore, whether we consider the Mae Enga of New Guinea, or the Dassanetch of East Africa, or the United States today, it is difficult to overstate the significance of educational systems, popular culture, and the media, among many mechanisms that each society uses to inculcate its young people, especially, into its norms. In this context it is crucial that warfare be described for what it is. Though it is thrilling for some of its participants, none would question that it is horrific for all involved, whether they survive or become one of warfare's millions of victims. If warfare is to be suppressed, it cannot be glorified in any society. Modern, liberal societies often make a determined—and praiseworthy—effort not to glorify warfare. It is lamentable that other societies and cultures have not matched liberal societies in this respect. Doing so might help avert some warfare, especially wars that result from aggressive ideologies or virulent nationalism, and thus spare some people from its horrors.

4 Implications of an Evolutionary Understanding of War

Building on the argument presented in chapter 3, here I explore the implications of an evolutionary understanding of warfare. I make four arguments. First, I discuss the implications of warfare for human evolution, in particular the growth of human intelligence and human society. I find that the threat of external attack, from both predators and other humans, provides a strong ultimate cause for the rapid growth of human intelligence. Moreover, external threat also explains the origins of human society. The impact of war on the creation of the state system in early modern Europe has been widely studied by such superlative scholars as Geoffrey Parker, Michael Roberts, and Charles Tilly.[1] Turning the clock back by several millennia, warfare was instrumental for the formation of larger human societies that in turn made larger civilizations possible.

Second, I argue that humans are not the only species to wage war. Warfare is also found among ants and chimpanzees. This fact strengthens my argument that natural selection is a deep cause of human warfare. While I will advance this argument, I also stress that I do not mean warfare is in any sense "natural" or currently serves any evolutionary purpose, as I discussed in chapter 3. In addition, recognizing that warfare is found elsewhere in the animal kingdom requires a broader understanding of what constitutes warfare. We must move beyond the famous definition of warfare as "the continuation of policy with other means" provided by the Prussian soldier and philosopher of war Carl von Clausewitz. This definition is accurate but incomplete, as war is waged for specific biological motivations as well as political ones.

Third, I examine how evolutionary theory helps explain the physically and emotionally stimulating effects that warfare has on combatants. I suggest that the stimulation often experienced in warfare may be a residue from group experiences of hunting and fighting in Pliocene, Pleistocene, and Holocene conditions. These activities in our evolutionary past may have caused or helped to cause many of the powerful emotions and stimulation that individuals still experience today while engaged in combat. In this section of the chapter I also argue that evolutionary theory explains why males, rather than females, are more likely to be warriors.

Finally, I explain the role of disease in international relations and argue specifically that disease will become an important aspect of international relations as new diseases and new strains of existing diseases emerge, making biological warfare increasingly viable. In addition, I submit that disease has played a significant, but neglected, role in the conduct of imperialism, particularly in the European conquest of the Americas. I argue that a critically important factor in explaining the cost of European imperialism is whether an "epidemiological balance of power" existed between conquering people and those conquered. To conquer the New World, Europeans had a great advantage, unknown to them, in the diseases they carried and in the vulnerability of New World peoples to smallpox and measles, among others. The epidemiological imbalance of power made the conquest significantly easier than it would otherwise have been. In Africa and Asia, the epidemiological balance of power tended to favor the native peoples, making European colonization more costly.

Implications of Warfare for Human Evolution

While warfare may have begun to gain resources in conditions of scarcity, I argue that once it started, like human language or culture, it had a profound influence on human evolution. In this section I will discuss two of these influences: I will explain why the threat of warfare had an effect on the evolution of human intelligence. Then I will address the origins of human society and submit that it resulted from a gradual and three-staged process. First, small groups developed early in human history to provide protection against predators; over time these groups began killing large animals for food; finally, ever-larger bands had to stay together to counter the threat posed by other groups of humans.

Warfare and Human Intelligence

In *The Dawn Warriors,* a classic study of war's effect on human evolution, Robert Bigelow suggests that the great growth of the human brain was the result of the constant pressure from conflict that yielded greater intelligence, communication, and the need to cooperate.[2] Bigelow begins by assuming that early humans needed their primitive social organization to defend against attacks by predators and then by other humans. He then postulates that the group or groups with "the most effective brains, and hence with the greatest capacity for effective cooperation in attack or defense, maintained themselves longest in the most fertile and otherwise desirable areas."[3] As a result, they produced more offspring who could better defend themselves and take resources from other humans who were less intelligent and less able to communicate. He suggests that warfare at relatively infrequent intervals, perhaps once a generation, would be a sufficient force to affect gene frequencies on average.[4]

Roger Pitt, building on Bigelow's argument, submits that warfare is the only evolutionary pressure that makes unlimited demands upon intelligence and so leads to greater intelligence. Pitt argues that once human ancestors "had mastered the effective use of weapons for defence against predators, and thence for hunting," warfare would result because the population explosion would in turn increase territorial competition: "the turning point from uneasy peace to war probably came at moments when a natural calamity, such as drought, caused starvation to become a serious threat to some of the groups."[5] Pitt suggests that human rudimentary intelligence gave rise to the use of tools, which in turn became important in driving intelligence. Weapons facilitate such a spiral for three reasons: first, weapons permit distance between combatants; second, they may take time to deploy, so some planning is required; and third, one directed blow can incapacitate even a stronger foe by stunning or killing.[6] In contrast, most animals cannot expect to disable another of nearly similar size with one strike. A man who succeeds in surprising a foe has a chance of stunning him and then moving to kill him with reduced risk of injury to himself. Should his attack fail, he might escape simply because he has time—because he attacked at a distance or because those he attacked had to gather weapons and pursue him.

Human need and basic intelligence explain warfare to gain resources, and this in turn fostered greater intelligence, which made weapons possible. Then, weapons were coupled with another product of intelligence—tactical ingenuity, such as surprise or ambush. This combination could reduce the risk of injury or death. But now the aggressor could be ambushed himself and thus needed greater awareness as well as good weapons. Thus, technological or quantitative arms races are not necessarily an aspect only of modern international relations; they might have been an aspect of intertribal politics in the Pleistocene, where greater intelligence led to several features of successful warmaking: better weapons, ability to communicate, ability to plan an attack and defend oneself from other humans, and ability to learn from mistakes. All of these helped individuals to gather resources for themselves, their families, and perhaps more distantly related individuals and other members of their group.

Darius Baer and Donald McEachron develop Pitt's insight by suggesting that the development of weapons made nonrelated individuals more dangerous because weapons made it easier to attack than to defend; the offense became dominant.[7] That is, weapons reduced the costs of attacking because they could be developed faster than physiological protection against them could evolve; and, as Pitt described, their destructive power increased because weapons might be thrown or used to quickly injure or kill another in a surprise attack.[8] Consequently, the danger to other groups increased, forcing them to adapt similar weaponry, flee, or be either killed or absorbed by the more advanced group.

Baer and McEachron suggest that aggression toward neighboring groups would increase as a result of the development of new types of weapons, mechanisms to deliver them, and increases in the speed of their manufacture. As a result of more sophisticated weaponry, a group would become more aggressive toward others to take their resources, and xenophobia would become more pronounced in humans due to the threat posed by other groups. Moreover, because weapons increased the lethality of intragroup hostility, "the need to reduce and redirect conflict within the group would increase as well," with the result that "redirecting ingroup aggression outward toward other groups was a method of lowering intragroup conflict."[9] Thus, the origins of the diversionary theory of war, such as social imperialism, or "wag-the-dog" policies, might be very old indeed, existing long before Wilhelm II's Germany or the Clinton administration.

McEachron and Baer describe how the development and improvements in weaponry would introduce a positive feedback system affecting human intelligence. "Better weapons led to increased levels of group conflict. Conflict selected for . . . enhanced mental capacity in the form of increased learning capacity, improved communications, the emergence of the ability to plan"; in addition, "increased mental capacity . . . not only created better weapons through an improving technology, but made the group a better fighting unit, and thus a more dangerous adversary," which in turn "increased the selective pressure for conflict—and the cycle began again."[10]

They further argue that this feedback system increases pressure for a strong social structure that ultimately results in larger human groups. "A superior social organization, armed with an improved technology, would have also had the effect of increasing the hunting success of the hominid groups"; increased success, coupled with greater pressure from neighboring humans, "will select for larger groups."[11] While they believe the trend between weapon development and intelligence is a positive one, they agree with William Hamilton that this might occur in what Hamilton calls a "stepping-stone" manner.[12] Hamilton proposed that a hominid group might expand into the territories of neighboring groups, enlarge in size and consolidate, and then expand again. This description recognizes, first, the value of taking the territory and resources of other tribes; and second, that the capture and integration of women from these tribes might be used to assist reproduction.

The argument that a relationship exists between intelligence and warfare is interesting. However, the argument makes an assumption that a constant pressure, such as from the interaction of technology and warfare, is necessary for the evolution of human intelligence. While this might be true, to determine its origins is beyond the scope of this book. Moreover, in this book I cannot assess the strength of this argument versus other potential causes of human intelligence. These include language, the use of tools, and the need to forge coalitions, perhaps for warfare or hunting, in addition to "Machiavellian intelligence," the need or ability to manipulate others for personal gain that is found among many primates, especially those who live in dominance hierarchies.[13]

Warfare and the Creation of Societies

Evolutionary theory can explain the key role of warfare in the creation of societies. Richard Alexander's "balance of power" theory of human social organization is useful here.[14] I introduced this argument in the last chapter. In sum, he argues that human society originated in three stages: first, small groups developed early in human history for protection against predators; second, over time these groups began killing large animals for food; and third, increasingly large bands had to stay together to counter the threat posed by other groups of humans. Alexander believes that the threat of war and the need for protection through balancing the power of neighboring groups gave rise to human society.

Physiologist Jared Diamond builds on Alexander's theory in his explanation of this process: "The amalgamation of smaller units into larger ones has often been documented historically or archaeologically. Contrary to Rousseau's conception, such amalgamations never occur by a process of unthreatened little societies freely deciding to merge."[15] Rather, such amalgamation occurs in one of two ways: "by merger under the threat of external force, or by actual conquest."[16]

Population density is critical for understanding why the external threat from neighboring groups increased around thirteen thousand years ago. Diamond argues that where population densities are low—for example, in hunter-gatherer societies—groups may migrate away from the external threat posed by others, as many did in the Amazon, New Guinea, and North America.[17] Compared to farmers, hunter-gatherers can move more easily because they are less tied to a specific territory. Where densities are moderate, as in simple agricultural tribal societies, there are fewer places for survivors to flee. Furthermore, Diamond submits, with little food surplus in these societies, the victors have little use for any survivors. Thus began the pattern that defined war in the ancient world: Men are killed, women and children become captives, and territory is occupied. Finally, where population densities are high, the survivors have nowhere to go, but the victors can more effectively exploit the defeated by making them pay tribute or absorbing them into their society.

Diamond gives an example of a tribe that merged under external force: the Cherokee Indian confederation formed in the eighteenth century to counter increasing pressure from white settlers.[18] Later the Sioux, facing a comparable pressure from settlers, adopted the same response. Similarly, the United States faced pressure from external foes, the British, as well as internal foes, as Shays' Rebellion illustrated. Both experiences led early U.S. leaders to replace the Articles of Confederation with the Constitution, which provided for a strong central government to address both internal and external foes. In Europe, Germany unified in the nineteenth century in response to increasing threats from the larger, unified states around it, despite the strong opposition of some Germans.

The second cause of population aggregation according to Diamond, i.e., by conquest, has been studied more thoroughly.[19] Charles Tilly, perhaps the foremost contemporary scholar of war's effect on state formation, succinctly observed that "war made the state, and the state made war."[20] Political scientists also recognize that the competitive pressure of the international system forces socialization.[21] The system forces states, or prestate political entities, to socialize or run the risk of losing sovereignty. The greater power made possible by the state apparatus created an external pressure for those prestate political entities to merge and create their own states or risk the consequences. But as Robert Carneiro and others have demonstrated, war created complex societies long before European states formed in the sixteenth and seventeenth centuries, and in non-European cultures as well.[22] States formed by conquest include the Zulu state in southeastern Africa, the Ashanti state of west Africa, and the Aztec and Inca empires, as well as Rome and the Macedonian empire under Alexander.

Warfare in the Animal Kingdom and Broadening Clausewitz's Definition of Warfare

In 1904, William James famously described "man" as "simply the most formidable of all the beasts of prey, and, indeed, the only one that preys systematically on its own species."[23] A half-century later, Erich Fromm echoed James when he argued that only humans are capable of some forms of aggression.[24] Both James and Fromm are correct, in part. Humans are "formidable beasts of prey," but it is perhaps only in their technology that they are unique in the animal kingdom.

I say this because we now know that James was wrong: warfare is not a unique human form of behavior. Ethological studies of warfare in the animal kingdom have shown that warfare is not unique to humans, but is conducted by ants and chimpanzees as well. This fact is significant because it suggests that warfare has contributed to fitness in several animal species, not just humans. Thus evolutionary theory may have dealt one more blow to human pride.[25]

My analysis in this section of the chapter is divided into three parts. I first discuss warfare among ants. I then address the evidence that warfare is found among chimpanzees. Finally, having shown that warfare is not solely human, I argue for broadening Clausewitz's famous definition of warfare.

The idea of warfare in animals may be a difficult one for people to accept. Indeed, aggression is common among animals, but coalitionary, or polyadic, aggression is not, and warfare is a behavior few animals practice.[26] However, having observed those species that engage in coalitionary aggression, ethologists agree that humans are far from being the most violent animal. Wolf packs occasionally invade neighboring packs' territory and attack residents.[27] In fact, in one area, intraspecies conflict among wolves accounted for 43 percent of deaths not caused by humans.[28] Among spotted hyenas, subclans patrol the territory of larger clans and may attack intruders.[29] This behavior is also found among prides of lions.[30] Ethologists Alexander Harcourt and Frans de Waal find that "in dolphin communities, as well as among humans, separate groups will cooperate in attacks on rival groups."[31] Bernard Chapais, who has studied alliance behavior among nonhuman primates, found that they commonly form alliances to attack conspecifics.[32] Studies of numerous animals, including cheetahs, hyenas, langur and rhesus monkeys, and lions, reveal that individuals engage in lethal fighting, infanticide, and even cannibalism much more frequently than do humans.[33] Other studies have shown ethologists that animals commit, in actuality, not perhaps in motive, murder, rape, deception, theft, torture, and genocide.[34] As Edward O. Wilson has documented in his study of insect societies, "alongside ants, which conduct assassinations, skirmishes, and pitched battles as routine business, men are all but tranquilized pacifists."[35] He believes, in fact, that an unbiased Martian observer might conclude that *Homo sapiens* is rather peaceful when compared to other species.[36] Table 4.1 provides a list of animals in which intergroup aggression has been documented between members of two or more spatially separate, distinct, and identifiable groups; such aggression is defined as an interaction in which members of one social group cooperate in threatening, chasing, striking, injuring, or killing at least one member of another group.[37]

Just as humans do, many animals engage in polyadic aggression. "At the simplest level of comparison," Harcourt and de Waal argue, "the phenomenon of cooperation during conflict, far from being a human trait, occurs in a wide array of animal societies," from prides of African lions to colonies of Arizona ants.[38] In these species and others, all "compete with one another, and in all, individuals within groups cooperate in the conflict."[39] Moreover, in their summary of cooperation for conflict among animals they submit that: "lethal intergroup conflict is not uniquely, or even primarily, a characteristic of humans."[40]

In contrast to aggression, which is common in the animal kingdom, warfare among nonhuman animals has only been documented in some social insects, such as ants, and in chimpanzees.[41] The term warfare is used deliberately: organized violence conducted to force others to yield to the attacking group. Ethologists and social scientists have long thought that the most important distinguishing characteristic of human intergroup aggression was the deliberate killing of conspecifics by organized groups. Although many species are violent, and will aggress against conspecifics, at this time we know that only a handful of species—some social insects, chimpanzees, and humans—wage war.[42]

Table 4.1. Documented Intergroup Aggression in Animal Species	
Animal	*Common Name*
1. *Acynonyx (Cynaelurus) jubatus*	cheetah
2. *Alouatta palliata*	mantled howler monkey
3. *Alouatta seniculus*	red howler monkey
4. *Alouatta fusca*	brown howler monkey
5. *Anoplolepis longipes*	ant
6. *Ateles belzebuth*	long-haired spider monkey
7. *Brachyteles archnoides*	muriqui or woolly spider monkey
8. *Callicebus moloch*	dusky titi monkey
9. *Callicebus torquatus*	yellow-handed titi monkey
10. *Callithrix jacchus*	common marmoset
11. *Canis lupus*	wolf
12. *Cebus albifrons*	white-fronted capuchin
13. *Cebus apella*	brown, black-capped or tufted capuchin
14. *Cebus capucinus*	white-faced capuchin
15. *Cebus olivaceus*	wedge-capped capuchin
16. *Cercocebus [Lophocebus] albigena*	crested or grey-cheeked mangabey
17. *Cercopithecus [Chlorocebus] aethiops*	vervet
18. *Cercopithecus ascanius*	redtail monkey
19. *Cercopithecus mitis*	guenon or blue monkey
20. *Cercopithecus neglectus*	De Brazza's monkey
21. *Cercopithecus talapoin*	talapoin
22. *Colobus [Procolobus] badius*	red colobus
23. *Colobus guereza*	black-and-white colobus
24. *Crocuta crocuta*	spotted hyena
25. *Eciton buchelli*	army ant
26. *Eciton hamatum*	army ant
27. *Erythrocebus patas*	patas monkey
28. *Formica omnivora*	ant
29. *Gallinula mortierii*	Tasmanian native hen
30. *Gazella gazella gazella*	mountain gazelle
31. *Gorilla g. berengei*	mountain gorilla
32. *Hapalemur griseus*	gentle lemur
33. *Helogale undulata*	dwarf mongoose
34. *Heterocephalus glaber*	naked mole rat
35. *Hylobates agilis*	agile or dark-handed gibbon
36. *Hylobates klossii*	Kloss's gibbon
37. *Hylobates lar*	whitehanded gibbon
38. *Hylobates moloch*	moloch or silvery gibbon
39. *Hylobates pileatus*	pileated or capped gibbon
40. *Hylobates [Symphalangus] syndactylus*	siamang

Table 4.1. Documented Intergroup Aggression in Animal Species (cont'd)	
Animal	*Common Name*
41. *Helogale undulata*	dwarf-mongoose
42. *Indri indri*	indri
43. *Lagothrix lagothricha*	(Humboldt's) woolly monkey
44. *Lemur catta*	ringed tail lemur
45. *Lemur [Eulemur] fulvus*	brown lemur
46. *Linepithema humilis*	Argentine ant
47. *Lycaon pictus lupinus*	Cape hunting dog
48. *Macaca fascicularis*	kra or long-tailed or crab-eating macaque
49. *Macaca fuscata*	Japanese macaque
50. *Macaca mulatta*	rhesus monkey
51. *Macaca radiata*	bonnet macaque
52. *Macaca sylvanus*	barbary macaque
53. *Macaca thibetana*	Tibetan macaque
54. *Mungos mungo*	banded mongoose
55. *Myrmica scabrinodis*	ant
56. *Oecophylla smaragdina*	ant
57. *Otaria byronia*	sea lion
58. *Pan paniscus*	Bonobo or pygmy chimpanzee
59. *Pan troglodytes*	chimpanzee
60. *Panthera leo*	lion
61. *Papio anubis*	olive baboon
62. *Papio cynocephalus*	yellow or savanna baboon
63. *Papio hamadryas*	hamadryas or desert baboon
64. *Papio ursinus*	chacma baboon
65. *Pheidole megacephala*	common ant
66. *Presbytis aygula*	Sunda Island leaf monkey
67. *Presbytis cristata [Trachypithecus cristata]*	silver leaf monkey or lutong
68. *Presbytis [Semnopithecus] entellus*	gray or Hanuman langur
69. *Presbytis [Trachypithecus] johnii*	Nilgiri langur
70. *Presbytis pileata [Trachypithecus pileatus]*	capped langur
71. *Presbytis potenziani*	Mentawai langur
72. *Presbytis senex [Trachypithecus retulus]*	purple-faced langur
73. *Presbytis thomasi*	Thomas' [leaf] langur
74. *Propithecus verreauxi*	white or Verreaux's sifaka
75. *Rattus norvegicus*	brown rat
76. *Saguinus imperator*	emperor tamarin
77. *Saguinus fuscicollis*	saddleback tamarin
78. *Saguinus mystax*	moustached tamarin

Table 4.1. Documented Intergroup Aggression in Animal Species (cont'd)	
Animal	*Common Name*
79. *Saimiri sciureus*	squirrel monkey
80. *Solenopsis invicta*	fire ant
81. *Solenopsis saevissima*	fire ant
82. *Suricata suricatta*	slender-tail meerkat or viverrid
83. *Tetramorium caespitum*	common pavement ant
84. *Turdoides squamiceps*	Arabian babbler
85. *Tursiops truncates*	bottlenose dolphin

Sources: Johan Matheus Gerradus van der Dennen, *The Origin of War: The Evolution of a Male-Coalitional Reproductive Strategy,* 2 Vols. (Groningen, Netherlands: Origin Press, 1995), Table 3.1, pp. 155–156; van der Dennen, "Animal Intergroup and Intercoalitional Agonistic Behavior," correspondence with the author, September 2002; Edward O. Wilson, *The Insect Societies* (Cambridge, Mass.: Belknap Press of Harvard University Press, 1971), pp. 55–66, 451–452; and Wilson, *Sociobiology: The New Synthesis* (Cambridge, Mass.: Harvard University Press, 1975), p. 50.

Note: This list is representative of the broad spectrum of intergroup aggression among animals but is necessarily incomplete.

Warfare among Ants

The fact that warfare may be found among ants may surprise some: how can insects possess the intellectual ability to wage war? Preeminent entomologists, however, say that ants do engage in warfare; they even employ elaborate tactics that demonstrate their mastery of attrition warfare. In addition, their battles result in heavy casualties for both winner and loser, and the loser often faces extermination. As Bert Hölldobler and Wilson describe, the warlike and aggressive nature of ants suggests to an international relations theorist that ants are the embodiment of offensive realism. Ants, they say, are "arguably the most aggressive and warlike of all animals. They far exceed human beings in organized nastiness; our species is by comparison gentle and sweet-tempered."[43] They sum up the foreign policy of ants: "restless aggression, territorial conquest, and genocidal annihilation of neighboring colonies wherever possible. If ants had nuclear weapons, they would probably end the world in a week."[44]

According to Wilson, territorial fighting among mature colonies of both same and differing species is common but not universal among ants, including the following fifteen genera: *Pseudomyrmex, Conomyrma, Myrmecocystus, Pogonomyrmex, Leptothorax, Solenopsis, Pheidole, Tetramorium, Linepithema, Azteca, Anoplolepis, Oecophylla, Formica, Lasius, Camponotus.*[45] Wilson says that territorial wars between colonies of different ant species occur only occasionally in the cold temperate zones. "In the tropics and warm temperate zones," however, "intense aggression is . . . very common. . . . Certain pest species, particularly *Pheidole megacephala,* the fire ant *Solenopsis saevissima,* and the so-called Argentine ant *Linepithema humilis* (formerly called *Iridomyrmex humilis*), are famous for their belligerency and destructiveness of their attacks on native ant faunas."[46]

Entomologist E.S. Brown provides the following account of a war at Lunga on Guadalcanal in the Solomon Islands between colonies of the introduced African ant, *Anoplolepis longipes,* and defending colonies of the native ants, *Oecophylla smaragdina* and *Iridomyrmex myrmecodiae.* From Brown's description, the conflict on one tree seems to occur on a scale equivalent to World War I. The story begins as the invading *Anoplolepis* ants move to the base of the tree trunk and "large numbers of *Oecophylla*" descend, "to ring the trunk in defensive formation just above them."[47] Brown continues, "It then becomes a ding-dong struggle, the dividing line between the two species sometimes moving up or down a few feet from day to day; any ant wandering alone into the other species' territory is usually surrounded and overcome. Eventually one species will get the better of the other, but this may not happen for several days or weeks."[48]

He describes another struggle between the attacking *Anoplolepis* and the defending *Iridomyrmex:* "*Anoplolepis* had advanced on to the base of the trunk of a palm occupied by *Iridomyrmex,* which had descended in force from the trunk and formed a complete phalanx of countless individuals, almost completely covering the trunk over about 2 ft. of its length. After a few days this defensive formation was still intact, but had retreated higher up the trunk; eventually it was driven from the trunk altogether, and later *Anoplolepis* took possession of the crown."[49]

The war Brown witnessed is not atypical in length. The nineteenth-century entomologist Henry C. McCook observed campaigns that lasted for weeks.[50] Indeed, Wilson argues that one can easily witness these wars "in abundance on sidewalks and lawns in towns and cities of the eastern United States throughout the summer."[51] All a casual observer must do is look down to see "masses of hundreds or thousands of the small dark brown workers locked in combat for hours at a time, tumbling, biting, and pulling one another, while new recruits are guided to the melee along freshly laid odor trails."[52]

Ants wage warfare for some of the reasons humans do: to gain resources such as food and territory. As Hölldobler and Wilson explain, "warfare among the ants is all about territory and food."[53] If victorious, attackers remove stored food, grain by grain, to their own nest. According to Hölldobler, ants are sensitive to the balance of power between colonies, with mature, larger colonies often raiding younger ones: "when one colony is considerably stronger that the other, i.e., when it can summon a much larger worker force, the tournaments end quickly and the weaker colony is raided."[54] Then, "foreign workers invade the nest, and the queen of the resident colony is killed . . . or driven off" and the workers killed or enslaved.[55]

Ants that engage in warfare avoid a major threat faced by individuals of other species: the risk of serious injury or death that would affect their ability to reproduce and raise offspring. This is because warriors are a sterile caste. The colony can lose workers without hindering the reproductive potential of their genotype, since reproductive ability lies solely with their relatives.

While ants and, according to Pitt, some termites wage war, other eusocial insects, such as bees, do not.[56] This might be because bees—with their much greater range—face less territorial or population pressure than the earth-bound ant or termite, although there is evidence that some wasps and hornets wage war as well. Additionally, why has the relationship between warfare and intelligence—suggested by Bigelow, Pitt, and Baer and McEachron—not affected eusocial insects that wage war? There are two reasons why. First, biological factors such as an exoskeleton seriously limit the growth of insect brains; second, predation has a greater effect on ants than on hominids and so population pressure is lessened. As a result, these insects face no selection pressure for greater intelligence.

Warfare among Chimpanzees *(Pan troglodytes schweinfurthii)*

In perhaps the most famous of all ethological studies, the leading authority on chimpanzee behavior, Jane Goodall, extensively observed the chimpanzees of the Gombe National Park in Tanzania and showed that aggression and warfare are part of chimpanzee behavior.[57] In the course of her studies, she watched not only violence associated with the struggle for dominance among males but also significant intercommunal violence, including attacks, murder, and a four-year war between rival communities.[58] These conclusions may be rather startling on their own, but she draws one even more unsettling one, reached after years of observing chimpanzee society: "If they had had firearms and had been taught to use them, I suspect they would have used them to kill."[59]

As with ants, evolutionary theory can explain why warfare would contribute to fitness among chimpanzees. Richard Wrangham provides an explanation of chimpanzee warfare, arguing for an "imbalance of power" hypothesis.[60] That is, unprovoked intercommunity aggression provides successful males with increased access to reproductive females and/or material resources. Both may be acquired directly and immediately, for example by seizing territory and attempting to hold it. However, Wrangham's insight is that these males may also acquire these assets over time by reducing their rivals' power to form or maintain coalitions.[61] This behavior contributes to both absolute and relative fitness. As Wrangham argues, "by wounding or killing members of the neighboring community, males from one community increase their relative dominance over the neighbors. . . . this tends to lead to increased fitness of the killers through improved access to resources such as food, females, or safety."[62]

Thus conflict may arise over access to resources but also to reduce the power of neighbors by eliminating males and incorporating females into the group. Indeed, the warfare that Goodall documents is of the latter type: One chimpanzee community attacks another but not over resources; in each case a group of males, occasionally accompanied by an anestrous female, attacks and kills a single male in the neighboring group. Over time, the competing group is eliminated. Goodall's evidence suggests the group is attacking the individual to injure or kill him because he is seen as a competitor or potential threat.

If an attack leads to the wounding or death of a male, then that community's strength is substantially reduced. As Wrangham explains, "if a neighboring community has ten males, its fighting power is reduced by 10%. This reduction lasts for a considerable time, because the system of male philopatry means that a dead male can be replaced only via births within the community."[63] This obviously is a slow process but does increase the aggressors' probability of winning subsequent encounters and reduces, perhaps significantly, competition for resources.

Wrangham's argument provides the motivation for warfare and also explains why the means or mechanism of attack is almost always a group attack. He submits this is because unprovoked aggression will result only if one gains by acquiring resources or reproductive opportunities, but also if the opportunity to attack is "economical." That is, can it be conducted at relatively low risk to the individual?[64] This is possible in a group attack. If a group encounters an individual or significantly smaller group, creating an imbalance of power should lead to success for the attacking, larger group.[65] A group also clearly provides greater defense from attack. This provides one explanation of why chimpanzees often move in groups, for defense as well as for offense if the opportunity for a successful attack presents itself.

In my discussion I aim to determine if Wrangham's imbalance of power argument can explain chimpanzee warfare. I argue that the ethological evidence does support Wrangham's argument. Ethological research was conducted at the major continuous, or almost continuous, chimpanzee observation sites at Budongo Forest in Uganda, Gombe, Kibale Forest in Uganda, Mahale Mountains National Park in Tanzania, and Taï National Park in Ivory Coast. This research shows that chimpanzees often act defensively and offensively toward other chimpanzees. They are aware that other chimpanzee males may be a threat to an individual, particularly if he is male, and they almost always seek to avoid them.[66] Moreover, males most frequently participate in actions directed at "strangers," chimpanzees unknown to human observers, or neighboring groups.

Specifically, five types of chimpanzee behavior are relevant for confirming Wrangham's argument: (1) territorial defense; (2) border patrols; (3) deep incursions; (4) coalitionary attacks; and (5) coalitionary kills and warfare. With respect to the first four, these behaviors are common, having been documented consistently at most, if not all, major chimpanzee research sites. Concerning the fifth, coalitionary kills and warfare are less common forms of behavior, as with humans, which is expected given the costs and risks associated with both forms of behavior. Through 1998, ethologists at the research sites mentioned have documented twenty-four confirmed (eleven) or suspected (thirteen) interspecific kills among chimpanzees and two wars between neighboring groups.

In the following discussion, I will examine each form of behavior. Table 4.2 includes the type of territorial behavior documented among different chimpanzee subspecies (*Pan troglodytes schweinfurthii* and *Pan troglodytes verus*), and table 4.3 includes the evidence of lethal violence among chimpanzees at the principal research sites.

The first type of behavior, territorial defense, has been documented in every study that has described intercommunity relationships.[67] This behavior is common in the animal kingdom among tens of thousands of territorial species, including chimpanzees, ants, bees, lions, wolves, and humans. Evidence of it is based on counter-calling between neighboring males; the rapidity with which male chimpanzees move toward a site where an opposing party has been detected; avoidance of larger opposing parties; and charging displays directed toward an opposing party.

Second, in addition to forming groups for hunting and foraging trips,

Table 4.2. Territorial Behavior in Chimpanzees: *Pan troglodytes (P.t.) schweinfurthii* and *P.t. verus*

	P.t. schweinfurthii				*P.t. verus*
	Gombe	Mahale	Kibale	Budongo	Taï
1. Territorial Defense	+	+	+	+	+
2. Border Patrols	+	+	+	?	+
3. Deep Incursions	+	+	+	?	+
4. Coalitionary Attacks	+	+	?	?	+
5. Coalitionary Kills	+	+	+	+	—

Source: Richard W. Wrangham, "Evolution of Coalitionary Killing," in *Yearbook of Physical Anthropology,* Vol. 42 (New York: Wiley-Liss, 1999), Table 2, p. 6. Wrangham's data includes all long-term studies of chimpanzees in typical, natural surroundings. "Coalitionary Kills" refers to adult victims only. "Coalitionary Attacks" means nonlethal attacks by several males on a single victim.
Symbols: "+" = positive evidence, "?" = insufficient data, and "—" = no evidence.

Table 4.3. Lethal Chimpanzee Violence by Site

Site	*Subspecies*	*Adult Deaths*	*Infanticides*	*Years of Study*
Gombe	*P.t. schweinfurthii*	7 (5)	6 (3)	38
Mahale	*P.t. schweinfurthii*	1 (6)	4 (6)	33
Kibale	*P.t. schweinfurthii*	2 (2)	1 (0)	11
Budongo	*P.t. schweinfurthii*	1 (0)	1 (1)	8
Taï	*P.t. verus*	0 (0)	0 (0)	19
Total		11 (13)	12 (10)	109

Adapted from: Richard W. Wrangham, "Evolution of Coalitionary Killing," in *Yearbook of Physical Anthropology,* Vol. 42 (New York: Wiley-Liss, 1999), Table 3, p. 6. For "Adult Deaths" numbers include kills recorded on the basis of direct observation and/or fresh bodies as well as, in parentheses, those animals that disappeared suspiciously. For "Infanticides" numbers include kills observed directly or, in parentheses, inferred from context. "Years of Study" is number of years from beginning of continuous study until 1998. Wrangham reports one suspect death at Budongo, but current research did not confirm this finding.

Goodall has found that adult males form groups to patrol their territory.[68] A group of males, perhaps with an occasional female, will visit peripheral areas of a territory. Goodall describes a typical patrol as "typified by cautious, silent travel during which the members of the party tend to move in a compact group"; they are quieter than normal, carefully watch and listen for evidence of trespass, and may "climb tall trees and sit quietly for an hour or more, gazing out over the 'unsafe' area of a neighboring community."[69] During a patrol, group members are "very tense" and "may sniff the ground, treetrunks, or other vegetation" for evidence of neighboring chimpanzees and "may pick up and smell leaves, and pay particular attention to discarded food wadges, feces, or abandoned tools on termite heaps."[70] Furthermore, if they see a "fairly fresh sleeping nest . . . one or more of the adult males may climb up to inspect it and then display around it so that the branches are pulled apart and it is partially or totally destroyed."[71]

Goodall was impressed the most by the silence of the chimpanzees on patrol. "They avoid treading on dry leaves and rustling vegetation. On one occasion vocal silence was maintained for more than three hours"; in addition, males do not make typical calls and "copulation calls are suppressed by females."[72] In stark contrast to this behavior, when they return to familiar areas, there is often "an outburst of loud calling, drumming displays, hurling of rocks, and even some chasing and mild aggression between individuals."[73]

Having studied the Taï chimpanzees, ethologists Christophe Boesch and Hedwige Boesch-Achermann also describe a "resolute and silent progression" among the chimpanzees on patrol; "they advance rapidly in a line one behind the other and stop regularly to listen and search for signs" of other chimpanzees.[74] The farther they intrude into another community's territory, "the more carefully they progress and the more regularly they stop to listen. . . . They remain silent during the whole patrol, very occasionally eating a leaf or two."[75] At specific geographical points, such as a ridge, the chimpanzees may stop and listen for some time before going on. They "cover large areas within the stranger territory, and normally come back into their territory at a different point. They regularly smell at tree trunks and leaves, even more so if they come upon fresh traces of chimpanzees, such as a wadge of fruit, a nut-cracking atelier, or a nest."[76]

Border patrols appear to be quite common. In Gombe, patrols occurred during 28 percent of the 134 boundary visits that parties from the Kasakela community of chimpanzees made from 1977 to 1982.[77] In Taï, border patrols along or within the territory of neighboring chimpanzees occurred in 29 percent of 129 territorial actions.[78] Their frequency also seems to vary depending on the status of relations between communities. During and after the 1974–1977 period in Gombe in which the Kasakela community killed the males of the rival and neighboring Kahama community, Kasakela border patrols were disproportionately directed toward the Kahama territory. These patrols were most frequent in 1972–1973, when thirteen patrols occurred in fifty-eight days, for a rate of eighty-two per year. In contrast, in the five years following the extinction of the Kahama community (1978–1982), border patrols continued at a rate of eighteen per year.[79]

Third, deep incursions are characterized by substantial penetration—up to a kilometer or more—into the neighboring territory and silent and cautious patrolling. Deep incursions have been documented at Gombe, Kibale, Mahale, and Taï. All were led by males and lasted up to six hours. Boesch and Boesch-Achermann find that in 87 percent of their patrols, the Taï chimpanzees made deep incursions into neighboring territory.[80]

The fourth type of behavior is coalitionary attack. Wrangham defines these as interactions in which observers determine that the intent of those in the aggressive party is to hurt or kill one or more of the victims. To this definition I add that the aggression must be directed against one or more chimpanzees of a neighboring group. In Mahale, ethologist Toshisada Nishida reports attacks by the M-group toward K-group males. In one instance, three M-group males chased a K-group male for two hundred meters; when they caught him, they bit and stomped on him before the attack ended.

In Taï, Boesch and Boesch-Achermann report similar behavior. In what they term "commando" attacks, parties of males penetrate into the neighboring range and attack one or more of these chimpanzees, sometimes after hours of silent waiting and listening. "We twice saw the study community being victim of a commando attack, in one of which Macho [adult male] escaped with nineteen wounds."[81] They also report seeing six incidents of another form of coalitionary attack that they term a "lateral attack."[82] In this form of attack a party moved laterally while looking in the direction of strangers, then approached, chased, and on at least one occasion caught and attacked one of the opponents. This appears to be a tactic for isolating a victim from the rest of the party, for maximizing surprise, or for maximizing the chances for a community in decline that could not succeed in a direct, frontal assault against a larger community.

In Kibale, no complete coalitionary attacks have been seen, but twice parties of neighbors have silently charged toward isolated males of the Kanyawara study community and then stopped when they saw humans; this suggests the human presence prevented an attack. In a third case, five Kanyawara males attacked an adolescent male and a nulliparous female near the border, but later retreated when confronted by four adult males from the neighboring community who chased them for seven hundred meters.[83]

In Gombe, Goodall documents a series of coalitionary attacks. When chimpanzees on patrol encounter no other chimpanzees, they may return to their own range or trespass onto a neighboring one. Intruders who happen across a neighboring subgroup usually retreat after making a display. However, if they encounter one male or female or a consorting male-female pair, they will often attack or "hunt" through a coordinated effort of several males. If captured, the individual may be bitten or beaten until severely wounded or dead. Having described this behavior in several discrete chimpanzee communities, Goodall says: "it is clear that interactions between males of neighboring communities are typically hostile."[84] In her 1986 study, Goodall outlines the common aspects in the five attacks she witnessed:

> (a) the attacks were all long—the shortest lasted at least ten minutes, and three continued more than twice as long; (b) all were gang attacks, during which the aggressors sometimes assaulted the victim one after the other, or two to five assailed the victim simultaneously; (c) all the victims were, at some point, held to the ground by one or more of the aggressors while others hit and pounded; (d) all the victims, in addition to being hit, stamped on, and bitten, were dragged first in one direction, then another; . . . and (g) during each incident the observers, all thoroughly experienced in chimpanzee behavior, *believed* that the aggressors were trying to kill their victims. . . . because the attacked showed some of the patterns which, while commonly seen during the killing of large prey, have not been seen during *intra*community fighting—as when one of Goliath's [adult male] legs was twisted, when a strip of flesh was torn from Dé's [adult male] thigh, or when Satan [adult male] drank the blood pouring from Sniff's [adolescent male] nose. Moreover, in all cases the attacks continued until the victims were incapacitated.[85]

The fifth type of behavior is the coalitionary kill, which may become warfare. Warfare, defined in this instance as a systematic confrontation and killing between free-ranging chimpanzees, has been well documented at two research sites, Gombe and Mahale, and lethal coalitionary attacks have been reported from all four study sites of the eastern subspecies *Pan troglodytes schweinfurthii.*

Table 4.4 lists known and suspected cases of coalitionary killing. In the

Table 4.4. Intraspecific Kills and Possible Kills of Adult Chimpanzees to 1999

Result	*Site*	*Date*	*Aggressor's Community*	*Aggressor's Party*	*Victim's Community*	*Victim's Party*	*ID*	*Reference*
Death	Gombe	1972	Kahama	Unknown	Kalande?	1M or more	1F	Wrangham
Death	Gombe	1974	Kasakela	7M, 1F	Kahama	1M	Godi	Goodall
Death	Gombe	1974	Kasakela	3M, 1F	Kahama	3M, 1F	Dé	Goodall
Death	Gombe	1975	Kasakela	5M	Kahama	1M	Goliath	Goodall
Death	Gombe	1975	Kasakela	4M	Kahama	1F	Madam Bee	Goodall
Death	Gombe	1977	Kasakela	5M	Kahama	1M or more	Charlie	Goodall
Death	Gombe	1977	Kasakela	6M	Kahama	1M	Sniff	Goodall
Death	Kibale	1991	Rurama	Gang	Kanyawara	1M	Ruwenzori	KCP and Wrangham and Peterson
Death	Mahale	1995	M-group	Gang	M-group	1M	Ntologi	Nishida
Death	Budongo	1998	Sonso	Gang	Sonso	1M	Zesta (+2 injured)	Wrangham
Death	Kibale	1998	Kanyawara	Gang	Sebitole	1M or more	Unknown	KCP
Death?	Mahale	1970	M-group	Unknown	K-group	Unknown	Kaguba	Nishida 1985
Death?	Gombe	1973	Unknown, possibly Kasakela	Unknown, possibly Kasakela	Kahama	Unknown	Hugh	Goodall
Attack and Death?	Gombe	1974	Kahama	3M	Kalande	1M, 1F	M attacked, caught, both escaped	Goodall
Attack and Death?	Mahale	1974	K-group	3M	M-group	1M	1 on 1 fight	Goodall

Death?	Mahale	1976	M-group	7M	K-group	Unknown	Infant male	Nishida 1985
Death?	Gombe	1977	Unknown, suspected Kasakela	Unknown	Kahama	Unknown	Willy Wally	Goodall
Death?	Mahale	1979	Unknown, suspected M-group	Unknown	K-group	Unknown	Sobongo	Nishida 1985
Death?	Gombe	1980	Kalande	Unknown	Kasakela	1F	Passion killed?	Goodall
Death?	Gombe	1981	Kalande	Unknown	Kasakela	Unknown	Humphrey killed?	Goodall
Death?	Mahale	1982	Unknown suspected M-group	Unknown	K-group	Unknown	Kamemanfu	Nishida 1985
Death?	Kibale	1988	Unknown, possibly Wantabu	Unknown	Unknown, possibly Kanyawara	1M Unknown	Unknown	Wrangham and Peterson
Death?	Kibale	1994	Unknown, possibly Wantabu	Unknown	Unknown, possibly Kanyawara	1M Unknown	Unknown	Wrangham and Peterson
Death?	Mahale	1996	M-gp	Unknown	M-group	1M	Jilba	Wrangham

Sources: Richard W. Wrangham, "Evolution of Coalitionary Killing," in *Yearbook of Physical Anthropology,* Vol. 42 (New York: Wiley-Liss, 1999), Table 4, p. 9; and Wrangham and Dale Peterson, *Demonic Males: Apes and the Origins of Human Violence* (Boston: Houghton Mifflin, 1996), pp. 20–21. *Symbols:* "Death?" = cause of death is suspected attack by neighboring chimpanzees; "M" = male, "F" = female. Goodall = Goodall, *The Chimpanzees of Gombe: Patterns of Behavior* (Cambridge, Mass.: Harvard University Press, 1986). "KCP" = Kibale Chimpanzee Project. Nishida = T. Nishida, "The Great Chimpanzee Ruler Killed by Coalition of Previous Group Mates: Cruel Political Dynamics in Wild Chimpanzees," *Asahi-Shimbun*, 31 January 1996, p. 1; Nishida 1985 = Nishida, Mariko Hiraiwa-Hasegawa, Toshikazu Hasegawa, and Yukio Takahata, "Group Extinction and Female Transfer in Wild Chimpanzees in the Mahale National Park, Tanzania," *Zeitschrift für Tierpsychologie*, Vol. 67, (1985), pp. 284–301. Wrangham = Wrangham, "Evolutionary of Coalitionary Killing." Wrangham and Peterson = Wrangham and Peterson, *Demonic Males*.

inferred cases, observers believed the most likely explanation for a chimpanzee disappearing was its being killed by neighbors. They made these inferences for three reasons: those who disappeared were healthy and not senescent, other causes of death seemed improbable, and the disappearances occurred at a time and place where there were patently hostile relationships with a neighboring community. Although the number of reported episodes of coalitionary kills is still small, the episodes are noteworthy because they have been reported from four separate sites and appear to be demographically significant in proportion to the total number of observed adult deaths. For example, for adult males in the Kahama community in Gombe, of the adult males confirmed dead, about 70 percent were killed in intraspecific coalitionary aggression, and possibly up to 100 percent.

At Gombe, Goodall found evidence of coalitionary killing and of warfare. As briefly noted above, she documented how one group of chimpanzees, the Kahama (also known as "southern") community, had split from a larger one, the Kasakela (also known as "northern"), and was then exterminated over the four years 1974 through 1977.[86] The Kahama community consisted of six fully mature males. Of these, four—Charlie, Dé, Godi, and crippled Willy Wally—were in their prime, Hugh was past prime, and Goliath was old. An adolescent male, Sniff, was also part of this group.[87] The Kasakela community consisted of eight fully mature males, of whom six—Humphrey, Evered, Figan, Jomeo, Satan, and crippled Faben—were in their prime and two, Mike and Hugo, were old.[88] All five adult males in the Kahama community who were alive during the 1974–1977 period and one female (Madam Bee) were killed, while the remaining females (Little Bee, Honey Bee, Kidevu, and possibly Mandy and Wanda) were incorporated into the larger group, as often occurs in human warfare.[89] Border patrols by Kasakela males were directed mostly toward the Kahama community during this period, although Goodall also notes they attacked stranger females (presumed to be from a neighboring community) who were anestrous, not nulliparous, and sometimes carrying infants.[90] In the end, the Kahama community was eliminated and Kahama territory was incorporated into the Kasakela community. This leads Wrangham to suggest that unusual demographic and social conditions sparked this conflict.[91]

At Mahale, Nishida and his colleagues confirmed Goodall's insights about chimpanzee warfare and discovered that the chimpanzee community they called the K-group gradually became extinct over fifteen years as almost all of its adult males died in conflict with the neighboring and much larger M-group.[92] The evidence indicates that as many as six of these deaths were brought about by the M-group. As Nishida explains, "at least some adult males, particularly Sobongo and Kamemanfu, were killed by M-group chimpanzees. Severe fighting was occasionally witnessed between males of K-group and M-group in the overlapping area of [the] two unit-groups' ranges," and "M-group's males were sometimes seen to penetrate into the core area of K-group's range from 1974 onwards."[93] A third group, the B-group, also penetrated the K-group's range as the number of its males declined.[94] The males who disappeared were healthy and not senile or injured. In one case, M-group males were known to be near the K-group males, as there were many outbursts of calls; the next day another K-group male was missing.[95] As the number of males was reduced, the females were incorporated—at first gradually and then later en masse—into the invading group, as was most of the K-group's territory, where they mated with the victorious males.[96]

Kibale also provides evidence of coalitionary kills, but not of warfare. Ruwenzori, a male of the Kanyawara community, was found newly dead in a border area three days after males from the Kanyawara and Rurama communities had engaged in counter-calling. Ruwenzori's body showed clear evidence of a violent attack. In another incident, observers followed the Kanyawara males to the fresh corpse of an individual from a neighboring community (Sebitole) who had apparently been killed by chimpanzees the previous evening. He had many wounds on the front of his body, his trachea had been ripped through, and both testicles were removed. The nine Kanyawara males who had patrolled the border the previous evening were all present the next morning, when several beat the body and dragged it after they encountered it again.[97]

Among the western subspecies of chimpanzee, *Pan troglodytes verus,* there is no evidence of lethal intraspecific aggression toward adults or infants from the Bossou or Taï research stations. Since the Bossou community has few males and lives in an isolated forest with little or no connection to other wild chimpanzees, low rates of aggression are not surprising. The Taï community, however, is not as limited and has been studied for twenty-three years, leading Wrangham to estimate that it should have produced evidence of about five killings each of adults and infants if it conformed to the pattern of the *P.t. schweinfurthii.* To Wrangham, "the fact that Taï chimpanzees show all the components of lethal raiding but no coalitionary kills" or warfare suggests that aggressive relationships differ "between Taï and the eastern populations."[98]

Notwithstanding the findings at the Taï research facility, Wrangham's imbalance of power hypothesis is well supported by the evidence from all the research sites. The adage that God is on the side of the big battalions is not just true for humans. Both within and between primate groups, struggles tend to be won by the larger of the two parties, although variables such as dominance, rank, and geographic location are also important.[99] In the four longest studies of chimpanzees, the principal determinant of the nature of intercommunity interactions is not the geographic location but the relative size and composition of parties when they encounter each other.

This argument is based on direct observations at Gombe, Kibale, Mahale, and Taï, as well as recording playback experiments at Kibale. For example, Goodall found that conflicts between groups are usually determined by the relative size and composition of the groups rather than the geographic location of the encounter as is the case with most other territorial animals. As she explains, "a small patrol will turn and flee if it meets a larger party, or one with more males, even *within* its own range; whereas if a large party, traveling *out* of its range, meets a smaller party of neighbors, it is likely to chase or attack."[100] If the parties are approximately the same size "with similar numbers of adult males," the typical result will be "visual and auditory display exchanges without conflict."[101] Boesch and Boesch-Achermann found that small parties of males (one to three) mainly checked for the presence of strangers by drumming and listening to the response (67 percent of eighteen occasions).[102] Middle-sized parties (four to six males) tended to make incursions into the neighboring territory more frequently (37 percent of seventy-six observations), while large parties (seven to nine males) tended to attack the strangers (63 percent of thirty observations).[103]

At all sites, the likelihood that a party will advance, exchange displays, or retreat appears to be predicted well by whether it is larger than, equal to, or smaller than the opposing party. The evidence suggests that chimpanzees are bolder when their party contains relatively more males. Boesch and Boesch-Achermann summarize their findings:

> parties of between four and six males more commonly go on patrol inside the strangers' territory, which is the dominant strategy of parties of that size. Large parties of six or more males mainly carry out attacks. When considering adult males and females present in the party, no significant difference is observed. This suggests that Taï chimpanzees take into account the number of adult males and not the total number of chimpanzees that are present when deciding a strategy. At various times we saw parties of four males waiting, we presumed, for more males to join them, but as none came, they limited themselves to check or patrol the area.
>
> In the course of our study, the community declined in size and in the

> number of adult males. . . . As the males became less numerous, they changed their strategy . . . which is what one could expect if intrinsic power plays a role. When in 1988, the number of adult males fell below eight, they started to be more careful, investing more time in going on patrol and less in direct confrontation. Strikingly, from 1992 onwards, for the first time since it was possible to make observations, avoidance behaviour, in which the chimpanzees retreated in silence from the strangers, became the dominant strategy and avoidance replaced the checks used before. This suggests that they were aware of the imbalance of power and avoided taking risks.[104]

Furthermore, recording playback experiments at Kibale support the hypothesis that males are more likely to attack when a party of three or more males encounters a lone victim; this also supports the observational data from Gombe.[105] Thus, for aggressors, the low cost of lethal group raiding appears to be supported by the evidence and so strengthens Wrangham's imbalance of power hypothesis.

Finally, reflecting on the similarities between human and chimpanzee groups and the origins of warfare, anthropologist Keith Otterbein argues that war's roots lie in "the localization of related males, a condition that goes back to the common ancestor of humans, chimpanzees, bonobos (pigmy chimpanzees), and gorillas."[106] He submits that warfare has been present in human and pre-human ancestry for about 5 million years. Along with the great apes, we inherited it from our common ancestor. Because the common ancestor was patrilocal, organized into groups by territory and localized by related males, intergroup conflict likely resulted from conflicts over resources and territory.[107]

Wrangham concurs and says the argument is supported by two factors: the rarity of coalitionary lethal violence toward adult conspecifics in other primates, and evidence that after they separated from gorillas, chimpanzees and humans shared a common ancestor 5–6 million years ago. Thus, he argues that lethal raiding arose in an ape prior to the chimpanzee-human divide and has continued in both the chimpanzee and hominid lines.[108]

Indeed, chimpanzee (*Pan troglodytes schweinfurthii*) warfare shows similarities to tribal warfare and is one reason they are considered to be discrete from the vast majority of other animals. Goodall points to "the *violence* of their hostility toward neighbors" as the way that "chimpanzees, like hyenas and lions, differ most from the traditional territory owners of the animal kingdom."[109] This is because "their victims are not simply chased out of the owner's territory if they are found trespassing" but are "assaulted and left, perhaps to die."[110] She stresses that chimpanzees appear to be different from most other animals because they make raids into the heart of the territory of neighboring groups.[111]

Concluding her analysis of chimpanzee territoriality and warfare, Goodall writes: "as a result of a unique combination of strong affiliative bonds between adult males on the one hand and an unusually hostile and violently aggressive attitude toward nongroup individuals on the other," the chimpanzee "has clearly reached a stage where he stands at the very threshold of human achievement in destruction, cruelty, and planned intergroup conflict."[112]

For both chimpanzees and humans, Wrangham's imbalance of power hypothesis captures important similarities. First, small numbers of males make undetected incursions into neighboring territory, attack unsuspecting victims either alone or in small groups, and attempt to retreat without being captured or drawn into battle. Second, chimpanzees, like humans, often incorporate females from enemy or defeated groups into their own. Third, chimpanzees and humans can classify others as belonging to the group (in-group) or not (out-group).[113] Although we can only observe chimpanzee behavior, Wrangham suggests on the basis of his years of study that they view members of the out-group as equivalent to prey, or at best ambivalently.[114] As humans often dehumanize their enemies, it may be the case that chimpanzees also "dechimpanzeeize," if you will, their enemies and see them as not their own kind but as an inferior type. Fourth, both humans and chimpanzees are sensitive to power asymmetries between their own group and others, and can recognize when the asymmetry of power is favorable to one's group, not favorable, or equal. Fifth, both are ruthless when attacking members of the out-group when the perceived power asymmetry is favorable.

Broadening Clausewitz's Definition of War

Once we recognize that animals fight wars, we must broaden our definition of warfare. A definition of warfare informed by evolutionary theory permits us to recognize that warfare contributed to fitness for certain species by yielding, among other factors, more resources for survival; this occurred long before warfare was conducted for political ends. In the most famous and insightful definition of warfare, Clausewitz recognized the relationship between war and politics: "war is nothing but the continuation of policy with other means."[115] He repeatedly emphasizes the political component of warfare, and writes that warfare cannot be divorced from political reasoning: its "grammar, indeed, may be its own, but not its logic."[116] This definition has rightfully had great influence among political scientists and military and political leaders. Indeed, his recognition of the political primacy of war justifies both civilian control of the military and a dominant role for the civilian leadership in formulating a state's grand strategy and use of the military instrument.[117]

Despite Clausewitz's insight, an evolutionary understanding of the causes of war requires us to broaden Clausewitz's definition: it is conducted for resources as well as for political ends. An evolutionary understanding of the causes of war permits the recognition that war may be fought by animals or individuals in prepolitical conditions; in such circumstances it would be fought for territory, resources, or to gain security. As I described in chapter 3, evolutionary theory suggests that this is how war began. Any definition of warfare must recognize that war existed before organized political entities. Just as an individual would aggress against another to gain resources, so would an organized band of humans, perhaps an extended family group or a tribe. To be sure, war is conflict between autonomous territorial groups waged for political purposes. Nonetheless, it is also waged for purposes related to acquisition or defense of territory and resources necessary for the survival of oneself, one's family, and one's group or tribe.

The Stimulating Effects of Warfare and Why Males Are More Likely to Be Warriors

In this section of the chapter I intend to address two significant issues into which evolutionary theory provides insight. First, warfare often stimulates combatants and very frequently has a cohesive effect on populations. Second, males are more likely than females to be warriors.

With respect to the first issue I want to suggest that the stimulation often found in warfare may be the residue from such group experiences as hunting and fighting in the late-Pliocene, Pleistocene, and Holocene conditions. These activities in our evolutionary past may have caused or helped to cause many of the powerful emotions that individuals engaged in combat still experience today. I do not intend to demonstrate the point conclusively, but I do suggest that the intensity of the emotion, the sharpness of the mind, and the exhilaration may be anchored in the evolution of the human species.

I also begin this discussion with a caveat: war stimulates some individuals, not all. Clearly, the hierarchy of a modern military and the lethality of the modern battlefield do not resemble the human past; equally clearly, many combatants in modern warfare do not find it stimulating. Indeed, as many scholars have noted, most are probably reluctant to fight, and some are cowards.[118] Nor do I mean to diminish the impact of warfare on family and friends, civilians, animals, and other innocents. Notwithstanding these points, it seems that warfare is Janus-faced. At least some typical individuals, who are neither guilty of exceptional cowardice nor remarkable for spectacular feats of bravery, are both repelled by killing and yet thoroughly enjoy aspects of the experience of warfare. For the individuals cited below, their experiences in warfare left an indelible impression on their lives as members of what the American author and Civil War veteran Stephen Crane termed "a mysterious fraternity born out of smoke and danger of death."[119]

Warfare Is Janus-Faced: The Emotions of Warfare and Its Stimulating Effects on Combatants

At the end of the American Civil War, during which he cared for and listened to the accounts of hundreds of Union wounded, Walt Whitman declared: "The real war will never get in the books."[120] No doubt Whitman was referring to its brutality and terror. Yet he may have also been referring to its energizing effects. No matter how disturbing the horror of war may seem, individuals engaged in warfare often find it stimulating. William James wrote that warfare was a "thrilling excitement" and a "supreme theater of human strenuousness."[121] For him, talk of the horror and expense of war was beside the point. "The horror makes the thrill; and when the question is of getting the extremest and supremest out of human nature, talk of expense sounds ignominious."[122] The "thrill" that James identifies may take several forms. It may stimulate the emotions and senses of its participants and those who observe war from a distance, whether those on the homefront, or writers and poets.

James's "thrill" brings to mind Winston Churchill's comparison of war to champagne. "A single glass of champagne imparts a feeling of exhilaration," he opined, "the nerves are braced; the imagination is agreeably stirred; the wits become more nimble. A bottle produces a contrary effect. Excess causes a comatose insensibility. So it is with war, and the quality of both is best discovered by sipping."[123]

J. Glenn Gray, a former philosophy professor at Colorado College, who was first a U.S. soldier and later an officer in World War II, also notes the excitement of war: "Many veterans who are honest with themselves will admit, I believe, that the experience of communal battle . . . has been the high point in their lives."[124] He does acknowledge the pain: "despite the horror, the weariness, the grime, and the hatred, participation with others in the chances of battle had its unforgettable side, which they would not want to have missed."[125] But this feeling of war's excitement may be ineffable: "for anyone who has not experienced it himself, the feeling is hard to comprehend, and, for the participant, hard to explain to anyone else."[126]

For Gray, war's stimulation is inextricably linked to the danger of war; his sentiment is echoed by many others who have fought in combat. As James Jones, a veteran of the campaign for Guadalcanal in 1942 and 1943 explains in his fictional account of that struggle, "it was like facing God. Or gambling with Luck [or] taking a dare from the Universe."[127] Jones has one of his soldiers explain that warfare is more electrifying for him "than all the hunting, gambling and fucking he had ever done all rolled together."[128] During the invasion of Sicily in 1943, war correspondent Jack Belden recalled remembering as he plunged and rocked toward the shore how his senses were heightened: "all my senses were now altered to the straining point. A flush of thrill and excitement shot through me like a flame. It was wonderful. It was exhilarating."[129] Another World War II veteran and author, Harry Brown, also notes this exhilaration and a keen awareness: "in peacetime you could go into a store to talk to a clerk, but by the time you had left the store you had completely forgotten the clerk's face."[130] It was different in war, however: "you might see a man's face in the flash of an exploding shell or in the cab of a truck or peering out of a slit trench, and though you had never seen the face before and never would see it again, you couldn't forget it. War left impressions of unbelievable sharpness."[131]

But warfare also held other thrills. For Gray, war contains an "aesthetic appeal."[132] He recounts the stimulation of invading southern France in August 1944. For him, the first "aesthetic appeal" was a feeling of awe, a "fascination that manifestations of power and magnitude hold for the human spirit. Some scenes of battle, much like storms over the ocean or sunsets on the desert or the night sky seen through a telescope, are able to overawe the single individual and hold him in a spell."[133] However, the "chief aesthetic appeal of war surely lies in this feeling of the sublime," which must be understood "in the original meaning of the term, namely, a state of being outside the self."[134]

For Gray, another appeal of war was a "delight in destruction," or a "satisfaction that men experience when they are possessed by the lust to destroy and kill their kind."[135] He describes how, as a result of these experiences in World War II, "men who have lived in the zone of combat long enough to be veterans are sometimes possessed by a fury that makes them capable of anything"; this rage allows them both "to destroy" and to be "supremely careless of consequences" so that "they storm against the enemy until they are either victorious, dead, or utterly exhausted. It is as if they are seized by a demon and are no longer in control of themselves."[136]

Finally, Gray notes the delight in the comradeship war affords: "numberless soldiers have died, more or less willingly, not for country or honor or religious faith or for any other abstract good, but because they realized that by fleeing their posts and rescuing themselves, they would expose their companions to greater danger."[137] Such loyalty is critical for morale, but Gray submits that the "feeling of loyalty, it is clear, is the result, and not the cause, of comradeship. Comrades are loyal to each other spontaneously and without any need for reasons."[138] A great source of strength for individual soldiers and units arises "when, through military reverses or the fatiguing and often horrible experiences of combat, the original purpose becomes obscured"; then "the fighter is often sustained solely by the determination not to let down his comrades."[139]

Of course, scholars across many disciplines have noted that warfare influences humans by causing a growth in group identity, or cohesion, and sense of comradeship.[140] In his study of the Mundurucú Indians of Brazil, Robert Murphy notes that warfare is "an especially effective means of promoting social cohesion in that it provides an occasion upon which the members of the society unite and submerge their factional differences in the vigorous pursuit of a common purpose."[141] Camilla Wedgwood argues that warfare in Melanesia serves a "double purpose of enabling people to give expression to anger caused by a disturbance of the internal harmony, and of strengthening or reaffirming the ties which hold them together."[142] To political scientists, these arguments are familiar. Many political scientists have demonstrated how war unifies societies and also how social imperialism, creating or exaggerating an external threat, facilitates the creation of domestic unity in states that face serious class or ethnic divisions.[143]

Eugene Sledge, a U.S. Marine mortarman who fought against the Japanese in World War II, also strongly remembers the comradeship, and describes the feelings of his comrades who left combat and returned to the United States: "Plenty of people were ready to buy a Marine combat veteran wearing campaign ribbons and battle stars a drink or a beer anytime. But all the good life and luxury didn't seem to take the place of old friendships forged in combat."[144] He concludes his war memoirs: "War is brutish, inglorious, and a terrible waste. Combat leaves an indelible mark on those who are forced to endure it. The only redeeming factors were my comrades' incredible bravery and their devotion to each other."[145] Their training "taught us to kill efficiently and to try to survive. But it also taught us loyalty to each other—and love. That esprit de corps sustained us."[146]

Sledge's feelings are echoed by Guy Sajer, a German soldier on the Eastern Front in World War II, who wrote one of the most remarkable memoirs of that war. After reflecting on his many horrible experiences on the Eastern Front, he writes, "but even now, looking back on everything that happened, I cannot regret having belonged to a combat unit. We discovered a sense of comradeship which I have never found again, inexplicable and steady, through thick and thin."[147]

Author William Manchester wrote of his U.S. Marine compatriots in the Pacific in World War II: "those men on the line were my family, my home. They were closer to me than I can say, closer than any friends had been or ever would be. They had never let me down, and I couldn't do it to them."[148] In fact, this bond was so powerful that "I had to be with them, rather than let them die and me live with the knowledge that I might have saved them. Men, I now knew, do not fight for flag or country, for the Marine Corps or glory or any other abstraction. They fight for one another."[149] John Ahearn, an officer with the 70th Tank Battalion in Europe, echoes this. Reflecting on why men fight heroically during bitter combat, Ahearn concludes: "Recently when people were getting up in Congress on the flag issue, saying that men died for the flag, for a symbolism, I said bullshit. First of all there were no flags around, there was no standard out there. Men didn't die for the flag, or even in the heat of combat for their country; they died for their companions. You don't know the broad picture, at least I didn't. I knew my particular assignment, my platoon's assignment, my company's assignment. They were my concern. All fighting men are concerned about their fellows."[150]

Retired U.S. Army colonel David Hackworth explains the importance of bonding and his motivation to fight in the Korean War: "Sure, I was fighting for America, for all that was 'right' and 'true,' for the flag, the national anthem, and Mom's Apple Pie. But all that came second to the fact that the reason I fought was for my friends. My platoon."[151] He argues that this "was why most other soldiers fought, too. The incredible bonding that occurred through shared danger; the implicit trust in the phrase 'cover me'—these were the things that kept me going, kept me fighting here in Korea."[152] Hackworth continues, "the most important thing was that I knew with other troopers' respect came their *trust:* they knew that I wouldn't let them down. And to the best of my ability I never would."[153]

Former U.S. Army general Harold Moore writes of his experiences fighting the North Vietnamese Army at the Battle of the Ia Drang Valley in November 1965, the first major engagement between U.S. and North Vietnamese forces in the Vietnam War: "we discovered in that depressing hellish place, where death was our constant companion, that we loved each other, and we wept for each other. And in time we came to love each other as brothers."[154]

William Broyles, a U.S. Marine who fought in the Vietnam War and was later an editor at *Texas Monthly* and *Newsweek,* recognizes that "war is horrible, evil, and it is reasonable for men to hate all that."[155] But for Broyles, like Gray, it is a more complicated experience: "I believe that most men who have been to war would have to admit, if they are honest, that somewhere inside themselves they loved it too, loved it as much as anything that has happened to them before or since."[156] Broyles's reflections dovetail with Gray's: this love of war explains why "men in their sixties and seventies sit in their dens and recreation rooms around America and know that nothing in their life will equal the day they parachuted into St. Lô or charged the bunker on Okinawa."[157] Broyles continues, emphasizing the uniqueness of the experience, "war . . . intensifies experience to the point of a terrible ecstasy" and "if you come back whole you bring with you the knowledge that you have explored regions of your soul that in most men will always remain uncharted."[158] Also, it is a "utopian experience" where "individual possessions and advantage count for nothing, the group is everything" and what you possess "is shared with your friends" and is "a love" that "transcends race and personality and education—all those things that would make a difference in peace."[159] Author and Vietnam veteran Philip Caputo admitted after the war that battle made him feel "happier than I ever had" felt.[160]

For many veterans of combat, warfare seemed to resemble a terrific sport that placed the utmost emotional and physical demands upon its participants. Joanna Bourke provides many accounts from soldiers who found warfare a thrilling and terrifying sport.[161] For example, James Hebron, a Marine sniper in the Vietnam War, recounted the sense of power and sport he felt during combat: "That sense of power, of looking down the barrel of a rifle at somebody and saying, 'Wow, I can drill this guy.'"[162] Hebron continues, "Doing it is something else too. You don't necessarily feel bad; you feel proud, especially if it is one on one, he has a chance. It's the throw of the hat. It's the thrill of the hunt."[163] Murphy also notes this aspect of warfare in his study of the

Mundurucú: "the enemy was looked upon as game to be hunted, and the Mundurucú still speak of the pariwat [any non-Mundurucú human] in the same terms they reserve for peccary and tapir."[164]

Of course, those left behind often find warfare stimulating as well. No doubt the state engaged in warfare intentionally uses propaganda to create excitement so it can mobilize popular support for war and maximize the extraction of resources, such as men, money, and labor, from society. But it is equally true that artists, writers, and poets have found war to be a rich source of insights into the human condition, from unknown artists' paintings of scenes from the Peloponnesian War on Athenian vases to Picasso's *Guernica,* and from Homer's *Iliad* to Shakespeare's *Henry V* and Hemingway's *For Whom the Bell Tolls.*

Warfare also stimulates hatred of enemy. Soldiers engaged in warfare often possess a depth of hatred for the enemy that is not caused by training or indoctrination but is a result of the very experience of fighting. As Sledge recounts, "The attitudes held toward the Japanese by noncombatants or even sailors or airmen often did not reflect the deep personal resentment felt by Marine infantrymen. Official histories and memoirs of Marine infantrymen written after the war rarely reflect that hatred."[165] "But at the time of battle," Sledge states, "Marines felt it deeply, bitterly, and as certainly as danger itself. To deny this hatred or make light of it would be as much a lie as to deny or make light of the esprit de corps or the intense patriotism felt by the Marines with whom I served."[166] His experiences on Peleliu and Okinawa convinced Sledge "that the Japanese held mutual feelings for us."[167] Moreover, these "collective attitudes, Marine and Japanese, resulted in a savage, ferocious fighting with no holds barred. This was not the dispassionate killing seen on other fronts or other wars."[168] It was, rather, "a brutish, primitive hatred, as characteristic of the horror of war in the Pacific as the palm trees and the islands."[169]

In his excellent study of the ground war in the Pacific, Eric Bergerud explains why Sledge's feelings were not unique. For the soldiers who fought in them, the difference between the European and Pacific theaters "derived from the intense fear, coupled with a powerful lust for revenge, that poisoned the battlefield in the South Pacific. Both emotions, so closely linked, arose from a series of local events that were never part of the public's perception of the war," but were rather "the private property of the men at the front or those who have studied the campaigns closely."[170] Overall, he says, "the incidents in question all created or reinforced a perception that the Japanese soldier was uniquely dangerous and uniquely cruel."[171] This in turn affected the behavior of Australian and American soldiers and Marines. Early in the war, on

Guadalcanal, Marine veteran Donald Fall recounted the hatred he felt toward the Japanese and the desire for revenge:

> On the second day on Guadalcanal we captured a big Jap bivouac with all kinds of beer and supplies. Thank goodness for that because we needed the food to make it through those first two weeks or so. But they also found a lot of pictures of Marines that had been cut up and mutilated on Wake Island. The next thing you know there are Marines walking around with Jap ears stuck on their belts with safety pins. They issued an order reminding Marines that mutilation was a court-martial offense. On New Britain a lot of guys who captured Japs tried to pry their mouths open and take the gold teeth out. They did that with dead ones too. [Sledge also notes this behavior.] You get into a nasty frame of mind in combat. You see what's been done to you. You'd find a dead Marine that the Japs had booby-trapped. We found dead Japs that were booby-trapped. And they mutilated the dead. We began to get down to their level.[172]

Fall's account of the war causing hatred for and dehumanization of the enemy is echoed in many other war narratives. Another combatant on Guadalcanal, Ore Marion, explained in an interview with Bergerud how deep his hatred was for the Japanese and how this influenced his own behavior and that of his comrades:

> We learned about savagery from the Japanese. Those bastards had years of on-the-job training on how to be savage on the Asian mainland. But those sixteen-to-nineteen-year-old kids we had on the Canal were fast learners. Example: On the Matanikau River bank after a day and night of vicious hand-to-hand attacks, a number of Japs and our guys were killed and wounded. At daybreak, a couple of our kids, bearded, dirty, skinny from hunger, slightly wounded by bayonets, clothes worn and torn, wack off three Jap heads and jam them on poles facing the "Jap side" of the river. All of a sudden you look up and those goddamn heads are there.
>
> Shortly after, the regimental commander comes on the scene. He can't believe the scene in that piece of jungle. Dead Japs and Americans on top of each other. Wounded all around, crying and begging. The colonel sees Jap heads on the poles and says, "Jesus men, what are you doing? You're acting like animals." A dirty, stinking young kid says, "That's right, Colonel, we are animals. We live like animals, we eat and are treated like animals—what the fuck do you expect?"[173]

These accounts from the Pacific War should not suggest that such hatred or behavior did not occur in Europe, or that civilians on the homefront were immune from such deep feelings of hatred. Bourke mentions a book published for the Home Forces in Britain during World War II. The author warned against feeling any "qualms" about killing Germans, and exhorted his readers to "remember the utter bestiality of the Nazi mind," for if they did not "when they come here, the most damnable barbarians since Attila will lay our country waste, indulge in mass executions, and turn our women into whores for themselves. . . . Have at him, because if you do not he will have at you."[174]

Naturally, such feelings were to be found among the combatants in smaller wars as well. In his memoir of the Vietnam War, Caputo explains how he "burned with hatred for the Viet Cong and with . . . a desire for retribution. I did not hate the enemy for their politics, but for murdering Simpson," a member of his platoon.[175] "Revenge was one of the reasons I volunteered for a line company. I wanted a chance to kill somebody."[176]

Such deep hatred carries a major implication: it helps dehumanize the enemy and thus contributes to such war crimes as the killing of civilians or enemy prisoners of war. During the Gallipoli campaign in 1915, Captain Guy Warneford Nightingale wrote to his sister: "we took 300 prisoners and could have taken 3,000 but we preferred shooting them."[177] British author and World War I veteran Robert Graves thought that the killing of prisoners of war was the major war crime committed by British soldiers. He named several reasons for doing it: revenge for fallen friends or relatives, jealousy of the enemy's potential comfort in a prisoner of war camp, fear of being overpowered by the prisoners, or laziness that kept the perpetrator from escorting the prisoner out of the combat zone. "In any of these cases the conductors would report on arrival at headquarters that a German shell had killed the prisoners; and no questions would be asked."[178] Graves justifies this in part by explaining, "We had every reason to believe the same thing happened on the German side, where prisoners, as useless mouths to feed in a country already short of rations, would be even less welcome."[179] An additional factor, according to Graves, was personal gain. He quotes an Australian soldier:

> Well, the biggest lark I had was at Morlancourt, when we took it the first time. There were a lot of Jerries [Germans] in the cellar, and I said to 'em: "Come out, you Camarades! [*sic*]" So out they came, a dozen of 'em, with their hands up. "Turn out your pockets," I told 'em. They turned 'em out. Watches and gold and stuff, all dinkum. Then I said: "Now back to your cellar, you sons of bitches!" For I couldn't be bothered with 'em. When they were all safely down I threw half a dozen Mills bombs in after 'em. I'd got the stuff all right, and we weren't taking prisoners that day.[180]

In World War II many atrocities against prisoners of war were motivated by similar causes. German atrocities against prisoners of war in World War II have been well documented. For Americans, the most infamous was the Malmédy massacre of approximately seventy American prisoners of war and even more Belgian civilians conducted by German SS troops during the Battle of the Bulge. This massacre led to considerable retaliation against German prisoners of war during and after that battle.[181] For example, a written order from the headquarters of the 328th Infantry Regiment of the U.S. Army, dated 21 December 1944, reads: "No SS troops or paratroopers will be taken prisoner but will be shot on sight."[182] In an exceptional memoir of his experiences as an ordnance officer with the 3rd Armored Division, Belton Cooper notes how few SS prisoners were taken by his division and others. Near Parfondroy, Belgium, his division "discovered the massacre of innocent civilians by the brutal SS troops of *Kampfgruppe Peiper*. Numerous bodies of old men, women, and children were scattered around the village; they had been shot by the SS."[183] Nearby, "30th Division troops in Stavelot told us that they had encountered similar massacres there. Word of these massacres undoubtedly contributed to the shortage of live SS prisoners taken by American GIs."[184] Clarence McNamee, an American soldier serving with the 70th Tank Battalion, recalled how the news of the massacre angered his comrades: "We were up, though the front was very fluid—really nonexistent. Our mail overran us, and the Germans had taken one of our mail trucks and killed all the men in it. We then engaged the same outfit and completely overran them. When we did, we saw that the Germans were eating our Christmas cookies. Several men went into a rage and started shooting Germans before they could fire back or surrender, whatever their intention. This was after we heard about Malmédy. It was a wild time during the Bulge."[185]

The Eastern Front in World War II was known for its great brutality on both sides. Sajer recounts several incidents like the following during particularly intense combat:

> We were mad . . . running on our nerves, which were stretched to the utmost, and which alone made it possible to respond to the endless succession of crises and alerts. We were forbidden to take prisoners until our return trip. We knew that the Russians didn't take any, and although we longed for sleep we knew that we had to stay awake as long as there were any Bolsheviks in our sector. It was either them or us—which is why my friend Hals and I threw grenades into the bread house, at some Russians who were trying to wave a white flag.[186]

The Soviets were just as brutal. The head of the British military mission to Yugoslavia, Brigadier Fitzroy Maclean, asked a Soviet officer how they dealt with prisoners. The answer was, "If they surrender in large groups, we send them back to base; but if . . . there are only a few of them, we don't bother."[187] Later Maclean observed a row of corpses, "like ninepins knocked over by the same ball. They had clearly not died in battle."[188] This led him to consider how the Soviets defined a "large group."

The Japanese also committed atrocities against prisoners of war, though they are less well documented than those by the Germans.[189] In fact, in their treatment of British and American prisoners of war, the Germans and Italians were considerably more humane than the Japanese. Of the 235,473 British and American prisoners of war captured by the Germans and Italians during World War II, the death rate reached about 4 percent, or 9,348.[190] According to the Tokyo War Crimes Tribunal, 350,000 prisoners were captured by the Japanese after the start of the Pacific War. Of the 132,134 from Australia, Britain, Canada, the Netherlands, New Zealand, and the United States, 35,756 died while detained, a death rate of 27 percent.[191]

For Americans, perhaps the most famous example was the Bataan Death March in the Philippines. This started on April 9, 1942, when 75,000 American and Filipino survivors, who were sick, starving, and dehydrated, were forced to walk sixty-five miles in sweltering heat to a railhead at San Fernando. One survivor of the Bataan Death March, Lester Tenney, describes some of the horror: "The Japanese called four men out of line, and had them dig a ditch. When they dug the ditch, they told them to bury this man. One of the fellows said I can't bury him, he's still alive, so the Japanese shot this one man, called four more men out, and said now dig two ditches. . . . You don't want to bury your friend, you don't want to bury a live person, but you know that if you say no, you're going to be buried."[192]

The allies also abused human rights. On July 14, 1943, the American 45th Infantry Division massacred around seventy Italian and German prisoners of war at Biscari in Sicily.[193] Major-General Raymond Hufft of the U.S. Army ordered his troops across the Rhine River in 1945 with instructions to take no prisoners. After the war, when he reflected on the war crimes he authorized, he admitted, "if the Germans had won, I would have been on trial at Nuremberg instead of them."[194] In his masterful account of the American soldier in the European theater of operations, Stephen Ambrose explains, "I've interviewed well over 1,000 combat veterans. Only one of them said he shot a prisoner. . . . Perhaps as many as one-third of the veterans . . . however, related incidents in which they saw other GIs shooting unarmed German prisoners who had their hands up."[195]

In the war in the Pacific, similar factors motivated soldiers to kill prisoners of war. U.S. Marine Louis Maravelas described how his fellow Marines decided not to take prisoners on Guadalcanal. "During a battle along the Matanikau three or four were going to surrender. There must have been a dozen of us with a bead on them. Sure enough, one bent over and there was a carbine or submachine gun slung on his back that his comrades tried to grab."[196] So "we shot them down instantly."[197] In a later incident he was on patrol with a large Marine force. "Suddenly, one Japanese officer comes charging out of the jungle screaming and waving his sword. We riddled him. What did he accomplish? He was only one man, what could he hope to accomplish?"[198] After such incidents, "It got to the point where we took no prisoners. It wasn't a written order, but a way to survive. No one should take a chance to take a guy prisoner who might try to kill him."[199]

Bill Crooks of the Australia Imperial Force (AIF) fighting in New Guinea seconded Maravelas's comments: "We knew" what the Japanese "had done in Nanking. We knew they had machine-gunned our nurses on Banka Island. We knew they had bayoneted AIF prisoners after they captured Rabaul. We knew they had bayoneted hospital inmates in their advance down the Malayan Peninsula."[200] As a result, "We just would kill every bastard soul of them the moment we come against them. We knew their bushido banzai code was to take no prisoners in battle and never surrender. So we killed them."[201]

Writer George MacDonald Fraser served as a noncommissioned officer with the British Army in Burma during World War II and recounts in his memoirs one particular incident in 1945 when soldiers from an Indian unit killed all of the Japanese prisoners of war in a military hospital. He admitted it was a war crime, but he explained how he thought of it: "I probably grimaced, remarked 'Hard buggers, those jawans [Indian soldiers],' shrugged, and forgot about it. If I had made an issue of it with higher authority, I'd have been regarded as eccentric. I'd have regarded *myself* as eccentric."[202] Fraser further explained: "in the year 1945, towards the end of a war of a particularly vicious, close-quarter kind, against an enemy who wouldn't have known the Geneva Convention if it fell on him, I never gave it a second thought."[203] He continues, "and if I had, the notion of crying for redress against the perpetrators (my own comrades-in-arms, Indian soldiers who had gone the mile for us, and we for them), on behalf of a pack of Japs, would have been obnoxious, dishonourable even."[204]

Understandably, the hatred expressed by these veterans toward the enemy could also spread to civilians. Indeed, civilians were massacred during the Second World War and the Vietnam War, most famously at My Lai in March 1968. Of course, while we may understand how soldiers might develop such feelings, we cannot excuse their actions.

Finally, warfare may also be sexually stimulating, as Broyles recalls of his experiences in Vietnam. "Most men who have been to war . . . remember that never in their lives did they have so heightened a sexuality. War is, in short, a turn-on."[205] Similarly, another Vietnam veteran wrote: "a gun is power. To some people carrying a gun was like having a permanent hard-on. It was a pure sexual trip every time you got to pull the trigger."[206] Mark Bowden studied the events of October 3, 1993, in Somalia, when supporters of Somali warlord Mohamed Farrah Aidid shot down U.S. Army helicopters and ambushed the Army Rangers and other forces trying to rescue the survivors. He reported the sexual aspect of the experience: some considered masturbating during combat.[207] But the effect may be a general one as well. Gray notes that military service during World War II made women more attractive: "when we were in uniform almost any girl who was faintly attractive had an erotic appeal for us."[208] The soldier also became more attractive to women, as "millions of women" found a "strong sexual attraction in the military uniform."[209] The impact continues today, despite some remaining anti-military sentiment that developed in the United States after the Vietnam War. For example, U.S. Army Green Berets famously brag about their ability to attract interest from women, and Marines often boast of the sexual appeal of their dress uniforms.

This section of the chapter has shown that individuals who fight war recognize its horrors but also often acknowledge its stimulating effects. Warfare, indeed, can be Janus-faced. Many who fight do indeed find their experience to be the high point of their lives: never before or since have they had to exert themselves physically and emotionally so much, often far beyond exhaustion. Never before or since have they had such power, responsibility, and friendship. No civilian job can offer a similar stimulation. The themes that leap from these narratives are those of profound love of comrades, the deepest hatred of the enemy, fear of death, and fear of disappointing the other men. These combatants have tapped these profound emotions that have long served humanity in our evolution in the thrill of the hunt and in the protection of one's family and comrades. They are a part of warfare even today, as inseparable from it as emotion is from sport or relationships. Evolutionary theory allows us to understand the power of these emotions—emotions that allow or even force men to risk life and death in combat against other humans—and thus understand why they may have served fitness in human evolution.

Why Males Are More Likely to Be Warriors

Among primates, males are much more likely than females to engage in aggression—aggressive coalitions are sexually dimorphic in mammals—because it gives males a greater chance of reproductive success.[210] Before I explain in detail why this is so, I want to stress two points. First, this discussion is relevant for many animals, such as most mammals, reptiles, and birds, but in many species, ants perhaps most obviously, only females aggress. Second, as I have stressed elsewhere in this book, evolutionary theory provides insights into the probability of a particular behavior. No doubt some females can be as physically aggressive or violent as some males. Moreover, women have a formidable history as leaders in combat—Boadicea and Joan of Arc—as well as individual combatants. David Adams studied warfare in sixty-seven tribal societies, and found that women participated, at least occasionally, in nine: these were the Comanche, Crow, Delaware, Fox, Gros Ventre, Maori, Majuro, Navaho, and Orokaiva.[211] Turney-High also reports that many Indian women accompanied men on raids against their foes, often providing moral or logistical support for the attack: the "Pima always took one or two women along in their revenge expeditions against the predatory Apache, always females who had lost relatives to the ferocious foe."[212] In his exceptional study of gender and war, Joshua Goldstein notes that the Dahomey Kingdom in West Africa (Benin) had a large number (five thousand at its peak) of female soldiers in the eighteenth and nineteenth centuries.[213] More recently, during World War II women served as combatants in the Soviet army and air force.[214] Indeed, Ludmilla Pavlichenko, a history student turned sniper, was officially credited with causing 309 German casualties.[215] American women have participated, sometimes directly as combatants, in all of our wars. Examples are Lucy Brewer in the War of 1812, Sarah Borginis in the Mexican War, and Sarah Edwards during the Civil War.[216] Moreover, the history of terrorism and guerrilla warfare is replete with women serving in combat roles, with the Red Army Faction in Germany, the Tamil Tigers in Sri Lanka, or the Irish Republican Army; or as guerrillas fighting in Spain, Algeria, Vietnam, Nicaragua, and Eritrea, or Jewish partisans fighting the Germans during and after World War II.[217]

Given these caveats, an evolutionary perspective allows us to recognize that the two sexes have different rewards for taking the risks involved with potentially lethal conflict. John Tooby and Leda Cosmides suggest that a successful coalition has a greater chance to gain access to reproductive resources.[218] For males, females are the limiting reproductive resource, and a benefit of aggression is increased access to females. Of course other factors, such as wealth or power, may also lead to increased access, although aggression may affect these, too. Mulay Ismail "The Bloodthirsty," a Moroccan emperor (1672–1727), is reported to have fathered 888 children, many times more than the most fecund woman might.[219] Nonetheless, this suggests that males may be willing to go to war despite the potential fatal consequences for some. Because females are rarely limited by access to males, similar behavior is unlikely to enhance reproductive success. Moreover, not only does the death of each female hinder the reproductive success of the group, but also, among mammals, a mother's death hinders the likelihood that an infant will survive. As Bobbi Low argues, "if the offspring is dependent on its parent, a live parent is crucial. Thus, within mammals, males will more often be in a position to gain than females from risky, possibly lethal fights."[220]

In sum, females are more likely to restrain from organized aggression such as warfare because it has little reproductive utility in either an absolute or relative sense. The restraint does not necessarily come from men being physically more powerful than women. Indeed, the differences in strength and speed between human males and females are smaller than in our closest animal relatives, such as the chimpanzee and gorilla.[221] Nor are males more likely to be warriors because women incur the constraints of pregnancy, nursing, and raising children. If that were true, Low reminds us, "sterile women and postmenopausal women might broadly be expected to engage in intergroup conflict."[222]

Reflections on Disease and International Relations

Although it traditionally receives little study in the field, disease significantly affects international relations. In this section of the chapter, I discuss how this is so first by analyzing biological warfare. Its use as a weapon of war may be the most obvious example of how disease can affect international relations, but it is not the sole one. After this discussion, I examine the role disease has played in European imperialism, and address the commendable efforts of the Clinton and George W. Bush administrations to raise the profile of disease in U.S. foreign policy. I advance two major arguments. First, populations maintain an epidemiological balance of power that assists or hinders imperialism.

Where an imbalance of power was present, such as in the early centuries of settlement of the New World, European imperialism was facilitated by disease. Where a balance of power was present or the imbalance did not favor Europeans, such as in Africa, European imperialism was hindered. Pathogens have played a centrally important, if traditionally neglected, role in the history of European expansionism. Second, while disease poses a real threat to the security of the United States, it is equally important to keep the threat in perspective and neither inflate nor diminish it.

Biological Warfare

Throughout history, warfare and disease have been closely associated because warfare caused population movements and destruction, which in turn promoted disease. This important link between disease and warfare is significant because it has led to the deaths of millions of people throughout history, for example, during the Thirty Years' War, or the epidemics of typhus and other diseases that ravaged Russia in the era of World War I and the Russian Civil War. I focus my discussion, however, on a second and equally important component: biological warfare, the use of disease as an explicit instrument of warfare.[223]

The historical record of biological warfare dates to the fourteenth century. In 1346, Tartar armies besieging the city of Kaffa, now Feodossia, in Ukraine, catapulted diseased cadavers over the city walls to force it to capitulate.[224] This siege probably contributed to the second plague pandemic in history.[225] Kaffa was an Italian trading colony, and some of the defenders managed to flee the city, spreading the plague to Constantinople and to Messina in Sicily. By late 1347, plague was ubiquitous along the Mediterranean coast, and the next year it spread through Italy, France, and England, having already devastated many cities in the Middle East.[226] Between 1347 and 1350 at least a third of Western Europe's population, around 20 million people, died in what contemporaries called "the pestilence."[227] These numbers are stunning even today. The "Black Death" was the worst epidemic in recorded history.

In the New World, the British saw how biological warfare might help pacify Indian tribes. In 1763, Sir Jeffrey Amherst, commander of the British forces in North America, suggested using smallpox to attack Indian tribes hostile to the British during the French and Indian War (1754–1763). This suggestion was implemented by one of Amherst's men, Captain Simeon Ecuyer of the Royal Americans, who gave the Indians blankets and a handkerchief infected with smallpox, after which an epidemic of smallpox occurred among the tribes of the Ohio River valley.[228] During the American Revolution, some said that the British used smallpox as a weapon against the American forces in 1775–1777.[229] In 1777, George Washington, who had himself survived an encounter with smallpox in Barbados in 1751, ordered the entire Continental army variolated (that is, inoculated) to prevent further crippling outbreaks of the disease.[230]

More recently, the Japanese used biological weapons extensively in China during the Second World War.[231] It is now publicly known that the Soviet Union and then Russia maintained an offensive biological weapons program after signing the 1972 Biological Weapons Convention that prohibited such a program.[232] North Korea and other states possess these weapons, and many others have sought to acquire them. In addition, some terrorist groups, like the Japanese cult *Aum Shinrikyo* (Aum Supreme Truth) and the *al Qaeda* organization led by Usama bin Laden, are also interested in acquiring biological weapons.[233]

While the use of biological agents has a long history, it is likely that the future will see greater use of these agents. This is the case for four reasons. First is the issue of motivation. States may use biological weapons, but this threat is relatively unlikely, especially against the United States, because of the threat of retaliation with conventional or nuclear weapons. As the October and November 2001 anthrax attacks against U.S. officials and the media showed, terrorists are becoming more likely to attack with biological weapons.[234] This is the result of the rise of new types of terrorists, who, because of their political, religious, or millenarian beliefs, are more interested in killing many people than terrorists traditionally have been.[235] As an example, in 1995 two members of a Minnesota militia group were convicted of possessing ricin that they intended to use to kill governmental officials. A year later, one man in Ohio, Larry Harris, acquired bubonic plague (*Yersinia pestis*) cultures from American Type Culture Collection, which supplies microbial cultures to biomedical researchers; he was arrested again in 1998 in Nevada for possessing a biological agent for use as a weapon.[236]

The efforts of *Aum Shinrikyo* to acquire biological agents, specifically anthrax and botulism, are particularly chilling because the cult would almost certainly have used them in a major attack had they been able to perfect a mechanism of delivery. They did use a chemical agent, sarin, in their March 1995 attack on the Tokyo subway; less famously, they conducted ten unsuccessful biological weapons attacks between 1990 and 1995.[237]

Second, the danger from biological weapons is significant because they are potent weapons, relatively well known, quite accessible, and reasonably easy to deliver.[238] The potency of biological weapons is captured in the stark fact that a millionth of a gram of anthrax is a lethal dose for humans, and a mere kilogram can kill hundreds of thousands in a metropolitan area. Knowledge about biological weapons is readily available. Small groups of people with advanced knowledge of biology and modest financial support can make biological weapons if they can acquire a sample or feedstock. Indeed, information and some recipes are available on the Internet. Many of the agents are accessible in the natural environment, and are also used for biological and medical research in private firms, government laboratories, and universities; this helps explain how Larry Harris could acquire plague. As Kathleen Vogel documents, the former-Soviet bioweapons complex with its decaying security apparatus is another possible source of feedstock.[239]

While these potential sources are worrisome, the 2001 anthrax attacks demonstrate an even greater potential problem: an individual or group may have greater knowledge of weaponization than expected.[240] Furthermore, much of the technology required to produce biological weapons is readily available because it is "dual use," also needed for the production of legitimate products like vaccines and antibiotics. Finally, the means of delivering biological weapons are relatively simple but not necessarily easy. Biological weapons can be effectively disseminated as an aerosol, using simple technology like agricultural crop dusters or backpack sprayers. To be most effective, however, the agent should be respirable—which can pose a significant problem. As biological weapons expert Jonathan Tucker describes, "to infect through the lungs, infectious particles must be microscopic in size—between one and five microns (millionths of a meter) in diameter."[241] Manufacturing an agent that size is a challenge. Equally difficult is removing the electrostatic charge of particles to aerosolize better the bacteria. States with bioweapons programs, like the Soviet Union and Iraq, did this by using additives or coatings, such as silica, but it is generally thought that such a step would be beyond an individual and perhaps too great a challenge for a terrorist group. Moreover, once these steps are achieved, effective delivery also requires specific wind and other atmospheric conditions.

The third reason is the natural evolution of bacteria and viruses. Natural selection alone ensures that old viruses mutate; we have seen a new strain of smallpox (*Variola major*), called "India-1967," and new viruses, such as AIDS, Hepatitis C, Severe Acute Respiratory Syndrome (SARS), and Marburg or

Ebola, and these present great problems for public health officials.[242] Moreover, diseases can be transmitted between species, and changes in the earth's climate will make viruses and bacteria more viable, ensuring that humans will continue to confront pathogens.[243] As the distinguished historian William McNeill writes, humans may not like to acknowledge the fact, but "we are caught in the food chain, eating and being eaten."[244]

Indeed, existing viruses that are not significant threats now may rapidly become so. The virus responsible for the 1918–1919 influenza outbreak evolved very quickly. It came in two waves—spring and fall—with the fall wave causing much higher mortality and lasting into the spring of 1919. The spring virus was already highly contagious, both to humans and other animals (tens of millions of horses, pigs, and primates also died); virologists and epidemiologists suspect that it mutated and became even deadlier.[245] In fact, it was worse than the plague in absolute terms. Twenty-five million people died worldwide. The influenza pandemic killed more people in a shorter period of time than any other known contagion, including the smallpox pandemics discussed below.

In addition, there is always the problem that previously unknown viruses may suddenly appear, perhaps through "species jumping." AIDS, influenza, SARS, and monkeypox are zoonotic diseases—they are transmitted from other species to humans. An example of this happened in 1999, when the previously unknown Nipah virus infected pig populations in Malaysia and then spread to humans. Perhaps most famously, AIDS may have originated in apes or perhaps monkeys and then spread to humans. One explanation for its origins is that a hunter infected himself, perhaps through a cut on his fingers as he was butchering a chimpanzee with Simian Immunodeficiency Virus (SIV). It may be that the Ebola virus originally infected humans in the same manner. The monkeypox outbreak in the United States in 2003 is one more example. This disease heretofore had only been found in central and western Africa. Now it is in the United States and may be expected to spread. Its arrival in the New World has been traced to the importation of at least one exotic animal—a Gambian pouched rat (*Cricetomys gambianus*)—for purchase as a pet. That industry is large and poorly regulated, and this hurts the animals themselves and serves as well as an "interstate highway system" for disease to move from place to place in comfortable conditions. When one realizes that there are thousands if not millions of viruses in the world capable of making such a leap and infecting humans, it is difficult not to conclude that we have been fortunate so far.

But humans are pressing their luck. Diseases make the jump from other species to humans with increasing frequency due to the increased human population and their activities, such as rapid worldwide travel or farming and hunting in new areas, e.g. rainforests in Asia or South America. Suburban sprawl into previously unpopulated areas is also a significant cause of new interaction between humans and other species. But it is not just human activity. Climate change forces more birds and mosquitoes into places well populated by humans, and thus these animals bring new microbes into contact with people.

Finally, advances in biotechnology also contain dangers. Biotechnology's great benefits include unraveling the mysteries of the human genome, providing new ways to diagnose disease and other ailments, and more effective therapy. Other advances are likely to facilitate biomedical engineering, permitting better surgery, and the use of unspecialized cells called stem cells that can replace damaged brain or other cells. But biotechnology has a hidden cost. It may give rise to what is termed "black biology," its misuse for warfare or terrorism. Biotechnology enables laboratory workers to create or modify the more dangerous microbiological agents, such as smallpox, with other diseases, such as Ebola. One goal of the Soviet and Russian biological weapons program was to develop a smallpox-based biological weapon containing virulence genes from Ebola.[246] Biotechnologists can design and build viruses from constituent parts of other viruses to make an especially lethal weapon. Or they could design viruses or bacteria to mutate rapidly, staying a step ahead of any potential vaccine. The former Soviet and Russian biological weapons designer Kenneth Alibek (born Kanatjan Alibekov) has revealed that the Soviet biological weapons program made several genetically engineered strains of antibiotic-resistant *B. anthracis,* the bacterium that causes anthrax, by inserting DNA snippets from drug-resistant bacteria into the anthrax microbes.[247] Moreover, "binary" biological weapons may be created in which two or more benign diseases are brought together shortly before their use to create a dangerous agent. This is advantageous for the perpetrator of an attack since such a weapon might be safer to handle and possibly to produce.

Perhaps most worrisome, biological weapons may be designed to combine virulence with the ability to target a certain individual or group—so-called "designer diseases." According to Tucker, a senior virologist involved with the Iraqi biological weapons program, Hazem Ali, has admitted to having studied camelpox as an "ethnic weapon" because Arabs are largely immune, having regularly lived with camels and thus been exposed to the disease, while Westerners have not and thus are more vulnerable.[248] These dangers may strike some as fanciful, and perhaps camelpox is unlikely to be an effective weapon. But that is not the point. What is important is that the science to make such weapons exists now; all that is required is intent. After the anthrax attacks in the United States, these dangers can no longer be dismissed; they must be studied to ensure the protection of the population.

The Epidemiological Balance of Power

While biological weapons have posed a threat in the past and still remain a danger, it is useful to reflect on the role of disease more broadly, not just as a specific weapon but as a force that affects international relations. Disease has indirectly assisted in conquest and imperialism, affecting both the conquered and conquerors. Disease can be a powerful ally for conquerors and also a terrible foe. Its powerful effect has not always been recognized by historians of imperialism or by political scientists interested in explaining the rapidity of European imperialism.

The eminent historians Alfred Crosby and William McNeill and the physiologist Jared Diamond are changing this perception.[249] These scholars, in particular, have shown the importance of disease for explaining European imperial expansion. This explanation begins with the recognition that many human diseases were acquired first from the agricultural lifestyle and second from domesticated animals. Diamond in particular has shown that agriculture was a "mixed blessing for humanity" despite being commonly perceived as a universal good.[250] He sees three reasons why agriculture was not wholly beneficial for human health. First, hunter-gatherers enjoyed a varied diet with adequate amounts of protein, vitamins, and minerals, while farmers obtained most of their food from starchy crops. Even today three high-carbohydrate plants—wheat, rice, and corn—provide more than 50 percent of the calories consumed by humans. Second, dependence on one or a few crops made farmers vulnerable to starvation through crop failure, while hunter-gatherers were more mobile and less susceptible to that danger. Finally, most of the infectious diseases and parasites that afflict humans could not become established until humans made the transition to agriculture as a dominant lifestyle. With this lifestyle came the rise of "crowded, malnourished, sedentary people constantly reinfected by each other and by their own sewage," which in turn enabled the spread of crowd epidemics like cholera and measles.[251] As Diamond explains, "tuberculosis, leprosy, and cholera had to await the rise of farming, while smallpox, bubonic plague, and measles appeared only in the past few thousand years with the rise of even denser populations in cities."[252]

Like agriculture, domesticated animals were a mixed blessing for humans. By approximately 4000 B.C. societies across Eurasia had already domesticated cows, horses, goats, pigs, and sheep; in lieu of cows, East Asians domesticated four other cattle species: the banteng, gaur, water buffalo, and yak. Diamond shows how these animals—as well as plants like barley, oats, millet, rice, rye, and wheat—are critically important for explaining the rise of civilizations in Eurasia.[253] In addition, geography was significant. The Mediterranean, the Middle East, and China are along the same latitude, which made dissemination of animals and plants easier in Eurasia than in the New World since they never left the temperate latitudes of the northern hemisphere. Thus they were equally well adapted to China, Europe, India, or the Middle East.

Domesticated animals provided Eurasian peoples with great benefits as sources of labor and food, but those benefits came with a cost. These animals also created great dangers for humans. For example, measles, smallpox, and tuberculosis probably came directly or indirectly from cattle, influenza from pigs, and falciparum malaria from chickens.[254] As people in Eurasia domesticated many animals, they had a great opportunity to acquire immunity from these diseases, although they often paid a high price as epidemics ravaged populations. Europeans thus carried a potent ally—disease—when they traveled to the New World, and this ally would greatly help them defeat indigenous peoples by decimating their numbers.

Indeed, any complete account of European imperialism must consider whether an epidemiological balance or imbalance of power existed between the European and indigenous populations. When power was imbalanced in favor of the Europeans, conquest would be relatively easy, as it was for the English, French, Portuguese, and Spanish in the New World, or against the Hottentots in South Africa. So many died of disease that it destroyed the fabric of these civilizations and societies, hindering their ability to resist conquest. However, McNeill observes, sometimes, as in Africa, Europeans collided with native peoples who were not susceptible on a mass scale to the diseases carried by European explorers and colonists. Then an epidemiological balance of power made conquest significantly harder. For example, the Boer conquest of "the native Hottentots, whose vulnerability to imported diseases resembled that of other isolated populations," was relatively easy.[255] However, when Boer pioneers encountered Bantu populations, "who were about as disease-experienced as the Dutch," their expansion was more difficult.[256]

European Expansionism Facilitated by an Epidemiological Imbalance of Power

Disease greatly assisted the European conquest of the New World by decimating indigenous populations rapidly, often even before many Europeans arrived, as disease spread along trade routes. When the Spanish and Portuguese arrived in the New World, the losses of the indigenous population were terrific, although we have no precise estimate of the indigenous population of North and South America around 1500.[257] While population estimates vary greatly, modern demographic techniques provide more precise estimates. The first Europeans to arrive estimated indigenous populations; then missionaries and colonial administrators took censuses, and travelers added their estimates. These data can be weighed against the carrying capacity of the environment and the density of pre-Columbian artifacts and ruins.[258] After carefully analyzing these historical records, anthropologist Russell Thornton finds "that the aboriginal population of the Western Hemisphere circa 1492 numbered at least 72 million and probably slightly more."[259] David Stannard concludes that the indigenous population of the Americas was "within the range of 75 to 100,000,000 persons, with roughly 8,000,000 to 12,000,000 living north of Mexico."[260] Geographer William Denevan concludes that the indigenous population in 1492 was almost 54 million.[261] In contrast, the population of the Eastern Hemisphere is estimated to have been between 300 million and 500 million in 1500.[262]

Documenting the effect that the arrival of Europeans had on indigenous populations in the sixteenth century, historian Steven Katz believes the decline in population may have been as great as 96 percent if the highest estimate of indigenous population is used: "from approximately 112 million to between 4 and 5 million, within one century as a consequence of this colonial enterprise."[263] Although the estimated population of pre-Columbian North and South America varies greatly, and Katz's estimate of 112 million is probably too high, he argues that "before the arrival of the Portuguese and Spanish" the major empires of Mesoamerica—the Aztec, Incan, and Mayan—might have had a total population of between 70 million and 90 million people, with about one-third of the total living in modern Mexico, and "only 3.5 million two centuries later."[264] Katz continues, "in Chile the Mapuche are thought to have lost 90 percent of their upwards of 1 million population as a consequence of contact," while in Hispaniola sources suggest "depopulation rates of 400:1 within fifty years of the conquest."[265] In Central America, Nicaragua "is conjectured to have seen its approximately 500,000 to 1 million Indian population reduced over the course of the sixteenth century to 8,000 by 1574," and "Costa Rica had a preconquest Indian population numbered at between 75,000 to 100,000, which declined to 7,000 by 1581 and to 1,600 by the year 1665."[266] For all of Mesoamerica, Katz notes that the pre-Columbian population declined by about 78 percent.[267]

Thornton's data on the indigenous population in North America indicate a similarly substantial, if less dramatic, decline in population. He estimates that over 2 million American Indians lived north of the conterminous United States in 1492, and their numbers had declined to between 125,000 and 150,000 by the beginning of the twentieth century, or to about 7 percent of the 1492 population, representing a decline of about 500,000 per century.[268] In the conterminous United States, the over 5 million in 1492 had declined to 250,000 by the last decade of the nineteenth century, or to about 4 percent to 5 percent of its size in 1492, representing a decline of about 1.25 million per century.[269]

In South America, the population decline was equally substantial. In Peru, a population of about 3.5 million in 1520 dropped by over 80 percent to about 600,000 by 1630; and one analysis of the Andean region estimates an indigenous population of 10 million in 1530 that had fallen 60 years later to 1.5 million.[270] Denevan believes the indigenous population of North and South America declined from approximately 54 million in 1492 to 5.6 million by 1650, meaning a decline of 90 percent of the population in just over 150 years.[271]

Whatever the precise indigenous population size of pre-Columbian America, scholars of the demographics of Indian populations recognize that the indigenous population of North and South America was greatly reduced. Reflecting on genocides throughout history, Katz concludes: "There is nothing in the world-historical record, for all its cruelties and bloodlettings, to rival this demographic decimation. . . . as a matter of quantity the Indian catastrophe is unparalleled."[272] The principal culprit was not the cruelties of the Dutch, English, French, Portuguese, or Spanish colonists, although they surely contributed to depopulation; it was, instead, the diseases they carried, particularly smallpox, measles, influenza, and typhus.[273]

As Crosby writes, "it has often been suggested that the high mortality rates of . . . post-Columbian epidemics were due more to the brutal treatment of the Indians by Europeans than to the Indians' lack of resistance to imported maladies," but this is not the case.[274] Disease had the greater aggregate effect because the "European exploitation had not yet had time to destroy the Indians' health."[275] In his survey of disease in the history of the Americas, historian Noble David Cook concludes: "more than 90 percent of the Amerindians were killed by foreign infection."[276] While Europeans committed significant atrocities in the New World, disease was by far the most important killer of the Indians.

Mortality was so high because these Old World diseases found ideal conditions in the New: they are termed "virgin soil" epidemics. Virgin soil epidemics are different from others because of, first, their tendency to affect adults as well as children; and second, the number of victims and the mortality of those afflicted.[277] Of course, the Western Hemisphere was certainly not disease-free before 1492. The diseases and parasites present included American leishmaniasis, bacillary and amoebic dysentery, viral influenza and pneumonia, arthritides, salmonella, American trypanosomiasis, roundworms, and other endoparasites.[278]

Nonetheless, in conjunction with genocide and barbarous labor conditions, the devastation brought by Old World diseases radically transformed New World populations. Europeans brought to the New World smallpox, measles, diphtheria, mumps, whooping cough, and typhus; from Africa came malaria and yellow fever and possibly dysentery.[279] The epidemics of these diseases came in waves, typically parallel to human generations of twenty-five to thirty years, with particularly severe outbreaks usually occurring only when a population contained enough susceptible individuals.[280] Unfortunately, indigenous peoples faced not one disease but waves of new illnesses in rapid succession that increased mortality.

In Europe, acute infectious diseases primarily affected children; adults suffered less due to their frequent contact with these diseases. In the New World, adults had not developed immunity and thus suffered as much as children. Indeed, they continued to suffer long after their first exposure to many of these diseases. Mortality from Old World diseases was often just as high in the early or middle eighteenth century as it was at first contact. If a population were continuously exposed to a disease like smallpox, mortality would decrease as the disease became endemic, so that nearly every generation would be exposed as children. The indigenous peoples of the New World, however, had not had enough time to acquire immunity. Luckily, though, mortality declined as science and more modern medical practices brought some relief to these populations.

Smallpox, measles, and influenza are among the most communicable of acute infectious diseases. They move largely by air, as infected people exhale; smallpox also spreads by contact with the characteristic pustules. Other diseases, such as typhus, bubonic plague, yellow fever, and malaria, require vectors, such as lice, fleas, and mosquitoes, in order to spread.

Given its virulence, smallpox may be the worst disease in the historical record because it spread quickly, made almost all whom it infected very sick, and often killed them. It was, as Crosby says, "a thug" that "mugged and murdered and forced its way" into the historical record.[281] It affected both princes and peasants, as Donald Hopkins's famous study of smallpox reveals. Most famously, Queen Elizabeth I and Abraham Lincoln survived it, but many other leaders throughout history did not, including Egyptian pharaohs, two Japanese emperors, kings of Burma and Thailand, Joseph I of Austria, Louis XV of France, and perhaps the Roman Emperor Marcus Aurelius.[282]

All serious diseases inflict horrible suffering upon their victims, but the pathology of smallpox, as Hopkins describes it, invokes a special sympathy for its victims. "Some patients appeared exactly as if they had been severely scalded or burned, and even less seriously affected victims said the skin felt as though it were on fire," he writes; "in addition to the skin, which sometimes sloughed off in large pieces, the virus attacked the throat, lungs, heart, liver, intestines, and other internal organs, and that is how it killed."[283] Moreover, most victims "reeked of a particularly sickening odor," and in some "the disease caused hemorrhaging internally and externally, the so-called black smallpox, which was almost always fatal. Overall, about one out of every four victims died."[284] Those who survived were immune, but often at the cost of pockmarked faces due to the destruction of sebaceous glands, or blindness in one or both eyes due to ulcerations of the cornea.

The first outbreak of smallpox in the New World was probably on the island of Hispaniola in 1507, with a second, longer-lasting epidemic at the end of 1518 or early 1519.[285] From that time until 1524 or perhaps until 1528 a smallpox epidemic continued in the Spanish New World territories. Although it was so virulent that it would likely kill a nonimmune person before he reached the New World, smallpox might have come across in the form of scabs from pustules, in which the virus can live for weeks, if they dropped into textiles or bedding. Or perhaps the virus was passed along a string of several people until it reached the New World. The Spanish and other Europeans were not generally affected, since the disease was endemic in Europe and they had probably had the disease as youths. However, smallpox seems to have first struck people in India about 1500 B.C., and the peoples of the New World crossed from Asia to North America about 12,000–11,000 B.C. Thus, for smallpox, they were a virgin population.

Measles and other ailments also had a terrific effect. Between 1520 and 1600, Mexico experienced fourteen epidemics, and there were up to seventeen in Peru, with an especially severe measles outbreak from 1532 to 1533.[286] But as an agent of death, nothing could equal smallpox at this time.

The explosive spread of smallpox in the New World helped to defeat both the Aztec and Inca empires by killing many people and thus hindering the ability of the population to resist Spanish conquest.[287] Anthropologist and demographer Henry Dobyns estimates that mortality from the first smallpox pandemic in the New World was as high as 75 percent in the populations it touched.[288] Perhaps most famously, in 1519 Fernando Cortés and six hundred Spaniards captured the Aztec's capital city of Tenochtitlán and seized their ruler, Montezuma, but soon they were forced to withdraw as other Aztec leaders mobilized against them.[289] The next year Pánfilo de Narváez and his troops were sent by the governor of Cuba to arrest Cortés; a member of this expedition carried smallpox. Cortés won over Narváez's group and used them as reinforcements to conduct a second attack against the Aztecs. The far greater ally, however, was smallpox. According to Dobyns, "in a few weeks the smallpox pandemic swept through all of the native peoples allied with or against the Spaniards in central Mexico," greatly facilitating the final Spanish conquest of Tenochtitlán and the Aztec empire in 1521.[290]

In 1532, Francisco Pizarro and 168 men invaded the Incan empire. They met little resistance, largely because two half-brothers, Atahuallpa and Huascar, were locked in a dispute over the throne and would not unite to fight the Spanish. Their father, the emperor Huayna Capac, and the crown prince, Ninan Cuyuchi, had died from smallpox when the epidemic reached the Incan capital, Tawantinsuyu, in 1524, well before the Spanish. Half of the senior generals had also died, further hindering their ability to resist the invaders, and contributing to a civil war.[291] The people Pizarro conquered were survivors of one of the worst periods of their history.

Soon after, in North America, disease accompanied the expeditions of Francisco Vasquez de Coronado and Hernando de Soto. Beginning in 1540, Coronado explored parts of present-day Mexico, New Mexico, Texas, and Colorado, bringing smallpox and bubonic plague to the Pueblo Indians. During the 1540s, de Soto carried the same diseases to the indigenous tribes of the lower Mississippi Valley.

The English and French carried diseases just as efficiently as the Portuguese and Spanish. Sir Francis Drake brought typhus to the Caribbean and Florida. French fishermen and fur traders brought disease to the Atlantic coast of Canada, and in 1616 and 1617 disease swept through New England and, in the words of Cotton Mather, cleared the woods "of those pernicious creatures, to make room for better growth."[292] The seventeenth century saw twenty-four epidemics among North American Indians in contact with European settlers: twelve of smallpox, four of measles, three of influenza, two of diphtheria, and one each of typhus, bubonic plague, and scarlet fever.[293] In New England, a major epidemic of 1616–1617 (or possibly 1616–1620) nearly exterminated the Algonquin tribe.[294] This epidemic might have been bubonic plague, smallpox, or yellow fever, and it may have been transmitted through European contact or from other Indian tribes as these diseases spread throughout the Indian population.

French fur traders brought European diseases to the Huron tribe of the Saint Lawrence valley and present-day Toronto. The tribe suffered a series of epidemics, particularly smallpox, from 1634 to 1640 in which they lost one half to two thirds of their population, or about ten thousand of an estimated twenty thousand to thirty-five thousand.[295] Already weakened, they suffered more when defeated by the Iroquois in 1649. The surviving Huron fled to Lorette near Quebec or other locations, intermarried with the Ottawa and other tribes, or were adopted by the Iroquois.

These epidemics continued in each region of North America through the eighteenth century and into the nineteenth. The Iroquois and the tribes of New England suffered horribly from smallpox epidemics, especially in the early 1660s, 1717, 1731–1732, and 1763. In the South, the Cherokee were greatly affected by epidemics in 1711–1712, 1738, 1759, and the late 1700s; the 1738 epidemic is estimated to have killed one-half of the Cherokee population. Reports from Spanish missions in New Mexico report smallpox epidemics in 1719, 1733, 1738, 1747, 1749, and in the 1780s; and from Texas, the Spanish reported seven major epidemics among Indian tribes.[296]

In the Midwest, the Sioux may have had smallpox epidemics in 1734–1735, 1779–1780, and 1780–1781, with the latter years known as the "Smallpox Used Them Up Winter," and the "Smallpox Used Them Up Again Winter." Among southern Californian tribes, epidemics in 1709–1710, 1729–1732, and 1763 had typically horrific effects, with the first one alone killing two thousand of an estimated population of eight thousand Californian Indians.[297]

Even the comparatively isolated Northwest Coast of North America did not escape the ravages of smallpox. Brian Ferguson notes that a brief Spanish visit to the Tlingit of the Canadian Pacific Northwest in 1775 left smallpox that claimed, according to some estimates, half of the population of about ten thousand.[298] The Russian arrival around 1800 began a constant European presence that decimated this tribe through disease. By one calculation, the total Tlingit population in 1835 was 8,650, and another smallpox plague in 1836 reduced the population by 40 percent.[299] In addition to smallpox, North American Indians suffered from other epidemics in the eighteenth century, including six of measles, three of influenza, two each of typhus, diphtheria, and scarlet fever, and one each of typhoid and plague.[300] Tuberculosis was also an ever-present threat.

In the nineteenth century, Dobyns documents twenty-seven epidemics of Old World disease among North American Indians: thirteen of smallpox, five of measles, three of cholera, two of influenza, and one each of diphtheria and scarlet fever (both exclusively in Canada), tularemia, and malaria.[301] The thirteen smallpox epidemics he notes are the largest number known in any single century. Two of the thirteen were especially devastating. In 1801–1802, an epidemic in the U.S. Midwest and Northwest killed many among the Omaha, Ponca, Oto, and Iowa. It moved up the Missouri River to devastate the Arikara, Gros Ventre, Mandan, Crow, and Sioux. It also traveled down the Mississippi to affect tribes in the southern plains, Texas, and the South. From 1836–1840, smallpox ranged in the northern plains of the United States, Pacific Northwest, Canada, and Alaska, in one of the best-documented epidemics in the nineteenth century. Spread by a steamboat traveling the Missouri River in 1836, it killed six thousand to eight thousand Blackfoot, Piegans, and Bloods; two thousand Pawnee; almost all of several thousand Mandan; a half of the forty-five hundred Arikara and Minnetaree; one-third of the three thousand Crow; four hundred Yanktonai Dakota; over one-half of eight thousand Assiniboin; and many Kiowa, Ojibwe, and Coeur d'Alene.[302]

To this day, diseases continue to take a heavy toll among indigenous peoples. According to Ferguson, "detailed information from the Orinoco-Mavaca area" reveals that "even in this violent region, deaths caused by disease run three or four times higher than those attributed to war."[303] The Yanomamö recognize that disease is a cost of association with the outside world but still seek the benefits of trade nonetheless: "all Yanomama [*sic*] recognize disease as the worst dimension of contact with whites, yet some actively seek contact and the mortal risk associated with it, while others actively avoid contact believing the material benefit (e.g. steel tools) are not worth the potentially deadly results."[304] In 1968, when measles struck the Yanomamö, 8–9 percent of the population perished even though some modern medicines and treatment were available. In the 1970s the Kreen-Akroroes of the Amazon Basin lost 15 percent of their population to influenza.[305]

In addition, disease affects the animals or herds so crucial to the tribes for agriculture or as sources of food. For example, rinderpest has occasionally devastated tribes in East Africa by killing their livestock. In the epidemic of 1890, according to estimates, some Maasai in Kenya lost nearly 80 percent of their cattle herds to the rinderpest epidemic and a drought the next year.[306] The epidemic of foot-and-mouth disease in Argentina, Britain, and the Continent in 2001 shows that while the costs of modern animal epidemics in advanced societies are not as great as for the Maasai or other tribes, they can still cause agricultural crises and have a major economic impact on markets.

Finally, Diamond poses an interesting question: Why did not the New World, and specifically large cities like Tenochtitlán, have a deadly pathogen, the equivalent of smallpox, waiting for the Spanish?[307] The New World held no such lethal crowd pathogen, with the possible exception of syphilis. The answer is that, first, the most densely populated American centers, the Andes, Mesoamerica, and the Mississippi Valley, were never connected by a regular and rapid trade route and thus never became a large breeding ground for microbes. Second, in the era before Columbus, the New World had relatively few domesticated animals because about 80 percent of its large mammals had become extinct at the conclusion of the last Ice Age, about ten thousand years ago. Only five animals were domesticated by Americans: the turkey in Mexico, the llama/alpaca and guinea pig in the Andes, the Muscovy duck in tropical South America, and the dog throughout the Americas. None of these animals are likely sources of crowd diseases.[308] The result was an epidemiological imbalance of power between the Old and New World that facilitated European conquest. However, this was not the case in Africa and Asia.

European Expansionism Hindered by an Epidemiological Imbalance of Power

European imperialism encountered great difficulty where it met either a balance of power with the indigenous population or an imbalance against Europeans. While disease favored the European conquest of the New World, the case of West Africa demonstrates how disease could hinder, but not stop, European imperialism. Three major diseases affected Europeans in West Africa: malaria, yellow fever, and gastrointestinal infection. These three accounted for nearly 94 percent of all deaths among Europeans in this region in the late eighteenth and nineteenth centuries.[309]

Historian Philip Curtin found that the earliest and most carefully documented cause-of-death records for the British Empire in West Africa cover British troops stationed in Sierra Leone between 1819 and 1836. He writes, "deaths from 'fevers' (combining malaria and yellow fever, among others) were 84.93 percent of all deaths. Those from gastrointestinal infections were another 8.55 percent."[310] During this period, British troops perished at a high annual rate, 483 per thousand. The British recognized in West Africa they incurred an especially high "relocation cost": the cost in deaths from disease as the result of movement from a native environment to another.[311] Early in the nineteenth century this cost "was around 200 to 300 percent for movement to India, and it rose to 600 percent for movement to West Africa."[312]

The French also suffered. In Senegal from 1819 to 1838, French deaths were about 165 per thousand. Even eighty years later, with significant improvements in medicine and sanitation, death rates were high. The most scientific cause-of-death sample for the French Empire in West Africa is for 1909 to 1913. "At that time, 56.1 percent of all deaths were from malaria, 2.64 percent from yellow fever, and 21.71 percent from gastrointestinal infections."[313] Table 4.5 illustrates the deaths from disease for British, Dutch, and French soldiers in major colonies from 1817 to 1838. In all cases, disease extracts a significant toll from these soldiers, but the death rates from disease in the Dutch East Indies, Senegal, and Sierra Leone are the worst.

Disease not only affected troops in garrisons but almost always was worse during campaigns. During the Anglo-Asante war of 1824–1826, all-European units lost 638 soldiers per thousand annually before these units were withdrawn. Disease also affected Africans. For example, in 1824 during the Anglo-Asante war, the Asante besieging Cape Coast withdrew after losing many men to smallpox.[314] In Europe, because smallpox was endemic and infection during childhood was relatively common, the adult population was largely immune. In Africa, however, smallpox was a serious threat because many more communities were isolated. Thus, because many individuals in those communities were not immune, they suffered greatly.[315]

As Europeans continued their colonial campaigns in Africa in the nineteenth century, they learned to reduce deaths from malaria in the second half of the century by using quinine as a prophylactic. Deaths due to gastrointestinal diseases declined because troops used Pasteur-Chamberland water purification filters in garrison and even occasionally during campaigns. However,

Table 4.5. Mortality of European Troops in Major Colonies

Region	*Nationality of Troops*	*Date*	*Deaths per Thousand*	*Relocation Cost or Benefit (percent)*
Europe and North America				
France	French	1820–1822, 1824–1826	20.17	
Great Britain	British	1830–1836	15.30	
Northern U.S.	American	1829–1838	15.00	
Canada	British	1817–1836	16.10	
Southern U.S. and Malaria-Free Islands				
Bermuda	British	1817–1836	28.80	-88.24
Southern U.S.	American	1829–1838	34.00	-126.67
Mauritius	British	1818–1836	30.50	-99.35
Réunion	French	1819–1836	32.12	-59.25
Mediterranean Climate				
Algeria	French	1831–1838	78.20	-287.70
Gibraltar	British	1818–1836	21.40	-39.87
Malta	British	1817–1836	16.30	-6.54
Ionian Islands	British	1817–1836	25.20	-64.71
Cape Colony	British	1818–1836	15.50	-1.31
Pacific Islands				
Tahiti	French	1845–1849	9.50	50.01
New Caledonia	French	c. 1848	11.40	21.14
New Zealand	British	1846–1855	8.55	44.12

typhoid fever (more accurately, enteric fever) became a serious threat during this time; by the 1890s it was the most important single cause of death for garrisons in Algeria, Tunisia, Egypt, and India.[316] It also extracted a great cost during the Anglo-Boer War of 1899–1902, as Sir Arthur Conan Doyle wrote in his history of the war: "bad water," enteric fever and other ailments, cost the British "more than all the bullets of the enemy."[317] In fact, of the twenty thousand British soldiers who died in the Anglo-Boer War, fourteen thousand died from disease, and six thousand from enemy action.[318] Table 4.6 lists the major colonial campaigns of the British and French from 1860 to 1897 and

Table 4.5. Mortality of European Troops in Major Colonies

Region	*Nationality of Troops*	*Date*	*Deaths per Thousand*	*Relocation Cost or Benefit (percent)*
South and Southeast Asia				
Bombay	British	1830–1838	36.99	-141.76
Bengal	British	1830–1838	71.41	-366.73
Madras	British	1829–1838	48.63	-217.84
Ceylon	British	1817–1836	69.80	-356.21
Coastal Burma	British	1829–1838	34.60	-126.14
Straits Settlements	British	1829–1838	17.70	-15.69
Dutch East Indies	Dutch	1819–1829	170.00	-1,011.11
West Indies				
Jamaica	British	1817–1836	130.00	-749.67
Windwards and Leewards	British	1817–1836	85.00	-455.56
Guadeloupe	French	1819–1836	106.87	-429.85
Martinique	French	1819–1836	112.18	-456.17
French Guiana	French	1819–1836	32.18	-59.54
Tropical Africa				
Sénégal	French	1819–1838	164.66	-716.36
Sierra Leone	British	1819–1836	483.00	-3,056.86

Source: Philip D. Curtin, *Death by Migration: Europe's Encounter with the Tropical World in the Nineteenth Century* (New York: Cambridge University Press, 1989), Table 1.1, pp. 7–8.

indicates the terrific cost diseases imposed on campaigns in Africa and Asia. In almost every instance those costs were greater than the casualties incurred in fighting the enemy, and disease made being in those locations significantly more dangerous than staying in Britain or France.[319]

Of course, there was often an epidemiological blowback from colonization because European peoples had no monopoly on epidemiological protection: diseases also moved from the New World to the Old.[320] Beginning in the mid-1490s, a highly contagious venereal disease, ancestral to syphilis if not the disease itself, first affected port cities and towns in Spain, southern France, and

Table 4.6. Comparison of Deaths in British and French Campaigns in Africa and Asia, 1860–1897

British Army			French Army		
Date	*Campaign*	*Deaths from Disease per Thousand*	*Date*	*Campaign*	*Deaths from Disease per Thousand*
1860	China Field Force	14.9	1861	Cochinchina	140.0
1860	China (Talienwan)	5.4	1862	Cochinchina	117.0
1867–1868	Ethiopian	12.1	1862	China	118.0
1870	Zululand	24.8	1862–1863	Mexico	71.0
1874	Asante	17.4	1863	Cochinchina	107.0
1877–1878	Galeika Gaika	14.0	1881	Tunisi	61.0
1878–1879	Afghanistan	93.7	1883–1884	French Sudan	280.0
1882	Egypt	5.7	1884	Tonkin	60.0
1884	Suakin	nil	1884–1885	French Sudan	225.0
1884–1885	Nilotic Sudan	36.4	1885	Tonkin	75.0
1885	Suakin	2.2	1885–1886	French Sudan	200.0
1885–1886	Nilotic Sudan	4.1	1886	Tonkin	99.0
1885-1896	Asante	5.6	1886–1887	French Sudan	220.0
1889	Nilotic Sudan	6.0	1887	Tonkin	106.0
1895	Chitral	25.1	1887–1888	French Sudan	116.0
1896	Matabeleland	16.5	1888	Tonkin	133.0
1896	Dongola	46.6	1892	Dahomey	87.0
1896–1897	Mashonaland	2.0	1895	Madagascar	302.0

Source: Philip D. Curtin. *Disease and Empire: The Health of European Troops in the Conquest of Africa* (New York: Cambridge University Press, 1998), Table A8.2, p. 239.

Italy before spreading rapidly throughout Europe. It is probable although not certain that sailors on Columbus's first voyage to Hispaniola contracted the disease, or a close variant such as yaws, from the native Taino peoples.[321] Cholera was first reported in India in 1817 and arrived in Britain in 1831. In the nineteenth century, Britain lost an estimated 130,000 people to five cholera epidemics. Its population was 17.5 million in 1800 and 41 million in 1900, so even with significant population growth it was a major cause of death in nineteenth-century Britain.[322] In the nineteenth century and the first quarter of the twentieth, India lost 25 million people to cholera. This is a significant factor in explaining why India's population was nearly stagnant from the arrival of the British until after World War I. In 1891, India had a population of 150 million, several million fewer than when Charles, the Marquis of Cornwallis, arrived in 1786 to become governor-general.[323]

Disease and National Security

Disease was an important aspect of international relations in the classic age of imperialism, and it remains significant today, when modern transportation systems and globalization increase the number and speed of people traveling the world. The Clinton administration called disease a major national security priority, and disease has been termed the principal "unarmed threat" faced by the United States today. Richard Holbrooke, the Clinton administration's ambassador to the United Nations, argued in January 2001 that the AIDS epidemic in Africa was "the most dangerous problem in the world today" and the "biggest health crisis in 600 years."[324] In a major address on the national interests of the United States in October 2000, President Clinton's national security advisor, Samuel Berger, stated that stopping global infectious diseases is a major national security priority of the United States: "how can we say we are protecting our people if we fail to stop the spread of diseases like AIDS, malaria, and tuberculosis, which account for 25% of all deaths in the world?"[325] Berger continues, "flat earth proponents may not see disease as a national security priority. But a problem that kills massively, crosses borders, and threatens to destabilize whole regions is to me the very definition of a national security threat."[326]

Indeed, when we consider the number of people infected we see clearly that disease is a significant threat to the quality of life and even stability of some African states. About 30 million Africans are infected with HIV, and more than two million died of HIV-related diseases in 2000. In eight countries in southern Africa, more than 15 percent of the adult population is infected. In Botswana the situation is the worst—nearly 40 percent of the adult population has HIV, and projected life expectancy has fallen more than thirty years on average due to the disease. About 40 million people worldwide are living with HIV/AIDS, three million of whom are children under fifteen, and it continues to spread. Indeed, the fastest rate of HIV infection in the world is in Central and Eastern Europe and Central Asia, and by 2010, epidemiologists expect that the Asia-Pacific region could surpass Africa in the number of HIV infections. According to Nicholas Eberstadt, estimates of new HIV cases in China during the 2000–2025 time period range from 32 million to 100 million; in India, from 30 million to 140 million; and in Russia from 4 million to 19 million. Deaths in the same period may range from 19 million to 58 million in China; 21 million to 85 million in India; and 3 million to 12 million in Russia.[327]

Tuberculosis is especially deadly because it spreads relatively easily. About 1.75 billion people, almost a third of the world's population, are infected with the bacillus that can cause it. About 100 million of these people have active cases, and about 8 million people develop active tuberculosis every year. Although a cure was developed more than fifty years ago, the disease kills approximately 3 million people a year.[328]

Like its predecessor, the Bush administration is focused on the problem of HIV/AIDS, tuberculosis, and malaria in Africa, and in May 2001 announced a global fund to fight the spread of AIDS in Africa, with a founding contribution of $200 million to promote prevention, treatment, and training of medical personnel.[329] In January 2003, the administration dramatically expanded its support for stopping the disease by offering $15 billion over five years, beginning with $2 billion in 2004, in support to the most afflicted states in Africa and the Caribbean, as part of the Emergency Plan for AIDS Relief. This plan provides funds to prevent up to 7 million new infections, to treat 2 million people with advanced antiretroviral drugs, and to care for 10 million infected individuals and orphans. It is a notable step in fighting HIV/AIDS, but still falls short of the $7 billion to $8 billion annual spending epidemiologists believe is needed to retard its spread.

Is disease a national security problem? This is certainly questionable, despite Berger's insistence that it is. Critics might suggest that the physical security of the United States, as well as its other vital national and economic interests, are not threatened by diseases in Africa, Asia, or South America. A discussion of how disease threatens the United States must be more circumspect than Berger described. Disease is likely to threaten the national security of the United States in two ways. First, if a given disease were to spread to the United States by natural means (that is, not through biological warfare), as the West Nile virus has, then it would pose a national security threat because it could affect public health. Combating it would largely be the responsibility of the Department of Homeland Security, the Centers for Disease Control and Prevention, and state and local officials. Thus, in the modern world, the spread of disease is a national security threat that falls outside the responsibilities of the national security organs of the United States, such as the Department of Defense and the intelligence community. To address this threat adequately, these bureaucracies must coordinate with the Public Health Service and other agencies concerned with epidemiological and health issues.

Second, if the disease were to generate a refugee crisis in the state concerned or in neighboring states that threatened to destabilize these states or even an entire region, a case might be made for intervention. However, even in this case intervention might be unlikely given the checkered history of U.S. interventions in areas of less than vital interests. So while such a situation should be a matter of humanitarian concern for states, nongovernmental organizations, and public health authorities, it is certainly questionable whether it is a matter to be solved by the U.S. military.

While experts may disagree over what constitutes a national security threat and while the threat should not be exaggerated, Berger does point out a real danger of epidemics. That danger remains a significant national security concern, given the rise of drug-resistant bacteria, such as the new strain of cholera (*Vibrio cholerae 0139*) detected in Madras in 1992 and a puzzling outbreak of plague there in 1994, as well as viruses like SARS that could directly affect the health of the American population.[330] These diseases have the potential to be as great a threat as the Spanish influenza epidemic of 1918–1919, the last great pandemic before AIDS. Clearly, even if mortality were to fall far short of that epidemic, concern is warranted.

Contributing to the danger is complacency among governmental officials and the public in the belief that only a fixed number of infectious diseases need to be conquered. Because of natural selection, humans are locked in an evolutionary race with bacteria, fungi, and viruses, and humans are at some disadvantage in this race because pathogens have such short generation times. Thus, the problem of disease may actually increase for two reasons: first, the growing human population forces larger numbers of people into contact with animal reservoirs of disease organisms, and modern transportation allows more people to travel around the world more rapidly.

The end result is that disease will always be with us. It cannot be simply eliminated by advances in medicines and medical procedures. Without question, these are great assets for humanity. But natural selection, human population growth, and modern transportation ensure that diseases will continue to plague humans into the future, and Americans are not likely to be isolated from these diseases. Consequently, vigilance against disease should be a matter of concern for U.S. policymakers.

Conclusion

In this chapter I examined the implications of evolution for warfare. I argued that warfare contributed to the growth of human intelligence and human society. As with many aspects of human history and human evolution, we cannot say precisely how much warfare contributed to human intelligence. Nonetheless, given the conditions in which humans evolved, it is eminently reasonable to suggest that the threat of external attack, from both predators and other humans, provides a strong ultimate cause of the rapid growth of human intelligence and explains why humans would organize in ever-larger groups, ultimately developing into human society.

The recognition that warfare is not uniquely human but is also found among ants and chimpanzees may have been shocking and disheartening to some readers who conceive of warfare as a uniquely human pestilence or solely the product of human thought. In this, as in most other matters, humans are not singular creatures. Natural selection has produced organized violence in other animals as well for broadly similar purposes—to gain and defend resources. This requires us to broaden our conception of warfare beyond political causes to include the functional. War is conducted for political purposes but also to gain the resources necessary for life itself or to maintain a standard of living. This was true in our evolutionary past. However, it may also be true of the human future as natural resources decline or are depleted. Indeed, functional causes of warfare are likely to have increasing salience.

Of course, there is no doubt that humans are unique in the degree of violence and destruction they can create and the effect this violence and destruction has on other animals. Indeed, given the human development of nuclear and biological weapons, we might even be able to eliminate human life itself. In *Cat's Cradle,* Kurt Vonnegut memorably describes "ice nine," a fictional chemical that was intended to change mud to a frozen, solid surface. It was designed by scientists at the request of the U.S. Marine Corps, whose generals were tired of having to fight in the mud. However, once released it converted the entire planet—desert, water, forest, plains—to such a surface, thus eliminating most life, and all human life, on earth. Vonnegut drew the concept of "ice nine" from a genuine concern of physicist Enrico Fermi as his group prepared to conduct the first sustained nuclear chain reaction at the University of Chicago on December 2, 1942. Fermi's group had a small but significant concern that once fission began it would not stop, leading to the destruction of life on earth. The danger of unintended consequences applies not only to the world of fiction or physics, but also to biology and the growing risk of increasingly sophisticated biological weapons that could be specifically designed or genetically engineered to kill a specific individual or group, but, like Vonnegut's "ice nine," might have horrific consequences once released into the world.

Human evolution helps explain the physically and emotionally stimulating effects of warfare on combatants. As I argued above, the experiences individuals encounter in warfare are Janus-faced: the horror of war is linked with its stimulating effects on individuals, possibly a residue of group experiences of hunting and fighting in Pliocene, Pleistocene, and Holocene conditions. Our evolution caused or helped to cause many of the powerful emotions and stimulations that individuals engaged in combat still experience today. I have also discussed why males, rather than females, are more likely to be warriors. It may come as a surprise to some that this difference is rooted ultimately in the evolution of a female reproductive strategy and not, as an ultimate cause, in a discrete male culture or male bonding rituals. Of course in other genera, natural selection has created conditions that are radically different, as I demonstrated above in my discussion of warfare among ants. Any expectation that females are irenic by nature may be corrected if one pauses, perhaps on a summer or fall afternoon, and examines for a moment the warfare often conducted at one's feet.

Finally, the role of disease in international relations has been profound but neglected by scholars of international relations. Clearly, it is emerging from the darkness of neglect due to the efforts of Holbrooke and Berger in the final year of the Clinton administration and continued in the Bush administration, and scholars increasingly recognize that biological warfare is a significant threat to states at present and will become a greater threat in the future.

Disease is also relevant to our understanding of the human past. I demonstrated how the epidemiological balance of power played a significant role in understanding the relative ease of imperialistic conquest. Such conquest was relatively easy for Europeans in the New World. But it was significantly harder in Africa and Asia, where an imbalance of power against the Europeans obtained. Consequently, tens of thousands of Europeans died in the drive to colonize those continents, and hundreds of thousands more died as a result of syphilis and cholera—part of the cost of colonization.

5 Evolutionary Theory and Ethnic Conflict

In this chapter I explore how evolutionary theory increases our understanding of the causes of ethnic conflict. To do so, I examine the two major paradigms social scientists use to explain the causes of ethnic conflict. I then explain how evolutionary theory can improve our understanding of these causes. My central argument is that evolutionary theory allows scholars to understand better the origins of deep causes of ethnic conflict, such as the in-group/out-group distinction, xenophobia, and ethnocentrism. These behaviors contributed to fitness in the past, and so unfortunately remain ultimate causes or contributing causes of ethnic conflict.

At the outset of my argument, it is important to explain why ethnic conflict is an important issue to study. I suggest two major reasons. First, ethnic conflicts often generate great human rights abuses and humanitarian disasters. The slaughter of 1 million Tutsi and moderate Hutus in Rwanda and the widespread ethnic cleansing in the former Yugoslavia demonstrate the need to understand ethnic conflict.[1] The ethnic cleansing conducted in Rwanda and in the states of the former Yugoslavia has received significant attention from the world's media, but unfortunately the violence there is far from unique, even in its horrific effects on the population. Nonetheless, it gave rise to discussions of "compassion fatigue" in Western states, and forces one to recall Stalin's sardonic comment that one death is a tragedy, a million deaths is a statistic. In addition, since the end of the Cold War the frequency of ethnic conflict has increased, with conflicts in Bosnia, Chechnya, the Democratic Republic of Congo, Kosovo, Liberia, Macedonia, Rwanda, Sri Lanka, Sudan, and Turkey.[2] Despite the widespread ethnic conflict and genocide in the twentieth century, as documented in table 5.1, it is not a new phenomenon in world history.

Second, these conflicts often are of great interest to the creators of U.S.

foreign policy because they can affect neighboring states, as the conflict in Kosovo affected the Albanian minority in Macedonia, presenting the risk that the conflict would escalate to neighboring states, such as Greece and Turkey. This in turn might adversely influence NATO and the U.S. presence in Europe. Ethnic conflicts also can directly affect U.S. interests by undermining an important U.S. ally, such as Turkey in its conflict with the Kurds. Moreover, as a result, ethnic conflict is receiving greater attention from U.S. policymakers. In the spring of 1999, after the United States had intervened with Operation Allied Force in the conflict over Kosovo with the Republic of Yugoslavia, the Clinton administration declared a change of U.S. policy that became known as the "Clinton Doctrine."[3] The essence of the doctrine was that the United States and NATO will use force to advance and protect universal values, such as the prevention of ethnic conflict, instead of the narrower national interests for which states traditionally fight. In addition, the doctrine authorized military intervention in the internal affairs of sovereign states to protect those values. In September 1999, United Nations Secretary General Kofi Annan echoed this position when he announced that the U.N. would recognize the sovereignty only of states that supported human rights.

However, the Clinton doctrine was stillborn. After being announced it was quickly abandoned when the administration saw how little support it had among Congress, the Pentagon, or the U.S. foreign policy community. Nonetheless, the doctrine is important because it revealed the thinking of the Clinton administration, especially its motives and willingness to intervene in conflicts over less than vital national interests, particularly in ethnic conflicts. Despite some concern in Washington and among the Europeans, the administration of George W. Bush has not withdrawn U.S. forces from obligations incurred by his predecessor. It has deepened the commitment to Macedonia, and indeed has extended U.S. commitments with its involvement in Afghanistan, Iraq, and other states as part of the war on terrorism. It is following Clinton's willingness to intervene in deed, if not always in rhetoric.

For these reasons, ethnic conflict is a major issue in international relations and U.S. foreign policy, making it crucial to understand its origins in order to prevent or mitigate it. As with warfare, its origins are complex and often multifaceted. It is the result of cultural, economic, historical, political, social, and evolutionary causes. The last has received the least attention from scholars, but this does not mean that it is the least significant, only the least studied.

Referring to the Serbs' ethnic cleansing of Kosovo, President Clinton said that a bright future for humanity is threatened "by the oldest problem of human society: our tendency to fear and dehumanize people who are different from ourselves."[4] Clinton identified a central problem concerning the origins of ethnic conflict. Xenophobia—the fear of strangers—and ethnocentrism—the belief that members of one's own ethnic group are superior or preferable—exist among almost all peoples, and both contribute to the ubiquity and scale of ethnic violence. Evolutionary theory permits students of ethnic conflict to understand its causes in a more detailed and sophisticated way.[5] Specifically, it explains, first, why humans make in-group/out-group distinctions; and second, why humans can be xenophobic and ethnocentric. Comprehending the origins of these phenomena will assist scholars and policymakers as they seek to understand the fundamental causes of ethnic conflict, and as they try to stop or prevent it.[6]

Causes of Ethnic Conflict: Two Paradigms—The Primordial and the Modern

Although ethnic conflict has multiple causes, fundamentally its causes may be divided into two paradigms. Primordialists argue that ethnic conflict is the result of deep cultural, historical, and social causes, while modernists submit that ethnic conflict is the result of contemporary political and social forces.[7] Each paradigm is valuable because it offers important insights into the origins of ethnic conflict. Although the paradigms disagree about the causes of ethnic conflict, they do agree on two points. First, the risk of ethnic conflict in modern international relations is present when the state is biethnic or multiethnic, as are most states. Second, the causes of ethnic conflict are complex; adverse cultural, economic, or political conditions are likely to be the immediate causes of any specific ethnic conflict.[8]

In this section of the chapter I will describe the two paradigms in detail. My intent is to present the argument of each one and then to provide several criticisms of them. In the next section, I argue that evolutionary theory provides insights into the ultimate causes of ethnic conflict. My argument for the paradigms has two major implications: that the primordialist position will become more scientific than it is at present, and that modernists will also benefit because evolutionary theory permits some insights into why the people of a nation may be manipulated by leaders like Slobodan Milosevic. My effort here is only to demonstrate these insights. It is not to provide a complete account of the origins or conduct of ethnic conflict, or of elite manipulation of the masses. Furthermore, understanding the origins of ethnic conflict does

Table 5.1. Partial List of Genocides in World History, 1492–2000

	Deaths	*Victims*	*Perpetrators*	*Place*	*Date*
1.	XXXX	Caribbean Indians	Spanish	West Indies	1492–1600
2.	X	Beothuk Indians	French	Newfoundland	1497–1829
3.	XXXX	Indians	Spanish	Central and South America	1498–1824
4.	XX	Protestants	Catholics	France	1572
5.	XXXX	Indians	Americans	USA	1620–1890
6.	XX	Bushmen, Hottentots	Boers	South Africa	1652–1795
7.	XX	Aleuts	Russians	Aleutians	1745–1770
8.	XXX	Aborigines	Australians	Australia	1788–1928
9.	X	Tasmanians	Australians	Tasmania	1800–1876
10.	X	Morioris	Maoris	Chatham Islands	1835
11.	XX	Araucanians	Argentineans	Argentina	1870s
12.	XX	Hereros	Germans	Southwest Africa	1904
13.	XXXX	Armenians	Turks	Ottoman Empire	1915
14.	XX	Jews	Ukrainians	Ukraine	1917–1920
15.	XXXXX	Political Opponents	Soviets	Soviet Union	1929–1939
16.	XXXXX	Political Opponents	Nazis	Occupied Europe	1939–1945
17.	X	Polish Officers	Soviets	Katyn Forest	1940
18.	XXX	Serbians	Croatians	Yugoslavia	1941–1945
19.	XXX	Ethnic Minorities	Soviets	Soviet Union	1943–1946
20.	XXXX	Ethnic Germans	Central Europeans	Central Europe	1944–1950
21.	XXX	Hindus, Muslims	Muslims, Hindus	India, Pakistan	1947
22.	XXX	South Sudanese	North Sudanese	Sudan	1955–1972
23.	XX	Indians	Brazilians	Brazil	1957–1968
24.	XX	Tutsi	Hutus	Rwanda	1962–1963
25.	X	Arabs	Africans	Zanzibar	1964
26.	XXX	Communists, Chinese	Indonesians	Indonesia	1965–1967

27.	X	Ibos	North Nigerians	Nigeria	1966
28.	X	Aché Indians	Paraguayans	Paraguay	1970s
29.	XXXX	Bengalis	Pakistani Army	Bangladesh	1971
30.	XXX	Ugandans	Idi Amin	Uganda	1971–1979
31.	XXX	Hutu	Tutsi	Burundi	1972–1973
32.	XXXX	Cambodians	Khmer Rouge	Cambodia	1975–1979
33.	XX	Timorese	Indonesians	East Timor	1975–1976
34.	XX	Muslims, Christians	Christians, Muslims	Lebanon	1975–1990
35.	XX	Argentinean civilians	Argentinean military	Argentina	1976–1983
36.	XX	Political Opponents	Dictator	Equatorial Guinea	1977–1979
37.	X	Political Opponents	Emperor Bokassa	Central African Republic	1978–1979
38.	XXX	South Sudanese	North Sudanese	Sudan	1983–
39.	XX	Tamils, Sinhalese	Sinhalese, Tamils	Sri Lanka	1985–
40.	XXX	Croats, Muslims, Serbs	Muslims, Serbs Croats	Bosnia	1992–1995
41.	XX	Croatian Serbs	Croatians	Croatia	1993–1995
42.	XXX	Hutus, Tutsis	Tutsis, Hutus	Burundi	1993
43.	XXXX	Tutsis, Hutus	Hutus, Tutsis	Rwanda	1994
44.	X	East Timorese	Militias, Indonesian Military	East Timor	1998–1999
45.	X	Serbs, Albanians	Albanians, Serbs	Kosovo	1998–1999

Sources: Michael E. Brown, "Introduction" in Michael E. Brown, ed., *The International Dimensions of Internal Conflict* (Cambridge, Mass.: MIT Press, 1996), pp. 4–7; Jared Diamond, *The Third Chimpanzee: The Evolution and Future of the Human Animal* (New York: HarperCollins, 1992), pp. 284–286; R.J. Rummel, *Death by Government* (New Brunswick, N.J.: Transaction Publishers, 1997), p. 10; United Nations, "East Timor—UNTAET Background," <http://www.un.org./peace/etimor/UntaetB.htm>; and Alfred-Maurice de Zayas, *A Terrible Revenge: The Ethnic Cleansing of the East European Germans, 1944–1950* (New York: St. Martin's Press, 1994), p. xvi.

Notes: X = fewer than 10,000 people killed; XX = 10,000 or more; XXX = 100,000 or more; XXXX = 1,000,000 or more; XXXXX = 10,000,000 or more.

not excuse the atrocious behavior of individuals or groups in the past or, unfortunately, the present.

To advance this discussion, precise definitions of nation, nationalism, and ethnicity are particularly important in order to avoid confusion. Among scholars of ethnic conflict, political scientist Walker Connor has been particularly careful in his definition of terms, especially for explaining the differences between nation, nationalism, and ethnic group; he has also forcefully admonished scholars to maintain proper definitions of these words in order to advance scholarship and understanding. Too frequently, Connor argues, important differences between these words and between nation and state are not recognized, leading to needless confusion.[9] With Connor's admonition in mind, I will briefly discuss each term.

Nation

I define nation as a group of people who believe they share a common ancestry, often claim a homeland or specific territory, and possess shared historical memories as well as a distinct culture and often a discrete language.[10] A sense of shared blood or kinship among the group is also important. People hold the belief, however fictive it may be, of consanguinity or an unalloyed descent from a common tribe in the distant past: A nation or *ethnie* is defined by its members' belief that they are ancestrally related.

Each nation possesses common codes and shared symbols and myths of common descent. These codes, symbols, myths, and the associated historical memories of common past experiences are the main elements of collective identities in most societies and thus serve to differentiate the nation from other social bonds or groups. One of the first theorists of nationality, Ernest Renan, described the nation in his 1882 lecture "What Is a Nation?": "the nation, like the individual, is the culmination of a long past of endeavours, sacrifice and devotion. Of all cults, that of the ancestors is the most legitimate, for the ancestors have made us what we are. A heroic past, great men, glory . . . this is the social capital upon which one bases a national idea."[11]

The nation provides identity and a sense of pride to its members. It gives its members reasons why they are different from other nations—allowing them to create a sharp division between the nation and others—and describes the nation's history, emphasizing its trials and tribulations in relationships with other nations.[12] These memories may emphasize past injustices to that nation, such as massacres or economic, cultural, or political discrimination. These memories help forge the nation's identity, and may also promote bias or discrimination against other nations, perhaps by maintaining that they are untrustworthy, inferior, or a threat. As philosopher Yael Tamir argues, "drawing the boundaries of a nation involves a conscious and deliberate effort to lessen the importance of objective differences within the group while reinforcing the group's uniqueness vis-à-vis outsiders."[13]

When people identify with their nation, they are identifying with a wider kinship or with a fictive "super-family" linked by ties of ancestry, as seen for example in states like China, France, Japan, Germany, Greece, Israel, Korea, Poland, or Thailand.[14] People in a nation may know that the history of their nation is myth and that multiple ethnic groups have contributed to their bloodline, but these facts need not and often do not diminish the feeling that they are of a unique group. Nothing more is necessary, as Connor writes, for "all that is irreducibly required for the existence of a nation is that the members share an intuitive conviction of the group's separate origin and evolution."[15] In sum, a nation is a social group that shares a sense of homogeneity, even if that homogeneity is illusory, based on ethnic as well as linguistic and cultural factors, such as a common language, territory, religion, and customs. This homogeneity delineates them from other nations.

Finally, several scholars of nationality argue that a nation itself determines membership in the group. Put differently, the concept is self-defined, not other-defined. The eminent historian Hugh Seton-Watson suggests that this is one reason why the concept of nation will never be precisely defined but rather will be said to exist when "a significant number of people in a community consider themselves to form a nation, or behave as if they formed one."[16] Tamir echoes this, saying that a group may be thought of as a nation when it exhibits a sufficient number of "shared, objective characteristics—such as language, history, or territory—and self-awareness of its distinctiveness."[17] Similarly, political theorist David Miller argues "nations exist when their members recognize one another as compatriots."[18] Of course, in practice this can lead to complex political and social questions, like the "Who Is a Jew?" debate in Israel. Correspondingly, Germans wonder whether German identity is the result of blood, *ius sanguinis,* so that German parents make one part of the German nation, or of residence, *ius soli,* the fact that even an individual of a discrete nation who lives in Germany is German.

Recognizing the proper definition of nation allows us to understand that the word does not substitute for the citizenry of a state or "the people." Indeed, most states in international relations contain many nations; examples are Canada, India, Russia, and the United States. Nor is a nation a synonym for state. A state is a sovereign, territorial, and juridical unit of international relations and is, as sociologist Max Weber described, the "human community that (successfully) claims the *monopoly of the legitimate use of physical force* within a particular territory."[19] Seton-Watson distinguishes between a state as "a legal and political organization with the power to require obedience and loyalty from its citizens," and a nation, which is "a community of people, whose members are bound together by a sense of solidarity, a common culture, a national consciousness."[20]

Connor believes that writers on international relations tend to describe states as nations, because as nation-states like France arose, their borders coincided or almost coincided with the territorial distribution of a nation.[21] These homogenous nation-states are relatively infrequent in international relations, although in most states one single nation represents the majority, even the overwhelming majority, of the population. Of 202 states and major principalities in international relations, about 27 percent (fifty-four) may be described as nation-states. In 18 percent (thirty-six), one nation accounts for between 75 percent and 89 percent of the total population; in 20 percent (forty-one) the largest nation accounts for between 50 percent and 74 percent of the population; and in 35 percent (seventy-one) the largest nation accounts for less than half of the total population.[22]

Nonetheless, Michael Brown, a preeminent scholar of the causes of ethnic conflict, argues that national homogeneity is no guarantee of internal political stability. For example, he says, "Somalia is the most ethnically homogenous state in sub-Saharan Africa, yet it has been riven by clan warfare and a competition for power between and among local warlords."[23] Thus, homogeneity can assure us only that the state will not be plagued by ethnic conflict, not other types of conflict.

Nationalism

Nationalism is an ideological movement that aims to attain and maintain the political autonomy of an actual nation, such as the Japanese, or a potential one, such as the Kurds.[24] It is the ideology of the nation and clearly denotes a strong identification with and loyalty to one's nation, which can often serve to unify it or be a source of political action.[25] Nationalism implies fealty to one's nation; it is not the same as patriotism, which is loyalty to one's state. In the United States for example, an Irish-American might identify with the Irish nation, if one is a Chinese-American with China, but of course they can still be patriotic and give equal or greater loyalty to the United States. Connor argues that nationalism is often used improperly as loyalty to the state, and when so used it creates confusion in the study of ethnic conflict: "With nationalism preempted, authorities have had a difficult time agreeing on a term to describe the loyalty of segments of a state's population to their particular nation."[26] Therefore, various terms are used, among them "ethnicity . . . pluralism, tribalism, regionalism, communalism, and parochialism," with the result that "this varied vocabulary further impedes an understanding of nationalism by creating the impression that each is describing a separate phenomenon."[27]

To demonstrate his point, Connor argues that commentators often identify the conflict in Northern Ireland as a religious conflict because there is no racial or language difference between the two sides.[28] However, it is actually a conflict over national identity. Likewise, tension in Singapore is described as racial due to physical differences between Chinese and Malay people, when it is really a national conflict. Difficulties in Indonesia are often labeled regional because of the geography of that many-islanded state when in fact some of its myriad conflicts are principally national, as in Aceh, Borneo, and Irian Jaya. One need not agree wholly with Connor's characterizations of these conflicts to acknowledge that an accurate understanding of "nation" and "nationalism" allows scholars of ethnic conflict to determine whether conflicts described as religious, racial, or regional are really conflicts between and among nations.

Ethnicity

Like the terms nation and nationalism, ethnicity is often loosely defined, leading to confusion in discussions of ethnic conflict. Therefore its etymology is worth examining. "Ethnicity" is relatively new to the English language; the *Oxford English Dictionary* records it for the first time in 1953. It is derived from the word "ethnic," which dates to the Middle Ages. In turn, ethnic derives from the Greek word for nation, *ethnos,* a group characterized by common descent, which was used as a synonym for "gentile" in New Testament Greek.[29] The Greeks used *ethnos* in several ways: "In Homer, we hear of *ethnos hetairon,* a band of friends, [and] *ethnos Lukion,* a tribe of Lycians. . . . Aeschylus calls the Persians an *ethnos* . . . [and] Herodotus speaks of *to Medikon ethnos,* the Median people."[30] All these meanings reflect a group of people who share biological and cultural characteristics. Ethnicity certainly has other meanings in modern discourse, but this is the meaning I adopt in this book—ethnic group is synonymous with nation for present purposes. Thus my discussion of ethnic conflict should be understood as conflict between nations as I have defined that term here.[31]

My definition of ethnicity draws from Weber's conception of it. In his classic work, *Economy and Society,* he defined the term in the following manner: "We shall call 'ethnic groups' those human groups that entertain a subjective belief in their common descent because of similarities of physical type or of customs or both."[32] Weber continues, emphasizing that the belief in the common descent of the nation may be false or may underestimate the degree of intermarriage with other groups: "It does not matter whether or not an objective blood relationship exists. Ethnic membership (*Gemeinsamkeit*)" may be "a presumed identity" which does not diminish its effect on people.[33] Later, in discussing nationality, he stresses the importance of common descent for nation as well as ethnic group: "the concept of 'nationality' shares with that of the 'people' (*Volk*)—in the 'ethnic' sense—the vague connotation that whatever is felt to be distinctively common must derive from common descent."[34]

The discussion of nation, nationalism, and ethnicity now places us in a position to understand that "ethnic conflict," properly understood, is often conflict between or among competing nations. Thus "national conflict" would be an equally accurate term to describe conflict between nations. Nonetheless, in this book I will use the term ethnic conflict for three reasons: the term is accurate where ethnic is understood to identify or describe a specific nation; the term is widely used by scholars, policymakers, and the media; and using it will avoid the confusion with interstate war that might arise with the use of "national conflict."

The Primordialist Paradigm: Ethnic Conflict from Within

In the primordialist paradigm, the causes of ethnic conflict are anchored within the nation. The central argument is that people have strong primordial attachments to their nation, and when appeals or exhortations are made by leaders on this basis, for purposes of unity, defense, or aggression, they will respond vigorously. This attachment is a fundamental basis for ethnic conflict. In this section, I will provide the argument of the paradigm and then examine two critiques made by modernists.

Three suppositions unite the scholars who endorse this paradigm. First, human beings classify themselves and others in accordance with national identity; as this is almost always the most powerful source of identity, it strongly affects their behavior.[35] Second, this identity is often exclusionary, and contains misperceptions about one's own nation as well as other nations. Third, all the paradigm's scholars recognize that ethnic conflict stems from a cause deep within the nation and is directed against other nations, although, as I describe below, their exact explanations are different. In turn, this means that ethnic conflict is not principally caused by external forces (such as economic conditions or political ideology) but rather by a force found within the nation. These suppositions provide coherence to the paradigm.

The particular explanations used by scholars Clifford Geertz, Donald Horowitz, and Walker Connor are quite diverse, as they are drawn from anthropology, psychology, and political science, respectively. Nonetheless each recognizes that the nation is critical to understanding the problem of ethnic conflict. Before I go into the argument in detail, I want to stress three points. First, the explanations of this paradigm are noumenal. That is, they depend upon ideas or beliefs that are nonrational and rely on intuitive convictions or beliefs about the genesis and development of one's nation and why it is discrete from others. These convictions arise from multiple sources but often may be taught informally within each family, either by parents or grandparents, or by elders of the community, or in the educational system.

Second, advocates of this paradigm should not be dismissed as simply believing in "ancient hatreds" as the cause of ethnic conflict, although some certainly subscribe to such an argument, at least in their public statements. President George H.W. Bush argued in 1992 that the conflict in Bosnia was the result of "age-old animosities," and Clinton stated that the end of the Cold War "lifted the lid from a cauldron of long-simmering hatreds."[36] The primordialists recognize, as do the modernists described below, that the origins of a specific ethnic conflict are often multiple and may be unique; a specific ethnic conflict may be triggered by forces such as economic hardship or political unrest, not by historical beliefs or old hatreds between nations. Thus, it must be understood, this paradigm acknowledges that while many or even most ethnic conflicts are rooted in the nation, not all are.

Third, advocates of this paradigm recognize that individuals within a particular ethnic group may have a greater affinity or identity with groups other than the one into which they were born. Nothing in this paradigm negates the possibility that class, religious, or occupational cleavages will supercede ethnic ones. The supposition is that for most people national identity will be more important. Moreover, primordialists recognize that one's identity in one's own mind can take several forms. For some it may be tiered: an Iraqi may think of himself as first Arab, next Sunni, then Muslim, and finally Iraqi. For others, identity may be multiple at the same level: an individual living in Punjab may think it equally important to be a good Sikh, Punjabi, and Indian. Or, one's identity may be a combination; in one of the most famous and important examples of multiple and conflicting identities in American history, Robert E. Lee thought of himself first as a good Christian and a Virginian, seeing both as equally important, and then as a citizen of the United States or Confederate States of America.[37]

Now I address the central issue of primordialism: Why people have strong attachments or fealty to their nation that may result in ethnic conflict. I will examine the three major explanations in turn. The first draws from the work of sociologist Edward Shils and anthropologist Clifford Geertz. Shils first used the term "primordialism" when he sought to distinguish different types of social bonds in modern societies.[38] He argued against the intellectual current of the 1950s, which typically held that individuals in the advanced industrial democracies gave their loyalty to the state and participated in its political life only for self-interested reasons. In contrast, Shils submitted that individuals possessed personal, primordial, sacred, and civil ties in contemporary societies and that these bonds had a powerful effect on the individuals in them, strongly influencing their decisions.

In the early 1960s, in one of his early studies of non-Western societies, Geertz drew upon Shils's idea of primordial ties. In these societies, he argued, "primordial attachments" bind people and can lead to tension and conflict with other groups.[39] He defined "primordial attachment" as an "immediate contiguity" and "kin connection" among people that defines them as a nation. But for Geertz a definition of nation must include cultural elements, such as a particular religion, a language or specific dialect, and shared social practices.[40] The crux of his argument is that members of nations feel that their communities are primordial, extending into the distant past, a feeling that binds them and strongly influences their behavior.

In certain cultures, Geertz submits, such as India, Indonesia, and Nigeria, "these congruities of blood, speech, [and] custom . . . are seen to have an ineffable, and at times overpowering coerciveness in and of themselves," and become "primordial bonds" for a community or the natural affinity which unites it.[41] In contrast, Geertz maintains that in modern, Western societies national unity is maintained by a civil state, backed by its coercive powers and shared political ideology. In general, these societies regard identities based on primordial ties as illiberal and thus flawed. Nevertheless, if Geertz is correct, in non-Western societies at least the contiguities of the nation are a powerful source of social and political attachment, and conflict can result when these primordial ties are challenged by other nations or by cultural, economic, and political circumstances.

Sociologist Steven Grosby builds on the scholarship of Shils and Geertz. He points out a common theme running through all cultures, even contemporary Western ones: individuals have a powerful and deep attachment to specific biological and cultural relationships and objects. As Grosby explains, "the family, the locality, and one's own 'people' bear, transmit, and protect life," in all cultures.[42] "This is why human beings have always attributed and continue to attribute sacredness to primordial objects and the attachments they form to them."[43] People attach specifically to the parents who "give one life"; the locality "in which one is born," lives, and is nurtured; and finally the "larger collectivity in which one is born and in which one lives" that provides protection from the potentially dangerous outside world.[44] Thus the family, locality, and community are the source of primordial attachments and for most people will trump other attachments or loyalties.

Political scientist Donald Horowitz offers a second explanation for an individual's attachment to his nation and consequently for ethnic conflict. He uses a different terminology, that of "in-group" and "out-group," that is typical of psychology or political science rather than anthropology.[45] However, consistent with the explanations of Geertz or Grosby, he identifies a noumenal original position that keeps the nation together, distinguishes it from others, and may result in ethnic conflict. Horowitz argues that an ethnic group shares "a strong sense of similarity," which allows its individuals to "submerge their own identities in the collective identity, to favor ingroup members and make sacrifices for them."[46] He continues, "one sees oneself, so to speak, in other group members," which generates "empathy and in extreme cases even obliterates the boundary between one individual group member and another. . . . People assume ingroup members are similar to each other . . . and that assumption strengthens their attraction to" the in-group.[47] Horowitz emphasizes that ethnic groups exhibit "tendencies to cleave, to compare . . . to manifest ingroup bias, to exaggerate contrasts with outgroups, and to sacrifice for collective interests. . . . They appear frequently to engender more loyalty from their members than competing group-types do and to engage in severe conflict with other ethnic groups."[48]

Finally, like Geertz, Grosby, and Horowitz, Connor emphasizes the significant ineffable aspect of national identity and ethnic conflict. Indeed, Connor sees a nonrational belief in the nation: "logic operates in the realm of the conscious and the rational; convictions concerning the singular origin and evolution of one's nation belong to the realm of the subconscious and the nonrational."[49] Renan concurs, describing the nation as both a "soul, a spiritual principle" and "a grand solidarity" that is "constituted by the feeling of the sacrifices one has made in the past and of those one is prepared to make in the future" and thus binds the people of the nation.[50] To Connor, then, although the emotional essence of a nation may be inexpressible or unscientific, that does not in the least diminish its powerful effect on its people.

He notes that such a distinction may not be readily accepted by intellectuals, who typically (and often rightfully) discount nonrational explanations.[51] Moreover, for many Western intellectuals, identifying with one's nation or holding nationalist beliefs indicates a lack of enlightenment or even an illness. Einstein is quoted as having once said, "nationalism is an infantile sickness. It is the measles of the human race."[52] A little over a decade before the Cold War ended, political philosopher John Dunn wrote, "Nationalism is the starkest political shame of the twentieth century, the deepest, most intractable and yet most unanticipated blot on the political history of the world since the year 1900."[53]

Nonetheless, Connor is less interested in intellectual antipathy to nationalism and more in its power. The power of this explanation is not diminished by the rational explanations for ethnic conflict, further considered below, such as economic hardship or deprivation or elite manipulation of the masses. As Chateaubriand argued two hundred years ago, "men don't allow themselves to be killed for their interests; they allow themselves to be killed for their passions."[54] Or as Renan wrote, nations are not mere communities of interest or business partnerships. "Community of interest brings about trade agreements, but nationality has a sentimental side to it; it is both soul and body at once; a *Zollverein* is not a *patrie*."[55] According to Connor, these rationalist explanations may be "faulted principally for their failure to reflect the emotional depth of national identity: the passions at either extreme end of the hate-love continuum which the nation often inspires, and the countless fanatical sacrifices which have been made in its name."[56]

Primordialist scholars recognize how strongly people hold these beliefs and how easily political leaders may manipulate them. Primordialists are not surprised when political leaders of both ethnically homogenous and heterogeneous states emphasize common kinship to strengthen state solidarity. Think of Bismarck's exhortation to the German people to "think with their blood" or the metaphors or images of nations. Just as an individual has a home, so nations have homelands and the idea of a common blood and soil, as in "England," the land of the Angles, or a Kurdistan for Kurds. Recall the images of nations as maternal or paternal figures, the Emperor of Japan as the father of Nippon, "Mother Russia," the Fatherland, Uncle Sam, Marianne, and John Bull. Terms used to describe the nation in political or poetic exhortations include common blood, brothers, mothers, fathers, and ancestors. The British and Americans may think of each other as cousins, while Schiller famously demanded in *Wilhelm Tell* that all German peoples should be a "single nation of brothers, standing together in any hour of need or danger."[57] A century ago, the Chinese nationalist revolutionary Chen Tiannua perceived the Chinese to be a large family: "As the saying goes, a man is not close to people of another family. When two families fight each other, one surely assists one's own family, one definitely does not help the 'exterior' family. Common families all descend from one original family: the Han race is one big family. The Yellow Emperor is a great ancestor, all those who are not of the Han race are not the descendants of the Yellow Emperor, they are exterior families. One should definitely not assist them; if one assists them, one lacks a sense of ancestry."[58]

On a personal level, Sigmund Freud expressed his attachment to the Jewish nation not on the basis of a sense of shared religion or national pride. Rather, he was "irresistibly" linked with other Jews by "many obscure and emotional forces, which were the more powerful the less they could be expressed in words, as well as by a clear consciousness of inner identity, a deep realization of sharing the same psychic structure."[59] Writing decades later, Tamir describes how Jews and Israelis share a sense of belonging to a common nation despite significant cultural, religious, and ethnic differences.[60]

For the primordialists, then, the fundamental cause of ethnic conflict is grounded upon people's identification with and intense loyalty to their nation. The argument is that people have strong psychological attachments to their nation and will respond strongly when appeals are made on this basis or when it is criticized, challenged, threatened, or attacked. They will also respond just as vigorously when appeals are made to attack or demonize other nations. This attachment may be thought of as a stack of kindling, ready for a spark to ignite ethnic conflict. As political scientist Anthony Smith concludes, "wherever ethnic nationalism has taken hold of populations, there one may expect to find powerful assertions of national self-determination that . . . [may] embroil whole regions in bitter and protracted ethnic conflict."[61] Unless these assertions are solved through partition, mediation, or federation, "there can be little escape from the many conflagrations that unsatisfied yearnings of ethnic nationalism are likely to kindle."[62]

Of course, both of these paradigms of ethnic conflict offer critiques of the other. For modernists, primordialism is a flawed paradigm for studying ethnic conflict for two principal reasons. First, modernists argue, the primordialist argument depends upon a static conception of national identity; it discounts the possibility that national identity can be malleable.[63] It is obvious that individuals can have multiple identities, and may assume different ones as circumstances require. In addition, some modernists point out a fact freely acknowledged by primordialists: many individuals, particularly in multinational states, may be descended from two or more nations due to factors like colonization, intermarriage, and migration. These ensure that nations are not wholly discrete or unalloyed.[64] In response, primordialists advance the argument presented above: what is important is the belief, however false it may be, that a nation is descended from common ancestors. In the case of Icelanders and the Japanese, those nations are indeed descended from small groups of settlers, with comparatively little genetic input from the rest of the world, but for the Russian or Italian nations, for example, descent is clearly alloyed.

In response, primordialists do clearly recognize that identity is malleable. I have not discovered any primordialist who claims that identity is fixed and cannot change. Rather, their argument is that national identity is a powerful source of identification for many people and can thus contribute to ethnic conflict. Modernists are on stronger ground with a related criticism: that primordialists cannot clearly explain why national identity is such a powerful source of identification, trumping other sources of identity, such as class. The primordialist paradigm needs support to address this criticism; as I show later, evolutionary theory can help here.

A second principal criticism is advanced against the primordialists: they do not precisely identify the conditions in which ethnic conflicts are most likely to begin.[65] They postulate that national differences may explode into ethnic conflict, but of course many national differences do not result in ethnic conflict. Moreover, economic crises come and go, cultures change, political systems are overthrown, sometimes accompanied by ethnic conflict and sometimes not. So, how can we explain the variation?[66] This is a powerful criticism, because it seems that the primordialists are attempting to explain variation with a single constant, which hinders the coherence of their argument.

The primordialist position is not wholly convincing because its proponents offer few conditions delineating when ethnic conflict is likely to begin or to become worse. The modernists are correct that their own argument carefully describes how changing cultural, economic, and political conditions cause ethnic conflict. However, that said, I also see this criticism as anchored in part in a separate dispute over how to approach the very object of investigation—understanding ethnic conflict. In contrast to the modernists, who focus upon changing conditions, the primordialists prefer to focus on the problem over time; this position allows them to argue that national identity is responsible for much ethnic conflict. Thus, they take a different approach to the problem. What the primordialists are attempting to explain is similarity: why, they ask, do we have ethnic conflict over time in diverse locales, in all known times and places, given great variation in culture, economic system, political organization, etc.? Focusing attention on similarities, rather than differences, is an equally appropriate way to understand ethnic conflict. Here I stress again a point I made at the beginning of this section: both paradigms contribute to the study of ethnic conflict.

Indeed, if we compare the differences to the game of baseball, we might think of the primordialists as focusing upon the rules of the game, which give it structure and are similar for all teams, whereas the modernists are experts at understanding particular games and the differences between the teams in terms of league, pitchers, or even particular pitches. We need both elements to understand baseball in its totality but not necessarily to understand particular aspects of the game.

Nonetheless, what is not wholly convincing thus far is the explanation advanced by the primordialists. Again, evolutionary theory can assist this paradigm by providing it with its first scientific and categorical explanation for the tractability of national identity. This in turn can greatly improve the coherence of the paradigm's central argument.

The Modernist Paradigm: Ethnic Conflict from Without

In contrast to the primordialists, writers within the modernist paradigm of ethnic conflict view the nation and nationalism as recent phenomena in international relations because they result from the processes of economic and political modernization.[67] Modernists do not see ethnic conflict as the result of age-old animosities. Rather, they point to changes that led to our modern era: the ending of the feudal economy and political absolutism. With these changes, religion declined as a force to unite and control the population. Ethnic conflict, they say, was an unfortunate consequence of these liberating changes. As with my discussion of primordialism, I will first present the modernist argument and then offer critiques of the paradigm.

In preindustrial agrarian societies, nationalism did not exist; no one needed "to unify the tiny élite strata and the vast mass of peasant food-producers and tribesmen subdivided into their many local cultures."[68] Modernization brought with it both the need and the possibility of unifying a population. Nationalism, a major political component of modernization, began with the French Revolution, which introduced the idea that a people or nation were sovereign and should thus control the state. This created the first nation-state, and made possible the rise of mass armies, which were first seen during the French Revolutionary and Napoleonic Wars.[69] But modernization also has an economic component, industrialization, which eroded traditional society and caused mass migration to cities. As a result, traditional religious agrarian societies were replaced by the modern urbanized, secular society. As industrialization also required a literate workforce, the state supported standardized, public education to advance literacy and instill nationalism in children. One major consequence of widespread education was the rise of mass media. The media as well, whether state-owned or private, often advanced both nationalism and a fierce patriotism: examples are the Hearst newspapers clamoring for war with Spain, and the jingoistic press in Britain at about the same time. Thus, for those in the modernist paradigm, nationalism is a thoroughly modern phenomenon. Smith describes it as the "religion surrogate" of modern society; it is the force or the cement without which the society would disintegrate, as religion helped to bind agrarian societies in the past.[70]

Not surprisingly, then, modernists view ethnic conflict as the result of thoroughly contemporary forces. Among the modernists, Michael Brown and Stephen Van Evera provide thorough documentation of these forces. They argue that the causes of ethnic conflict should be classified as either "proximate" or "remote" (or "underlying"). Proximate causes immediately affect ethnic conflict, while its remote causes are deeper and produce the proximate.[71] Proximate causes of ethnic conflict are found at the mass and elite levels of politics, and result from developments both within and outside the state. Brown also says that four kinds of factors—structural, political, economic, and cultural—make some states or regions and some situations more likely to yield ethnic conflict. Relying on these explanations, modernists see no need to resort to "primordial attachments" or other arguments concerning ineffable ties or connections to the nation that may or may not result in ethnic conflict.

The first of the underlying causes in Brown's analysis are structural factors. These include the impotence of the state, for example in Africa, where many states were carved out of colonial empires and thus lacked political legitimacy, strong political institutions, or borders that matched the actual divisions among nations.[72] Another factor is intrastate security concerns. When a state is weak or expected to become so, nations and other groups within the state have incentives to provide for their own security and thus must be concerned about other groups. In this situation, Barry Posen argues, a security dilemma is present.[73] When a nation arms to defend itself in these circumstances, it threatens the security of the others, who then arm in response, thus diminishing the security of the first nation. This problem is "especially acute when empires or multiethnic states collapse and ethnic groups suddenly have to provide for their own security."[74] The ethnic geography of a state is the third structural factor identified by Brown and is also emphasized by Van Evera. This problem affects multinational states to various degrees. For example, states carved out of empires generally have complex demographics; their populations may be intermingled, or minorities might live in separate regions. Brown argues that countries with "highly intermingled populations are less likely to face secessionist demands because ethnic groups are not distributed in ways that lend themselves to partition."[75] Van Evera also sees that the demographic arrangement of national groups is an important cause of ethnic conflict, particularly if the populations have a large diaspora and are intermingled. He argues there is less danger of conflict if national groups are compact and homogenous: "The Czechs, for example, can pursue nationalism with little risk to the peace of their neighborhood, because they have no diaspora abroad, and few minorities at home."[76] These factors made the Czech-Slovak "velvet divorce" peaceful, unlike the partition of India in which the Hindu and Muslim populations were intermingled.

Second, political factors are important. The likelihood of ethnic conflict certainly depends upon whether or not the political system discriminates against one specific minority group or against all minorities. Furthermore, the state's ideology is significant, especially in states where one nation is the majority and political rights or privileges are anchored within that nation. In contrast, drawing on a distinction made by Smith, political scientist Jack Snyder argues that where "civic nationalism" is present, citizens enjoy equal and universal citizenship, institutions that give minorities a forum for their concerns, and laws to protect minority rights. In such states, ethnic conflict is less likely.[77] Brown is careful to note that a divisive state ideology need not be based solely on ethnicity; religion or other factors could also play a role.[78] Intergroup and elite politics are also important. The prospects for ethnic conflict are great "if groups—whether they are based on political, ideological, religious, or ethnic affinities—have ambitious objectives, strong senses of identity, and confrontation policies" in their domestic politics.[79] Concerning elite politics, Brown argues that ethnic scapegoating and other abusive tactics employed by "desperate and opportunistic" politicians, particularly in times of turmoil, may generate ethnic conflict.[80] Both Milosevic in Yugoslavia and Franco Tudjman in Croatia were accused of such behavior in the 1990s.

The third source of ethnic conflict is the economy. Economic downturns, stagnation, or crises can destabilize society, leading to ethnic conflict. Transforming a socialist command economy to a market one can worsen, or even cause, economic problems; this happened in certain states in Africa and in Eastern and Central Europe. As Brown summarizes, "Unemployment, inflation, and resource competitions, especially for land, contribute to societal frustrations and tensions, and can provide the breeding ground for conflict."[81] Then, the process of economic transition or reform "can contribute to the problem in the short run, especially if economic shocks are severe and state subsidies for staples, services, and social welfare are cut."[82]

In addition, discriminatory economic systems or practices may lay the foundation for ethnic conflict. Unequal access to resources or capital, or to economic opportunity or professional advancement for minorities, as well as great disparities in wealth or standard of living can generate resentment. This is an element of the struggle between the Tamil minority and Sinhalese majority in Sri Lanka, where the Sinhalese discriminate against the Tamils.[83]

Moreover, the very processes of economic development and modernization can lead to societal instability and conflict. As many scholars have noted, industrialization and technological change introduce adjustments for both workers and suppliers of capital.[84] Workers with specialized skills are required, as are new capital and legal arrangements for the society. Modernization brings equally important changes to the social arena. Increasing literacy generates mass media and a greater awareness of class differences and social inequalities, placing strains upon the traditional economy, and the state's social and political institutions. As Samuel Huntington writes in his classic study of modernization, "social and economic change—urbanization, increases in literacy and education, industrialization, mass media expansion—extend political consciousness, multiply political demands, broaden political participation."[85] But, just as importantly, these changes also "undermine traditional sources of political authority and traditional political institutions."[86] Many states experience a "lag in the development of political institutions behind social and economic change," resulting in political instability and disorder.[87] Economic development also raises expectations of higher standards of living and political liberalization that often exceed the economic or political system's ability to meet them. Of course, such unfulfilled expectations may cause discontent and unrest in the state, increasing the chance of conflict, especially if one class or ethnic group benefits disproportionately, or is seen as doing so.[88]

Finally, cultural factors can cause ethnic conflict. The first such factor is cultural discrimination against minorities. This may take several forms, including restrictions on minorities' religious freedom and educational opportunities, as well as on speaking, writing, or teaching of their languages. For example, Stalin introduced harsh cultural assimilationist policies in the Soviet Baltic states and the Caucasus after World War II to suppress minority cultures, and the Chinese in Tibet still pursue similar efforts to advance Chinese culture there at the expense of the indigenous one.[89] The second cultural factor is the self-image of ethnic groups, most often captured in their group history, their myths, and their perceptions in those histories of themselves and others.[90] Van Evera argues that such a self-image is almost always based on myth, transmitted through the group's politics and culture, and the myth largely takes three forms, which he calls self-glorifying, self-whitewashing, and other-maligning.[91] Together, these form what Brown terms the "ethnic mythologies" of the nation, the oral or written stories passed from one generation to another and so weaving themselves into the cultural tapestry and collective identity of the ethnic group.[92] Having studied the origins of ethnic conflict in Croatia, Georgia, Moldova, and the Karabagh region of Azerbaijan, Stuart Kaufman argues that these myths can be quite complex. Combining memories, values, and symbols, they help determine who is a member of the group and how people will respond both cognitively and emotionally to the myths.[93] He documents how political and religious leaders have used myths to mobilize ethnic groups, and to justify hostility toward and fear of other ethnic groups.

The first category of myths, the self-glorifying, includes claims of unique virtue or false claims of beneficence toward other groups. Ethnic groups often celebrate their own history, perhaps recalling the difficulties in their collective experience or the heroism of their cultural icons. They can be pernicious if rival ethnic groups have reciprocal perceptions of each other, which is frequent. For example, Brown says, Serbs "see themselves as heroic defenders of Europe and Croats as fascist, genocidal thugs," whereas their ethnic neighbors the Croats "see themselves as valiant victims of Serbian hegemonic aggression."[94] The second group, self-whitewashing myths, contains denials of wrongdoing against other groups, as the Turks continue to deny responsibility for the Armenian genocide during World War I, and in the early 1990s Croatian academics and politicians denied the Croatian Ustashi's massacres of Serbs during World War II. For Van Evera these myths are particularly dangerous because a denial "implies a dismissal of the crime's wrongness, which in turn suggests an ominous willingness to repeat it."[95] The third group, other-maligning myths, falsely blames other groups or states for an ethnic group's present problems. The myths also contain claims that others have negative intentions against the group and thus help defame others. Thus they may be directed against ethnic groups outside as well as within the state. These myths support other arguments that minorities should be denied equal rights because they "will appear to pose a danger if they are left unsuppressed; moreover, their suppression is morally justified by their (imagined) misconduct, past and planned."[96]

Van Evera argues that the degree of conflict that a national group's self-image or myth can generate depends upon three conditions. The first is the size of the crimes the ethnic group committed in the past: "past sufferings can also spur nations to oppress old tormentors who now live among them as minorities," perhaps generating conflict with the minorities' home countries.[97] The second is the object of blame for past crimes; using the example of Ukrainian suffering under Stalin, Van Evera notes that what is critical is how that suffering is interpreted today. Was it the sole product of Stalin's mind? That would require no lasting hostility because he is dead. Was it the result of communism? Again, there need be no residual tension because of the regime change. Or was it the result of traditional Russian imperialistic policies? In this case the Russians, perhaps especially those Russians living in Ukraine, ought to be held accountable.[98] Finally, Van Evera notes that mythmaking is an important element of nationalism and is present, in varying degrees, in all nations.[99] However, myths are likely to be inflated or strongly propagated when elites need them most, for example, to generate legitimacy for the regime during an economic crisis or when the elite must place more demands upon the population during a war. In these circumstances as well, the population might be more receptive to myths advanced by the regime. Also, a regime's myths may become more pernicious when opposition to them is weakest. In a state with no robust tradition of free press and free speech, and no strong and independent universities, scholars can rarely conduct and publish research that opposes governmental policies. While they are certainly not immune to nationalist mythmaking, Van Evera argues that democratic regimes are less susceptible to it because "such regimes are usually more legitimate and free-speech tolerant; hence they can develop evaluative institutions to weed out nationalist myth."[100] In contrast, the regimes most prone to this danger are those that depend to some degree upon popular consent but are governed by elites who are unrepresentative of the masses.

Building on the remote causes of ethnic conflict, Brown provides a solid analysis of the proximate causes of ethnic conflict.[101] He submits that conflicts may be sparked by developments both within the state and external to it, and by the elites of a state as well as the population. First, conflict may be triggered by internal, mass-level developments that he calls "bad domestic problems"; they include rapid economic modernization and political and economic discrimination.[102] The former triggered ethnic conflict in Punjab and the latter the conflict in Sri Lanka.[103] Second, external, mass-level problems may be catalysts for ethnic conflict. Brown notes that "swarms of refugees or fighters crashing across borders, bringing turmoil and violence with them, or radicalized politics sweeping throughout regions," may spark ethnic conflict.[104] For example, the expulsion of radical Palestinians from Jordan in 1970 helped destabilize Lebanon. Third, ethnic conflict may result from external, elite-driven causes, such as a government's decision to destabilize a neighboring state or states to advance its own political goals. "This only works," Brown cautions, "when the permissive conditions for conflict already exist in the target country; outsiders are generally unable to foment trouble in stable, just societies."[105] Examples include the efforts by the Rhodesian and later South African governments to undermine the government of Mozambique, Russia's efforts to destabilize Georgia, Pakistan's support of Kashmiri rebels in India, and India's support of insurgents in the Sindh province of Pakistan.[106] Finally, internal, elite-driven policies may also ignite ethnic conflicts. The leaders of a state may deliberately choose to demonize a minority ethnic group, for example, in order to stay in power in the face of growing dissension, or to augment the legitimacy of the regime.[107] Examples of such behavior are numerous, and have played a role in recent conflicts in Azerbaijan, Chechnya, Nigeria, Sudan, Togo, and Zimbabwe.

Just as the primordialist paradigm may be criticized, so too may the modernist paradigm. The modernists are much better at describing the conditions and mechanisms by which ethnic identity may escalate into conflict, as cultural, economic, and political changes make ethnic conflict more or less likely. Thus, modernist arguments are immediately useful for governmental and nongovernmental policymakers who must develop strategies for conflict prevention, management, and resolution. Nonetheless, I suggest two criticisms of the modernists. First, they offer an incomplete explanation for mass manipulation by the elites. Such manipulation certainly contributes to ethnic conflict, as Brown and Van Evera correctly argue. Yet, they fail to develop their explanation of why masses can be manipulated by elites. There is more to this issue than elite coercion or duplicity, or assumptions about the gullibility of the masses. Evolutionary theory generates insights into this issue and permits modernists to understand this phenomenon in a richer and more profound way.

Second, the modernist paradigm may be criticized for defining national identity largely in material terms and for failing to recognize the power in an individual's understanding of his national identity.[108] National identity clearly has significant and long-lasting emotional dimensions. Leaders of states certainly use emotional appeals based on community or family, especially when they need significant sacrifice from the people, as during war or crises.[109] Consequently, we cannot discount this emotive dimension as we seek to understand the origins of ethnic conflict. Upon reflection, we see that ethnic identity has clearly aroused the greatest passions, the deepest love, the most intense loyalty, and the most transcendent hatred. It has moved the greatest poets and artists. Its power seems to be so great that its only rival may be religion. So far, however, its origin has been intangible, and thus often discounted or ignored by scholars studying ethnic conflict. Fortunately, evolutionary theory will allow scholars to understand the origins of ethnic conflict and, through better understanding, work to prevent its occurrence. It also can help both paradigms answer some of the criticisms I have presented here. For the primordialists, it explains why people have deep attachments to their ethnic group. For the modernists, evolutionary theory allows a deeper understanding of how a population may be manipulated by political leaders.

Evolutionary Theory, the In-Group/Out-Group Distinction, Xenophobia, and Ethnocentrism: Insights for the Paradigms Provided by Evolutionary Theory

In this section of the chapter I describe how evolutionary theory contributes significantly to both paradigms by illuminating how evolutionary theory explains the ultimate causes of ethnic conflict. Specifically, I determine why three behaviors—making in-group/out-group distinctions, xenophobia, and ethnocentrism—contributed to fitness in the course of human evolution.

My first argument is that evolutionary theory explains why humans create in-group/out-group distinctions among themselves. This argument supports some of the insights of both paradigms. For the primordialists, the fact that humans can use ethnic identity to make strong in-group/out-group bifurcations, and that these are difficult to eliminate or supplant, assists their paradigm. These points should not be dismissed or discounted by the modernists. Yet it is equally true that in-group/out-group distinctions can be changed to a considerable degree if a state will make determined and consistent efforts, perhaps through public education, or if opinionmakers, such as the elite media or Hollywood, will do so. This is an important lesson for modernists and for all those who seek to stop ethnic conflict. If the distinctions can be removed or modified, people can place less emphasis on a fixed determinant of in-group status, such as ethnicity, and more on a less-fixed or more variable one, such as culture, political ideology, or language. Then new in-groups may be formed and the danger of ethnic conflict lessened.

The second argument I make in this section is that evolutionary theory explains why xenophobia would evolve in human ancestors; thus evolutionary theory provides an anchor for the primordialist school of thought. It also assists the modernist paradigm by helping to explain why people can be manipulated by elites and why the appeals of devious politicians who demonize other national groups can resonate powerfully in the proper circumstances.

My third argument is that evolutionary theory also explains ethnocentrism and its power as a source of ethnic conflict. This argument supports both the primordialist and modernist paradigms because it explains why people may be willing to respond when elites incite hatred of other nationalities. These and similar appeals will echo compellingly in the right conditions because of the force of the in-group/out-group distinctions based on nationality.

As I begin this discussion, I stress that my objective is to analyze how evolutionary theory can explain the ultimate causes of in-group/out-group distinction, xenophobia, and ethnocentrism. Two points are important here. First, as I have stressed throughout this book, we must always bear in mind the naturalistic fallacy: while the ultimate cause of a given behavior may be explained by evolutionary theory, it does not follow that the behavior is acceptable or the right way to behave in today's environment. Evolutionary theory allows us to study these behaviors, understand their origins by explaining how they contributed to fitness in the past, and apply these insights to the debate over ethnic conflict. Thus it is significant for understanding such conflict. However, the process of evolution by natural selection does not control the social actions of humans. Nor should this inherently legitimize governmental policies. To be sure, it may inform an understanding of human behavior and thus, if appropriate, governmental policies, but we must always consider the influence of the environment when we discuss the behavior of the phenotype. As writers in both paradigms of ethnic conflict recognize, many factors in the environment can mitigate as well as promote ethnic conflict.

Second, as I have argued in previous chapters, we must bear in mind the impact of the environment on human behavior. Human evolution has caused individuals to engage in certain behaviors. Few would doubt that eating, sleeping, and engaging in sexual intercourse are the product of evolution. I argue that the in-group/out-group distinction, and ethnocentric and xenophobic behavior, are also the products of evolution. Assuming that evolutionary theory can explain these behaviors, the critical point here is that the environment can suppress or exacerbate them. One's culture, politics, religion, and ideology all influence one's behavior, and just as an individual can suppress hunger or sexual desire at certain times, on occasions dictated by culture, for example, so too can an individual often suppress these three behaviors.

Understanding In-Group/Out-Group Distinctions

Evolutionary theory provides important insights for the in-group/out-group distinction commonly made by anthropologists, political scientists, psychologists, and sociologists.[110] The fundamental point of this division is that humans divide the world into an "Us," the in-group, versus "Them," the out-group, worldview in order to reduce the complexity of our mental information processing. Psychologists refer to the in-group as one's own group, to which one is positively biased. They argue that in-groups develop from a need for self-definition. The in-group identity helps to define one both positively and negatively. It provides one with meaning and purpose, knowing one is a part of a community. One knows what one is not—the out-group. In contrast, the out-group is stereotyped and homogenized as the "Other." Among the many different categories of an in-group, the most common and significant ones are family, friendship, age, race, sex, class, nationality, and citizenship.

Psychologist Henri Tajfel's famous in-group/out-group experiments demonstrate the force of these distinctions. Tajfel used as his subjects unrelated individuals to whom he assigned casual, trivial, or random categories; almost all of them formed groups on the basis of the categories, and discriminated against other groups on the basis of their new group identity.[111] Tajfel's further experiments had interesting results. He gave his subjects three choices. They could maximize the joint profit of both in-group and out-group; second, maximize the total profit of the in-group; or maximize the difference between the profit of the in-group and the profit of the out-group. He found that the outcome that appealed most was the maximal differential between groups, which might also be called relative gains, even if this meant less in absolute terms for the in-group.[112] Tajfel's findings bring to mind the parable about the peasant and the genie. When a peasant finds a bottle and uncorks it, a genie appears and promises the peasant a wish. He replies, "Well, my neighbor has a cow and I have none. So, I wish for you to kill my neighbor's cow." Individuals may value or weight relative differences or gains much more than absolute differences or gains.

The ubiquity of the in-group/out-group distinction in human cultures and across time suggests that it is an evolutionary adaptation. But rather than simply relying on ubiquity of behavior, evolutionary theory provides an ultimate causal explanation of the in-group/out-group distinction made by humans: it explains why such a complexity-reducing mechanism would evolve. There are three major reasons why this is so.

First, as I explained in chapter 2, evolutionary theory explains the egoism of humans. Egoism results from the need of the phenotype to feed and care for oneself in order to continue living so that genes may propagate. Inclusive fitness recognizes that one who has the means or ability will provide for his relatives; as William Hamilton recognized, this is another mechanism for ensuring that one's genes survive. It is certainly possible, as Robert Trivers argued, that reciprocal altruism may occur: one who has the means might share resources with the larger social group of humans in the expectation that at a future point they will reciprocate.[113] Political scientist Vincent Falger nicely summarizes the distinctions people make: the "'normal' evolutionary heritage of mankind" is that "the individual will first see to his own interest, then to those of his close relatives (and pseudo-relatives like friends), and then to the interests of the larger group."[114] Unfortunately, given the paucity of resources in the Pliocene, Pleistocene, and Holocene environments, attention to a larger group would probably be rare, given the necessity of satisfying one's own needs as well as those of one's relatives.

Second, in the Pliocene, Pleistocene, and Holocene environments, our human ancestors faced varied and great threats, in the form of other animals and other humans, as well as natural dangers, such as disease, complications from injuries, and the terrain. Even something as simple as falling or slipping on a rock could cause a fatal injury. Rivers posed the risk of drowning, and weather could cause freezing or dehydration. For example, it appears that "Ötzi," described in chapter 3, was first wounded and then froze to death as he rested in the Italian Alps. As a result of these dangers, humans and other animals need the ability to rapidly assess threats and react quickly. The in-group/out-group distinction may be thought of as the human mind's immediate threat assessment: in sum, no threat/threat. Is the outsider a threat to oneself or to one's family? As psychologist Robert Bolles writes, "What keeps animals alive in the world is that they have very effective *innate* defensive reactions which occur when they encounter any kind of new or sudden stimulus."[115] These reactions vary, he says, "but they generally take one of three forms: animals generally run or fly away, freeze, or adopt some type of threat, that is, pseudo-aggressive behavior."[116] These reactions are elicited by the appearance of a predator, and also by innocuous but unfamiliar objects or animals. Bolles continues, "These responses are always near threshold so that the animal will take flight, freeze, or threaten whenever any novel stimulus event occurs. . . . The mouse does not scamper away from the owl because it has learned to escape the painful claws of the enemy; it scampers away from anything happening in its environment."[117] Likewise, he argues, the gazelle "does not flee from an approaching lion because it has been bitten by lions; it runs away from any large object that approaches it, and it does so because this is one of its species-specific defense reactions."[118] He concludes, "The animal which survives is the one which comes into the environment with defensive reactions already a prominent part of its repertoire."[119]

Third, a conspecific might pose an important immediate threat. Given scarce resources, our human ancestors would have used their intelligence and ability to reason into the future to assess outsiders. Any outsider would be judged fairly quickly to determine whether his presence was a threat to their future resources. Would he compete for the scarce resources they needed to survive? Would his presence threaten their position in the extended family or tribal group?

Consequently, as with the great majority of animals, the human animal can also assess threats posed by conspecifics or other animal and natural threats. Of course, the human ability to assess such threats is much more complex than it is in other animals. Thus we may consider other variables, such as the possibility of immediate trade or trade and cooperation in the long run, given the constraints of Trivers's reciprocal altruism argument. As discussed in chapter 2, Trivers says humans cooperate with unrelated individuals, expecting reciprocal behavior from the party that benefits at present. Animals have the "flight, freeze, or fight" instinct, as do humans, but human intelligence gives us a greater repertoire of behavior: flight, freeze, fight, or cooperate (expecting reciprocal behavior).

But the option of cooperating with outsiders is relatively new in human evolutionary history, and so the central point remains: humans and other animals need to be able to assess the threats posed by outsiders quickly, so they know how to behave to save and protect themselves and their relatives. Do they cooperate, flee, or attack? The in-group/out-group distinction is one solution that developed through evolution.

Evolutionary theory provides considerable insight into the issue of nepotism. As evolutionary theorists widely recognize, nepotism evolved because it aided reproduction. Investment in nonrelated individuals is wasted, so it could never be as powerful a force in evolution. Of course, as I explained in chapters 1 and 2, this is a logical consequence from the perspective of inclusive fitness, since assisting genetic relatives is really assisting one's own genes. Recall that for almost all of their evolutionary history humans have lived in bands of genetically related people. Also, humans are obviously intelligent and opportunistic animals who use many clues of relatedness, especially those that let them easily distinguish between family and those more distant. Having acknowledged these facts, we see the advantage to an individual who is nuanced in his treatment of others, favoring those who are directly related over those who are not. He will leave more offspring, on average, than those who do not act in this way. Perhaps more than any other animal, humans make many fine distinctions and gradations of kinship that inform their decisions on how to treat others. A mother will entertain treatment from her son that she would not tolerate from the neighbor's boy.

Naturally, the fact that most humans feel a powerful bond to their genetic relatives does not guarantee it will prevail in extraordinary circumstances. Brothers do murder each other. But killing genetic relatives is uncommon, and seldom gratuitous. The motive may be some great gain for the murderer, like a throne or an inheritance. Richard III supposedly ordered his young nephews, Edward V and Richard the Duke of York, murdered in the Tower of London so he could inherit the throne; five hundred years later, the Menendez brothers killed their parents in Beverly Hills to inherit their wealth.

Anthropologist Pierre van den Berghe suggests that the willingness to sacrifice for genetic relatives, advanced by inclusive fitness, helps explain the political mobilization of the in-group. Consciously or not, we favor those we perceive as closer to the genetic family. Such identification with people from the same ethnic and racial backgrounds can have significant political consequences for domestic politics—for example, the "Irish vote," or the "Jewish vote"—as well as for international relations. As van den Berghe argues, "ethnic and racial groups can be politically mobilised, even on a huge scale, with greater ease and rapidity, than other social groups, especially under external threat from an enemy who is himself defined in ethnic or racial terms."[120] Of course, people identify themselves in many other terms than race, and one person may have quite intricate reasons for voting for a particular candidate. Nonetheless, it does make sense for political analysts to speak in terms of a "Hispanic vote" leaning toward the Democrats even while the Cuban-American vote is solidly Republican, because more Hispanics vote for the Democratic Party. Van den Berghe suggests this type of common political behavior is in part an extension of identification with the family in-group.[121]

For the study of international relations, van den Berghe's insight suggests a reason why racially distinct adversaries in warfare may be easily demonized: they are more readily seen as the out-group than enemies who are not. A classic example is the conflict between the United States and Japan in World War II. It was relatively easy to generate greater animus toward the Japanese during World War II than against the Germans or Italians, as the distinguished historian John Dower has carefully documented.[122] While noting that the reasons why the Japanese were more hated than the Germans is complicated, Dower says the difference "is surely in large part racial."[123]

Paul Fussell, literary critic and veteran of World War II, also notes the differences between the common American perceptions of the Japanese and the Germans during World War II.[124] The Japanese, he argues, "were typecast as animals of an especially dwarfish but vicious species," while "the Germans were recognized to be human beings," even if some of them were "of a perverse type."[125] The surprise attack on Pearl Harbor meant that the Japanese were the major enemy of World War II: "Americans detested the Japanese the most, for only they had had the effrontery to attack the United States directly, sinking ships, killing sailors, and embarrassing American pretenses to alertness and combat adequacy."[126] As a consequence of the attack, "For most Americans, the war was about revenge against the Japanese, and the reason the European part had to be finished first was so that maximum attention could be devoted to the real business, the absolute torment and destruction of the Japanese."[127] Echoing this sentiment is playwright Arthur Miller, who worked in the Brooklyn Naval Yard during the Second World War and describes how his coworkers showed a "near absence . . . of any comprehension of what Nazism meant."[128] To them, "we were fighting Germany essentially because she had allied herself with the Japanese who had attacked us at Pearl Harbor."[129]

More recently, in 1999, several political commentators argued that Americans were more willing to assist Kosovo Albanian refugees streaming into Macedonia and Albania during the war than to assist the victims of the equally horrific civil war in Sierra Leone. The Kosovo refugees were white, and among them were boys whose baseball caps and tee shirts made them indistinguishable from many American boys, as well as mothers whose headscarves marked the only noticeable difference from most American women. While the civil war in Sierra Leone, as portrayed on American television screens, was all too easily dismissed as a continuation of almost seamless African misery and conflict, it also received less media attention. A *Washington Post* editorial writer noted that race was an important factor in explaining why popular opinion in the United States was more willing to help Kosovo Albanian refugees: as Caucasians they possess a strong physical resemblance to most Americans, whereas the residents of Sierra Leone do not. Nor do Tutsi and Hutu refugees from the Rwandan genocide.[130]

Van den Berghe stresses that only a few inherited phenotypes are often used to indicate race, not because they are innately meaningful—they are not—but rather because of their visibility. In turn, this can determine whether a person associates or identifies a stranger as being in-group or out-group, and thus how he reacts to the stranger. Skin pigmentation is the most widespread because it is the most visible from the greatest distance and subject to a limited range of environmental variation. Other features, such as those of the face, specifically the eye, lip, and shape of the nose; hair textures; and physical stature, are used where the differences are sharp and result from greater intergroup than intragroup heterozygosity. Racism is a consequence of these visible differences; they have proven to be more difficult to combat than cultural differences, which are easier to adopt or abandon. As van den Berghe argues, "You can learn a language, convert to a religion, get circumcised or scarified, adopt a dress style, but you cannot become tall or white."[131] One reason racism has become such a significant problem in modern times is the acceleration of mass migrations across genetic climes in circumstances in which individuals had to compete for scarce resources, thus facilitating racist efforts to dehumanize those of other races. Of course, this was made worse by other environmental factors, such as the exclusionary cultural, economic, political, and social practices identified by Brown and Van Evera.

Anthropologists and sociologists have long recognized the importance of physical differences in drawing in-group/out-group distinctions. For example, in *Economy and Society,* Weber notes the importance of consanguinity and the physical similarity that is common among ethnic groups: "Those who are obviously different are avoided or despised or, conversely, viewed with superstitious awe."[132] That is, "persons who are externally different are simply despised irrespective of what they accomplish or what they are, or they are venerated superstitiously if they are too powerful in the long run."[133] "Antipathy," he continues, "in this case is the primary and normal reaction."[134] For Weber, little is needed to trigger either antipathy or attraction: "Almost any kind of similarity or contrast of physical type and of habits can induce the belief that affinity or disaffinity exists between groups that attract or repel each other."[135] These habits, or more broadly differences in custom, and religious and linguistic differences, are just as important as consanguinity in sparking affinity or antipathy. As he writes, they "play a role equal to that of inherited physical type in the creation of feelings of common ethnicity and notions of kinship."[136]

Insights from the In-Group/Out-Group Distinction for the Ethnic Conflict Paradigms

For primordialists, inclusive fitness theory and van den Berghe's arguments mean that ethnic and racial variables for determining in-group/out-group identification are powerful and should not be discounted by state or nongovernmental policymakers seeking to prevent ethnic conflict. Specifically, these arguments strengthen the central tenets of the primordialist paradigm. The first tenet is that people do classify themselves by national identity. The nation is an in-group that helps to orient an individual's worldview. The nation is likely to be exclusionary as well, identifying in-group and out-group in the form of other nations, thus supporting a second tenet of the paradigm: that national identities are often exclusionary. This argument also assists the third tenet: ethnic conflict results from deep causes within the nation. If we recognize that the in-group/out-group distinction helps define identity, and that this process is important for every individual, then we see how this fuel may be exploited by politicians, such as Milosevic, or other elites for their own gain. When these elites frame another nation as an out-group, they make it quite easy to demonize the others in order to advance their own political agendas. The "us" versus "them" distinction is a common thread of ethnic conflict, as both paradigms acknowledge, and evolutionary theory allows us to understand why such in-group/out-group appeals often resonate powerfully.

An obvious implication for multiethnic states is that they will be vulnerable to appeals for division made along ethnic lines, a point well demonstrated in the 1990s with the breakup of both the Soviet Union and Yugoslavia, and widespread ethnic conflict from Rwanda to East Timor. The significance of this insight increases when we realize that most states in international relations are multiethnic and many are multiracial. In sum, the primordialist paradigm is correct to emphasize the danger posed by aspects of identity, such as ethnicity or race, which are difficult to change. This insight is not diminished if we also recognize that in order to get the in-group/out-group string to resonate, someone must pluck it: as the modernist paradigm readily acknowledges, individuals and social forces play major roles in causing ethnic conflict.

The modernist paradigm also benefits from the evolutionary explanation of in-group/out-group behavior. Using it, the modernists can understand how devious politicians or political parties can manipulate people time and again. The fuel is the human ability to generate readily in-group/out-group distinctions. This fuel may aid political parties or leaders to gain or maintain power through social imperialism—dividing the body politic into "us" against "them." Evolution's explanation of the in-group/out-group division allows modernists to understand why the masses are receptive to such manipulation by these elites. This situation, of course, is made worse when no countervailing pressures come from the domestic or international media, other states, or international society.

When the out-groups are defined along national lines, and when the appeals of political parties or political leaders are further linked with xenophobia or ethnocentrism, then humans are quite capable of demonizing others. The combination of in-group/out-group distinctions and xenophobic and ethnocentric messages makes the masses further receptive to such messages or manipulation by political parties and elites. The modernist paradigm should recognize these points because it reinforces a fundamental truth concerning successful propaganda or elite manipulation of the masses. Good propaganda requires not just that the leaders or elites advance the message, as the modernist paradigm recognizes; they must also transmit a message that resonates with the masses. Messages based on "us" versus "them," and those that inflate one nation at the expense of others or of minorities all too often resonate very well, especially in the cultural, economic, political, and structural conditions ascertained by Brown and Van Evera.

For those who seek to stop ethnic conflict, two points are important here. First, evolutionary theory makes an important contribution for understanding the force of the in-group/out-group division. Acknowledging the force of this distinction, or of xenophobia or ethnocentrism, does not prevent us from recognizing other sources of in-group identity or cooperative behavior, such as friendship, a genuine desire to assist strangers, fealty to one's *alma mater,* or identification with a person's economic and social class. Thus, if the in-group can be broadened to include ethnic outsiders, then ethnic identity will have less power as a cause of ethnic conflict.

This might be accomplished in several ways. Within multiethnic states, one way is to broaden the definition of the in-group to include ethnic outsiders, perhaps by re-making state identity. This might be done using the media, and film and television industries.[137] This has been done in the United States, where identity has been changing over the past generation; for example, blacks rarely appeared on American television before the 1970s.[138] In Germany, the Social Democrats are making similar attempts to change the common conception of "Who Is a German" to include the Turkish and other minorities. It may also be accomplished by using the public education system. For example, in December 2001, Great Britain's Home Office released the "Cantle Report" on racial cohesion, which stressed the increasing polarization of Britain's white and ethnic minority communities following ethnic disturbances that year in Bradford, Burnley, and Oldham. It noted that whites and minorities were leading "parallel lives" without significant interaction between communities, and suggested ensuring that schools be ethnically mixed so young people would experience classmates of other ethnic backgrounds.[139]

Another way might be to change the size of the group. As economists Anthony Downs and Mancur Olson famously argued years ago in their analyses of what became known as the collective action problem, the size of the group makes a critical difference for the likelihood of cooperation.[140] In general, the larger the group, the more diluted is an individual's sense of solidarity with it. This is the case no matter what the source of solidarity, whether it is genetic or rooted in one's class or ethnic identity or one's sense of loyalty to the state. Few individuals are as concerned about their second cousins as they are about their siblings. This obviously is a limited solution, of course; states cannot break easily into smaller parts to promote community. But in certain circumstances, such as in civil society, it is possible to have smaller, integrated groups that promote cooperation, as Robert Putnam argues in *Bowling Alone*.[141]

Second, understanding the importance of morphological differences for the in-group/out-group distinction is not to argue that such differences are immutable. Certainly they are not. That is the good news for stopping ethnic conflict. The bad news is that this division may take considerable time to eradicate, and may never vanish completely. Nonetheless, ensuring adequate nutrition for people who are undernourished is a significant force in erasing some physical differences between peoples. Naturally, this takes time, but progress may be very rapid, at least when compared to evolutionary changes, as the growth in the average height of males and females in Western societies demonstrates. Furthermore, more important than nutrition is exogamy, or marriage outside of one's own racial or ethnic group. Exogamy for three or four generations of 25 percent or more will typically erode both the racial and physical attributes of nations—such as the height difference between Tutsis and Hutus in Rwanda.[142] These differences can only remain as long as interbreeding remains relatively infrequent, as in Rwanda, where a rigid caste system hinders exogamy.

In cases where morphological differences are weak, national groups are likely to stress cultural differences in order to differentiate themselves, both to identify their members and to maintain their unique identity. Precisely because most neighboring ethnic groups appear similar, they must depend upon cultural distinctions for these purposes.[143] Neighboring ethnic groups appear similar to each other because the lack of major physical barriers allowed them to practice exogamy. The greatest differences arise when different environmental conditions and great physical barriers, such as deserts and oceans, prevent exogamy.

Understanding Xenophobia

Evolutionary theory allows us to explain why xenophobia evolved in humans and other animals. As I documented in chapter 4, xenophobia is found in nonhuman animals. Indeed, most species aggress against conspecifics, and most are territorial, as many dog and cat owners know from observation.[144] The near universality of these behaviors suggests that they evolved in animals in the distant past. The empirical evidence for this is widely documented by ethologists and biologists and is so strong as to be overwhelming, as I showed in chapter 4.[145] In fact, the amygdala, one of the brain's most ancient neurological structures, controls much of the behavior related to fear.[146] David Barash has found in his studies of humans and other animals that "both . . . tend to reserve their most ferocious aggression toward strangers."[147] Biologist John Fuller concludes that "xenophobia is as characteristic of humans as of ants, mice or baboons."[148] Jared Diamond argues that "xenophobic murder has innumerable animal precursors," and humans are unique in being the only species to have developed the weapons necessary for killing at a distance.[149] After a comprehensive review of xenophobia in animals, Johan van der Dennen concludes that it "is a widespread trait throughout the animal kingdom," one that helps to "maintain the integrity of the social group" and "ensures that group members will be socially familiar."[150]

While xenophobia is present in many animals, my central question is why the trait would also evolve and be maintained in humans. Why would fear of strangers and perceptions of them as threats exist in the repertoire of human behavior? Using inclusive fitness as a theoretical foundation, I see four reasons why xenophobia would contribute to fitness in the environments in which humans evolved.

First, as I argued in chapter 3, many anthropologists, archeologists, and historians surmise that humans lived in extended family bands that fought to aggress against or protect themselves against rival human bands as well as against large carnivores or packs of them, such as wolves or hyenas. This behavior is well documented in humans and chimpanzees, as I showed in chapter 4. This suggests that it was present in our common ancestor before humans and chimpanzees divided some 5 million years ago, although this has not yet been demonstrated since such evidence concerning our common ancestor is absent at this time.[151] However, in the Pliocene, Pleistocene, and Holocene conditions in which humans evolved, strangers were unlikely to be related to others living nearby, and were likely to be competitors for scarce resources and perhaps also a threat to the group.[152] Having thoughtfully reflected on the problem of xenophobia and conflict in human society, Diamond argues that competition for territory or other scarce resources is a central cause of xenophobia in humans: "Humans compete with each other for territory, as do members of most animal species. Because we live in groups, much of our competition has taken the form of wars between adjacent groups, on the model of the wars between ant colonies."[153] He continues, "as with adjacent groups of wolves and common chimps, relations of adjacent human tribes were traditionally marked by xenophobic hostility," which was "intermittently relaxed to permit exchanges of mates (and, in our species, of goods as well)."[154]

Second, xenophobia would be a mechanism of defense against the communicable diseases that could spread into a virgin population during contact with strangers.[155] I do not argue that these humans understood modern epidemiology. Of course they did not. But they were intelligent. It is certain that they recognized the powerful effect of disease, judging from the history of European encounters with the rest of the world during the Age of Discovery and after, some of which I described in chapter 4. Indigenous peoples in Africa, Asia, and the New World understood quickly that disease affected the Europeans themselves. They could comprehend quite easily that Europeans died in considerable quantities in Africa and Asia. In the New World, Native Americans comprehended rapidly that Europeans and their African slaves brought pestilence even if they did not realize how diseases were transmitted. Moreover, the cultural histories of many societies are rife with stories about a stranger or groups of strangers bringing pestilence. In these traditions, a stranger or strangers were associated with harm or evil that affected the community. No doubt this was accurate on occasion because a stranger would bring disease to a virgin population, as occurs even today. Primitive peoples with no understanding of disease and its prevention would associate illness with malevolent spirits acting through the stranger, or else with witchcraft. Of course, strangers would also be affected by disease from the populations they encountered, but being new to a community and perhaps transient, their suffering would be less likely to be recorded or enter into an area's folklore or cultural history.

Third, because humans are the only species to kill conspecifics at a distance, fear of strangers may have been accelerated or become especially important as warriors introduced weapons of ever-increasing technological sophistication: the spear, the atlatl, and ultimately the bow and arrow. Armed with weapons, even a single individual, to say nothing of a group, would greatly increase the threat posed to another individual, especially if he were separated from his group. Weapons probably brought about the first revolution in warfare by allowing one human to kill another, even a stronger one, with less risk to himself since he no longer had to engage the other in close proximity with his fists or even a club. Given this technology and the necessity of hunting and foraging to survive, it would be beneficial to stay in one's group if possible so the members could help fight off an attack. It would be equally useful to be able to recognize group members and extra-group members quickly and from as far away as possible.[156]

Fourth, a stranger might also be a threat to one's position in the dominance hierarchy. As I explained in chapter 2, most social mammals organize themselves in such hierarchies. Ethologists argue that these hierarchies evolve because they aid in defense against predators, promote the harvesting of resources, and reduce intragroup conflict.[157] The ubiquity of this social ordering strongly suggests that it contributes to fitness. Donald McEachron and Darius Baer suggest that strangers would have to find a place in the dominance hierarchy, which might entail conflict, especially among those displaced by the inclusion of the new member.[158] Such readjustments are certainly possible, although cleavages or internecine conflict in the group might lead some to seek the stranger as an ally.

Given these conditions, humans would consider other humans a threat and thus would rarely tolerate strangers. Low tolerance of strangers, or xenophobia, contributed to fitness and thus spread. As human communities grew larger, multiple groups would have reproduced, some containing genotypes that resulted in an increased suspicion of strangers. These genotypes would increase fitness by increasing the survival of the group over time. Like warfare, however, and indeed like much of human behavior, xenophobia may be augmented or weakened by psychological and cultural forces.

For these reasons, xenophobia contributes to fitness and thus explains why humans may react negatively to people with different morphological features, such as facial traits or skin color. If the genetic difference is physical, then identification of difference is obvious, such as that between Africans and Europeans, but xenophobia can be triggered even by small differences between neighboring tribes or populations, as Richard Alexander and Falger suggest. They argue that the ontogenetic flexibility of humans, such as morphological differences, even slight ones, is sufficient to cause xenophobia.[159] While Alexander acknowledges that the causes of intense xenophobic reactions are complex, he explains, "it is possible that morphological differences alone make different countenances more or less communicative."[160] Then the "differences between individuals of populations that diverged" because of geographic isolation or other factors that prevented exogamy "could lead to xenophobic reactions."[161] For example, Slavs typically have broad faces and Anglo-Saxons narrow ones; Tutsis tend to be tall, and Hutus short. Both Gérard Prunier and Christopher Taylor studied the 1994 Rwandan genocide and found that the physical differences were tragically important because they allowed easier identification of Tutsis by the radical Hutus.[162] These differences facilitate a type of discrimination that could not exist between people who are morphologically very similar, such as Irish Catholics and Protestants or Norwegians and Swedes, all of whom would likely have to listen to each other to determine who is who. Still, visible or auditory cultural differences even between similar groups might provoke a xenophobic reaction. As Diamond explains, "xenophobia comes especially naturally to our species," and "because so much of our behavior is culturally rather than genetically specified, and because cultural differences among human populations are so marked," consequently "those features make it easy for us, unlike wolves and chimps, to recognize members of other groups at a glance by their clothes or hair style" and react negatively to them.[163]

Insights from the Evolutionary Explanation of Xenophobia for the Ethnic Conflict Paradigms

Recognizing the evolutionary causes of xenophobia is useful for scholars who research ethnic conflict using the primordialist paradigm. Evolutionary theory explains why xenophobia contributed to fitness, and then spread, becoming a common human trait that helps account for the scope and intensity of ethnic conflict throughout history. This recognition assists the primordialist paradigm in two ways.

First, as with the in-group/out-group distinction, the evolutionary explanation of xenophobia supports the primordialist paradigm. Some of the paradigm's central tenets depend on xenophobic beliefs, so evolutionary theory gives primordialist scholars a scientific foundation for advancing the argument that evolution has selected xenophobia in people, and that this trait remains in them today. The major consequence of this connection is that it gives the paradigm a scientific basis. Appeals to the amorphous concepts of "primordial animosities" or "ancient hatreds" are no longer necessary.

Starting with the evolutionary explanation for xenophobia, primordialists can argue that just as people possess the trait of xenophobia, an aggregate of those people—a nation—can be xenophobic too, given the appropriate stimulus or conditions. Recalling the in-group/out-distinction made above, the evolutionary origin of xenophobia supports the primordial paradigm's central tenets: it helps explain why people are exclusionary. The in-group/out-group distinction is an important part of the explanation of exclusionary behavior, but so is xenophobia. Nations are likely to have a strong sense of xenophobia where the state either encourages it explicitly, through the speech and actions of its politicians and other elites, or implicitly, by not opposing xenophobic arguments. In addition, civil society and the media can both encourage and discourage xenophobic beliefs or behavior, and so have a significant impact on the nation.

Xenophobic behavior serves as a foundation for ethnic conflict because evolution caused it. There is always the potential it may be triggered. However, when this potential escalates to become actual ethnic conflict, we must look to the myriad of environmental factors named by both primordialists and modernists: a history of conflict or discrimination, language differences, economic difficulties, and structural problems such as the security dilemma.[164] This is a lesson for the primordialists as well. So, while xenophobia is a product of human evolution, the right environment can mitigate its effects and keep open conflict from erupting.

Combating xenophobia is crucial to stopping ethnic conflict, and the argument above provides good news for this effort. One mechanism for suppressing xenophobia is reducing in-group/out-group distinctions. States can accomplish this through governmental policies; the media, film industry and opinionmakers can play important roles as well. The goal in essence is to look at strangers, the "Other," and make them the "Self." That is, make strangers part of the nation, the in-group. If this is not possible, then the state should work to increase public tolerance for strangers by making them equally valuable citizens of the state. Of course, I recognize this is difficult because xenophobia is a powerful trait, because it is inherently hard to create a consistent message, and of course because other groups in society may create countervailing messages. It is possible to suppress xenophobia, but the task should not be underestimated.

Understanding Ethnocentrism

Like xenophobia, inclusive fitness explains why ethnocentrism would contribute to fitness in human evolutionary history, and thus evolve in humans.[165] As I stated above, ethnocentrism is commonly defined as a belief in the superiority of one's own ethnicity. Yet, defined this way, ethnocentrism would seem to have no evolutionary foundation as a belief. After all, one may hold many beliefs—that the earth is flat or that it is round, or that Roman Emperor Augustus was poisoned or that he died of natural causes—and these beliefs have no effect on fitness. So it might seem that evolutionary theory has little explanatory power for ethnocentrism.

Now let me provide a more precise definition of ethnocentrism: it is a collection of traits that predispose the individual to show discriminatory preferences for groups with the closest affinities to the self. Now the contribution of evolutionary theory becomes clear.[166] Here evolutionary theory can explain why ethnocentrism is common among people: because it stems from inclusive fitness. Thus the phenomenon of ethnocentrism has it origins not merely in one's random beliefs or opinions, but in human evolution. Of course, even with its foundation in human evolution, ethnocentrism—like xenophobia and almost all behavior—is open to considerable environmental manipulation. It may be either suppressed or supported by cultural, religious, or political beliefs and authorities.[167]

I made this argument in the discussion above but will briefly restate the central arguments here. Since our genus *Homo* first evolved in the Pliocene, humans have favored those who are biologically related. In general, the closer the relationship, the greater the preferential treatment. The vast majority of animals behave in this way, and humans were not different in our evolution. In a world of scarce resources and many threats, the evolutionary process would select nepotism, thus promoting the survival of the next generation. This process is relative however. Parents are more willing to care for their own children than for the children of relatives, and they are significantly less disposed to provide for those of strangers.

The essence of an inclusive fitness explanation of ethnocentrism, then, is that individuals generally should be more willing to support, privilege, and sacrifice for their own family, then their more distant kin, their ethnic group, and then others, such as a global community, in decreasing order of importance. That people are more willing to sacrifice for their family than for strangers or a larger community is obvious. In contrast, an individual like Mother Teresa is saintly. Her willingness to suffer and sacrifice for strangers throughout her lifetime is both noble and lamentable because her actions illustrate what we already know: few people are willing to sacrifice to help strangers who require great care due to endemic poverty and debilitating illness.

The in-group/out-group division is also important for explaining ethnocentrism and individual readiness to kill outsiders before in-group members. Irenäus Eibl-Eibesfeldt draws on Erik Erikson's concept of "cultural pseudospeciation," and says that in almost all cultures humans form subgroups, usually based on kinship; these "eventually distinguish themselves from others by dialect and other subgroup characteristics and go on to form new cultures."[168] As an unfortunate result of this process, such in-groups tend to perceive the out-groups as increasingly remote, from distant cousins to foreigners with strange customs and language. Ultimately they may even see them as less than human, as another species: animals.[169] Edward Wilson also argues that ethnocentrism has a strong in-group/out-group component. Ethnocentrism is the "force behind most warlike policies," as he notes: "Primitive men divide the world into two tangible parts, the near environment of home, local villages, kin, friends, tame animals, and witches, and the more distant universe of neighboring villages, intertribal allies, enemies, wild animals, and ghosts."[170] This "elemental topography makes easier the distinction between enemies who can be attacked and killed and friends who cannot. The contrast is heightened by reducing enemies to frightful and even sub-human status."[171]

Of course, while few scholars would doubt that ethnocentrism is a powerful force, we should not overestimate its power. It obviously can be offset or mitigated by other environmental pressures. In many examples from military history, for example, men of the same ethnicity killed each other readily. Ethnic ties did not keep Confederate General Cobb's mostly Irish Georgia regiments (the 18th and 24th) from firing on and decimating the Union "Irish Brigade" (the 28th Massachusetts) at Marye's Heights during the battle of Fredericksburg in December 1862. Of course, it is equally true that ethnic hatreds may be suppressed or even reconciled. The animus between the Scots and the English is well known; their shared history is filled with great conflicts that still resonate today in calls for Scottish independence. Nonetheless, for much of modern history the Crown has repressed that hatred. Indeed, Scots have readily served the Crown even after repression. Perhaps the most famous example occurred in 1758—only thirteen years after the bloody British suppression of the Scottish Jacobite Rebellion—when British and Colonial forces assaulted the French at Ticonderoga. The 42nd Regiment (the Black Watch) made one of the most famous assaults in military history. The Black Watch was a Scottish regiment composed mostly of Highlanders; many of its members were strongly Francophile and had strong personal reasons to hate the British government. Nonetheless, fealty to the Crown's orders and military professionalism overrode other beliefs.

While it is important not to overestimate the power of ethnocentrism, it would also be a mistake to underestimate its contributions to ethnic conflict. When an individual considers whether to support a larger group, several metrics are available. One of these—and I stress only one—is ethnocentrism, a continuation of one's willingness to sacrifice for one's family because of the notion of common kinship. As I discussed above, the ways humans determine their relations with unrelated individuals are complex, but key factors are physical resemblance, as well as environmental causes like shared culture, history, and language.

Insights from the Evolutionary Explanation of Ethnocentrism for the Ethnic Conflict Paradigms

Evolutionary theory helps explain why people hold ethnocentric beliefs. Again, as with the in-group/out-group division and xenophobia, the evolutionary explanation of ethnocentric behavior improves the argument of the primordialist paradigm. Instead of depending upon the "ancient hatreds" argument to explain ethnic conflict, scholars in this paradigm will be able to anchor their arguments about ethnocentric behavior in evolutionary theory. As with the in-group/out-group bifurcation and xenophobia, this argument improves the paradigm.

A second conclusion is that the masses are well prepared by human evolution to have their ethnocentric strings pulled by the elite. They will more readily identify with a diaspora population because they see that community as being similar to themselves, part of an extended, albeit fictitious, family who resemble one another, as well as perhaps possessing shared culture, history, and language. This conclusion assists both paradigms. It gives the primordialist paradigm an explanation of why people identify with a nation and thus often see the nation as family. It gives the modernist paradigm an explanation of why elites in many cultures and eras can manipulate the masses by using the common theme of ethnic or national identity.

While students of ethnic conflict recognize that ethnocentrism is a potent force, as a concept it is troubling to those who use the modernist paradigm to analyze ethnic conflict. How can modernism explain the explosive and forceful nature of ethnic nationalism that remains despite increasing economic exchange, globalization, and the dominance of liberalism in Western societies that work to reduce its force? Millions of men and women have sacrificed their lives for their ethnic homelands: for France, the Russian motherland, the German fatherland, Vietnam, China, Israel, Japan, Italy, and Eritrea, among many others. Although the modernist paradigm provides insights into ethnocentric beliefs, it still cannot explain why so many millions of people respond to flags and anthems, national festivals, monuments and shrines. Ethnocentric beliefs continue to be ubiquitous, and Connor has found them in almost every political and economic condition.[172]

In this case, understanding the evolutionary origin of ethnocentrism also provides a partial solution to ethnic conflict. Exogamy is a powerful mechanism for reducing ethnocentric beliefs because intermarriage between nations reduces a strong in-group/out-group division and ethnocentrism. The United States is a strong example of this. Widespread intermarriage between nations, what Americans usually call ethnic groups, in the United States has lessened national, that is, interethnic, tensions that existed in the country's past—for example, with the Irish or Italian immigrants in the last half of the nineteenth century, or eastern European immigrants in the early twentieth century.

To be sure, exogamy is only a partial and imperfect solution to the problem of ethnic conflict. The messages of civil society, the media, and political leaders, as well as economic conditions and opportunity are also significant, but scholars already understand them as potential solutions to ethnic conflict. Exogamy is not perfect for two reasons. First, as the example of the former Yugoslavia shows, the high rates of interethnic marriage between Bosnian Croats, Muslims, and Serbs could not guarantee the state's stability. Other environmental forces clearly identified by the modernist paradigm can overwhelm even familial loyalties in the right circumstances. Second, new problems can arise or people can create new in-group/out-group divisions. As Bosnian Croats intermarry with Serbs, or Irish-Americans with Italian-Americans, new in-group/out-group bifurcations or ethnocentric beliefs may well surface. Tajfel's experiments discussed above suggest that people can draw almost infinite numbers of very fine in-group/out-group distinctions. Lamentably, there is no easy solution to the problem of ethnocentrism, given the human need to associate with an in-group, and by doing so to feel superior to the out-group.

Conclusion

In this chapter I have shown that evolutionary theory can significantly advance the scholarly understanding of the origins of ethnic conflict. Evolutionary theory permits students of ethnic conflict to understand two human behaviors: why humans make in-group/out-group distinctions, and why we can be xenophobic and ethnocentric. Moreover, by comprehending the origins of these behaviors, scholars and policymakers can better understand ethnic conflict in order to prevent it from occurring.

As a result, evolutionary theory makes a contribution to the primordialist and modernist paradigms of ethnic conflict and provides a deeper understanding of the pernicious problem of ethnic conflict. Armed with a better explanation for the fundamental causes of ethnic conflict, primordialists can better explain the origins of these conflicts, and why they are relatively common, without asserting that the cause is an ancient hatred. Indeed, I suggest the modernists also consider human evolution to be an underlying cause of ethnic conflict, one that is just as important as the structural, political, economic, or cultural factors they identify.

I would like to suggest that evolutionary theory provides a common foundation for both paradigms. The primordialist paradigm can use evolutionary theory to argue for deep causes within the individual, as I have shown above. The modernist paradigm can accept the evolutionary origins of the in-group/out-group distinction, xenophobia, and ethnocentrism as a structural cause of ethnic conflict. This would allow the modernists to accept what science shows: that the roots of ethnic conflict are well grounded in human history and that almost all peoples make in-group/out-group distinctions, and can be xenophobic and ethnocentric. Rather than ignoring evolutionary science, the modernist paradigm can embrace it, and without having to accept that ethnic conflict is the result of ancient hatreds.

While it is important to recognize the contribution of evolutionary theory to both paradigms, I also want to reiterate what evolutionary theory does not do. It does not offer proximate explanations of ethnic conflict. In order to understand the causes of a particular ethnic conflict, one has to use the analytical tools provided by primordialist and/or modernist paradigms. For example, evolutionary theory will not explain the specific ethnic conflicts in Afghanistan or Kosovo. It will not explain the actions of the Kosovo Liberation Army, or the Yugoslav national army in suppressing the KLA and ethnic

Albanians in Kosovo. Nor will it explain NATO's decision for war against Yugoslavia to stop the suffering of the ethnic Albanians.

That said, evolutionary theory still makes a serious intellectual contribution to the study of ethnic conflict. In chapter 1, I explained the general circumstances that make evolutionary explanations useful for social scientists. Evolutionary theory assists social scientists by explaining ultimate causation, not proximate causation. A medical analogy captures this point well. When a patient is suffering from pneumonia, a doctor treating the patient need not be concerned with human evolution—the ultimate causation—or the patient's liver, an irrelevant proximate causation—or the patient's skin disease. The doctor must focus upon the particular danger and not be concerned with ultimate causation, irrelevant proximate causation, or other less threatening ailments. So too a policymaker confronting a situation like Kosovo in March 1999 need not be concerned with the ultimate causes of ethnic conflict. The point is to stop the immediate progress of the disease.

It is equally true, however, that a doctor's complete understanding of human physiology allows him to keep more patients in good health. It allows the doctor to recommend preventative care that lets the patient avoid illness or trauma in the future. When encountering an ill patient, the doctor's knowledge allows him to diagnose and categorize a patient's illness so that effective treatment may follow and the patient may be returned to health.

So it is for scholars of ethnic conflict. Understanding the causes of ethnic conflict, both ultimate and proximate, helps the scholar better predict the circumstances in which ethnic conflict may occur. He can then recommend to the policymaker certain courses of action that ideally prevent ethnic conflict from occurring, or when it does appear, advance policies to minimize the suffering of those afflicted. Given the horror of ethnic conflict, we must understand its causes so it can be stopped wherever it is possible to do so. In sum, the problem is simply too important to ignore any of the variables that help us understand it.

Finally, no matter what the ultimate causes of ethnic conflict, the international system, individual states, and nongovernmental organizations, such as the United Nations, can work to suppress it. The bipolar international system of the Cold War helped to control ethnic conflict, as we see in the great rise in ethnic conflict since the Cold War ended.[174] State policies may also help prevent or ameliorate ethnic conflict. Having surveyed, with Šumit Ganguly, ethnic relations in sixteen Asian and Pacific states, Brown notes that government policies "are often decisive in determining whether ethnic problems, which are inherent in multiethnic societies, are resolved peacefully and equitably."[174] The United Nations had also made great efforts to stop and suppress ethnic conflict in the post–Cold War world, for example in East Timor and Rwanda. Nonetheless, given how much the in-group/out-group distinction, xenophobia, and ethnocentrism contributed to fitness during human evolution, each continues to be an ultimate cause of ethnic conflict. Thus, ethnic conflict will likely recur as a social phenomenon, remaining, like war and peace, part of the fabric of international relations.

Conclusion

The more we learn about human evolution, the more we recognize what makes us human. We are not separate from the natural world; it influences us in countless ways, from natural selection to the blowback effect of environmental destruction. Seeing this, we can better understand life in the natural world: what makes us unique as humans, and what makes us truly akin to other animals, from our "cousin" the chimpanzee, to our more distant relatives. As E.O. Wilson explains in his wonderfully titled *Biophilia:* "Humanity is exalted not because we are so far above other living creatures, but because knowing them well elevates the very concept of life."[1] Understanding what makes us unique as humans and how we are similar to other forms of life is important for ethologists, philosophers, and all who study the human condition. More narrowly, as I have shown in this book, evolutionary theory also helps students of international relations.

This book is a first step in illustrating how evolutionary theory can assist important theoretical and empirical issues in the discipline of international relations. To this end, in chapter 1, I explained what evolutionary theory is, the essence of Darwin's ideas, and especially what natural selection is and how it works. Given this, we are then in a position to understand what fitness means and why some adaptations will obtain over time in specific environments and why some will not. The subdivision of natural selection termed individual selection provides a lens through which to study much of human behavior, and indeed I used it to generate this book's theoretical and empirical insights. Its components are neoclassical Darwinism, Darwin's idea of natural selection wedded to a modern understanding of genetics, and William Hamilton's idea of inclusive fitness. I emphasized that individual selection is not the only mechanism of natural selection, but most evolutionary theorists consider it the best.

In chapter 2, I demonstrated the theoretical contributions that evolutionary theory may make to realism, offensive realism, and rational choice, by offering a scientific grounding for them. This allows the theories of realism and rational choice to use egoism without begging the question of why humans are egoistic. For offensive realists, evolutionary theory explains why leaders of states often will act as expected, preferring more resources to fewer, more power to less. When we view human behavior through the perspective of an animal struggling for tens of thousands of generations in the dangerous conditions of the Pliocene, Pleistocene, and Holocene epochs, we see behavior that makes perfect sense in the anarchic environments of evolutionary adaptation, though it is now considered Machiavellian, characteristic of power politics, and a hindrance to international cooperation.

Evolutionary theory makes an equally significant contribution to explaining the ultimate causes of warfare, allowing scholars to understand why warfare originated and remained in the human species as well as in other social animals. As I showed in chapter 3, the origin of warfare is anchored in a struggle to gain and defend resources; we see these motivations for warfare even today in primitive societies as well as states. Given these origins, and the mechanisms of natural selection, it is not surprising that warlike behavior continues among some species that have the necessary preconditions, such as social living, to make it a viable strategy to gain and defend resources and thus to survive in an often hard, competitive, and perilous world.

It may be difficult to accept that natural selection could produce warfare. It certainly is unfortunate. Since warfare is such a serious subject, however, we must explore its causes so that we may understand and stop it wherever it is possible to do so. Reflecting on human foibles, viciousness, and cruelty in *Following the Equator*, Mark Twain wrote: "Man is the only animal that blushes. Or needs to."[2] Twain certainly had a gift for capturing elements of the human condition in succinct and witty prose, and we can acknowledge his genius as a writer without bowing to his knowledge of ethology. The genocide and warfare of ants and chimpanzees and the widespread infanticide and aggression among other animals studied in chapter 4 require us to amend Twain's remarks: other animals should blush if they could.

Ethologists, like the rest of us, do not rejoice when they document such behavior in animals. It is painful to read accounts by Jane Goodall or Richard Wrangham of vicious chimpanzee warfare and to recognize that the scourge of warfare also afflicts other animals. However painful, this recognition allows social scientists to see how warfare contributed to fitness for certain species by yielding, among other factors, more of the resources they needed to survive in difficult conditions and how it contributed to origins of social living. It also permits scholars of warfare to understand that war's origins are anchored in survival, and that this motivation has obtained throughout human history. Warfare was conducted for this reason long before it was used to advance political ends.

I know that the idea of chimpanzee (or ant) warfare is a shock to many readers. It was to me as well when I first studied this issue. I literally could not believe it and thought these scientists were simply anthropomorphizing. However, after studying this issue in considerable detail, I soon realized it makes perfect sense. Natural selection explains warfare in humans and other animals. I cannot say that I truly comprehended the force of Darwin's argument until then. That is, until I understood: "Of course war would be present in other animals. It could not be otherwise." At that point, I grasped the force of natural selection and why humans are not discrete from the natural world but a part of it.

While it may be a shock at first, ants and chimpanzees do indeed fight wars, and they are as lethal as I document. But, of course, ant or chimpanzee warfare and human warfare are different in many important respects. Humans wage war for many more causes, in many more forms, and with infinitely greater complexity. Gettysburg, Verdun, and Stalingrad clearly are not chimpanzee warfare. Yet, in another sense, human and chimpanzee warfare can be the same. As I stressed in this book, the causes of chimpanzee warfare (to gain or defend resources) and their tactics (raid, ambush, and even battle) apply to primitive and advanced societies today as well as to the human past.

The similarity between chimpanzee and human warfare is a key insight provided by my study. This suggests that natural selection produces social behavior, such as warfare, and confirms the value of studying it through the lens of evolutionary biology. This is a significant contribution of this book. Established truths and common conceptions about warfare are shown to be inadequate. As a result, paradigms must shift to incorporate new knowledge. This is how science advances.

As I have emphasized, it is important to understand the causes of warfare, both the ultimate causes illuminated by evolutionary theory and the proximate ones, so that scholars and policymakers can predict better the circumstances in which it may occur. Such a comprehension of both the ultimate and proximate causes of warfare will help policymakers prevent war in some instances, or else to advance policies designed to minimize suffering. Given the crucial importance of warfare in international relations, scholars should explore every major explanation that contributes to understanding its origins.

I also argued in chapter 4 that evolutionary biology allows us to understand why warfare would be stimulating for individuals as the evolutionary residue of the experience groups had while hunting and fighting in the environments of the Pliocene, Pleistocene, and Holocene. The traits that served humans in our evolutionary past help to cause many of the powerful emotions, including excitement, that individuals in combat still experience today. Evolutionary biology also gives us perspective on how disease affects international relations. We will need this perspective as biological warfare becomes more viable due to the appearance of new diseases, novel strains of existing diseases, and the advances in biotechnology. It is also important for understanding the past. Disease facilitated and hindered imperial expansion. The epidemiological balance of power favored European colonialists in America, where smallpox and other Old World diseases decimated the native populations. However, particularly in West Africa, the epidemiological balance of power favored indigenous populations. Diseases produced high mortality among the colonists and hindered their ability to control their colonies.

This book has also helped explain ethnic conflict. Evolutionary theory allows scholars to understand better the origins of its deep causes, including xenophobia, ethnocentrism, and the human distinction between in-groups and out-groups. This understanding, in turn, benefits both the primordialist and modernist paradigms of ethnic conflict, as I showed in chapter 5. The causes of ethnic conflict—like those of war—are both ultimate and proximate, and we need explanations at both levels to understand why ethnic conflict occurs. Understanding the causes of ethnic conflict, both the ultimate causes illuminated by evolutionary theory and the proximate ones, allows scholars to predict better the circumstances in which ethnic conflict may occur. This has the potential to help policymakers prevent ethnic conflict in some circumstances, or at least advance policies to minimize suffering when it does occur. Understandably, given the horror of such conflicts, we must understand their causes so we can stop them wherever possible. As with warfare, the problem of ethnic conflict is so important that scholars should explore every major explanation that may help clarify its origins.

In the rest of this conclusion I want to address some new issues, and also speak to the nascent but growing interaction between evolutionary theorists, geneticists, ethologists, and social scientists. But there is still much work to do. To advance these aims, I will first present an agenda for further study, and then reflect on the Darwinian Revolution and the social sciences.

Evolution and Social Science: Notes on a Research Agenda

In this study I explored some of evolutionary biology's contributions to international relations. I see this as a first step, the beginning of the process of bringing in Darwin. Because it is a first step, further research needs to continue on the topics I have addressed. This is true for both the theoretical and empirical aspects of the book.

Theoretically, there is much to be accomplished, but at this stage I want to emphasize a single point. Social scientists should determine where evolutionary biology is best used in their respective disciplines. To make such a determination, in turn, requires expanding knowledge of evolutionary biology in the social sciences. Once this is accomplished, social scientists will be able to generate new insights into social issues. Recognizing the revolution of 1859 will assist our understanding of these issues.

It is immediately clear that social science theories that depend heavily on psychology and emotion—such as theories of decision making and of cognitive misperception—will benefit by incorporating both evolutionary theory and advances in genetics. We will understand more of human behavior as evolutionary theorists, biologists, and geneticists discover and map how evolution affected genes, how genes affect proteins, and how proteins affect the human mind. Great advances will be made in both evolutionary psychology and in psychiatry as the mind is mapped and we better comprehend both its physiology and the ways it evolved to solve problems our ancestors encountered in past environments. As one consequence of this research, theories of decision making will be greatly improved as scientists discover how the mind makes decisions, weighs preferences, and prioritizes data or information. These advances hold great promise for international relations, particularly for scholarship on the leaders' decision making: How do they form psychological images of threats? Why do they form these images? What are the cognitive origins of misperception?

In considering how evolutionary biology may generate new insights for international relations, this book has merely scratched the surface. Many topics remain to be explored. I have not been able to analyze how evolutionary biology, and specifically cognitive neuroscience and evolutionary psychology, may assist the theories of threat perception, misperception, and decision making that are commonly used in the discipline of international relations. One example is Irving Janis's idea of "groupthink," a psychological phenomenon that often plagues governmental or corporate decision making. Janis calls it a "deterioration of mental efficiency, reality testing, and moral judgment that results from in-group pressures"; as one consequence, "members tend to evolve informal norms to preserve friendly intragroup relations and these become part of a hidden agenda at their meetings."[3] Certain group dynamics cause intelligent and knowledgeable individuals to make enormous collective mistakes. Janis argues that various groupthink-caused misperceptions have escalated wars, produced major failures of foreign and military policies, and resulted in domestic political crises.

To support his argument, he reviewed several major American foreign and military policy disasters, such as the failure of warning at Pearl Harbor, the 1961 Bay of Pigs invasion, and the Johnson administration's 1964 and 1965 decisions to escalate the Vietnam War. In their susceptibility to groupthink, the very intelligent and sophisticated individuals in these administrations were no different from ordinary people: "Members of policy-making groups, no matter how mindful they may be of their exalted national status and of their heavy responsibilities, are subject to the pressures widely observed in groups of ordinary citizens."[4]

Evolutionary biology allows us to understand that groupthink, as described by Janis, is not necessarily the product of individual pathology or cowardice in the face of in-group pressures. Perhaps we should stop blaming the groups of supposedly weak-minded or pusillanimous individuals who failed in their moral duty to speak truth to power and so condemned the country to the agony of the Vietnam War. Rather, evolutionary biology lets us see more clearly how the decision makers' actions that Janis describes are ultimately the product of evolution. This is so because the human species, like many other species, evolved in a dominance hierarchy. Once we understand this insight from human evolution, then we appreciate why individuals, especially in conditions of stress, may react as Janis describes.

Evolutionary biology allows us to appreciate how difficult it is to produce an environment that results in good policies. Most individuals will challenge the leader only with reluctance, and so good decision making requires an environment in which subordinates are actively encouraged to challenge ideas and are protected from retaliation. Similarly, understanding dominance hierarchy allow us to grasp why leaders generally do not like to be challenged, and so why the decision-making environment must be structured so that a leader's probable first inclination to suppress dissent is checked.

Empirically, much work also remains to be accomplished. If the agenda were an alphabet, we are at the alpha and have a long way to go before reaching omega. Although the historical and ethnographic literatures are extensive, we need to understand in more detail warfare in primitive societies. Similarly, as ethologists continue to study warfare among chimpanzees and social insects, social scientists should monitor their findings in order to gain further insights into how other animals wage war. We also need more research about ethnic conflict. The primordialists and modernists have both contributed greatly, to be sure, and both paradigms have active research agendas. But important issues illuminated in this book require more research. In particular, we must continue interviewing people actually responsible for ethnic conflict to determine how their environment instilled in them their beliefs about outgroups, as well as heightening their xenophobia and ethnocentrism. With more knowledge, we are better able to counter and lessen these sources of ethnic conflict.

However, when evolutionary theory is applied to the social sciences I stress again a point I have made throughout this book: Evolutionary theory concerns ultimate causation, not proximate. Ultimate causal analyses can explain why proximate mechanisms occur and why animals respond to them as they do. An ultimate causal explanation like evolution is not a direct description of behavior; rather, it frames the parameters of a proximate causal explanation, explaining for example how a particular behavior permits or facilitates an animal's survival. In addition, when we consider behavior, the environment is equally important.

In the introduction, I introduced the metaphor of a cake to capture this point. Let me now broaden and deepen that metaphor. We can think of evolutionary causes as a recipe from a cookbook.[5] Through natural selection, we have inherited genetic recipes that tell the cook which ingredients to use, their quantity, and the order in which to introduce them—the foundations of an excellent meal. But the recipe is not the final result. Between the recipe and the final enjoyable meal lie ovens and grills, cooking tools like pots and pans, the cook's taste in seasoning, and occasionally flawed or unpredictable ingredients: a bad egg or stale spices or yeast that worked last week but not now. Nor can we predict in every circumstance how the meal will look in its presentation, its taste, and how the diners will react. So it is with animals. Human behavior is the result of both ultimate and proximate causes, of both genes and environment.

Moreover, social issues may be studied at different levels of analysis, each focusing on different aspects of the issue. Once familiar with evolutionary theory, a social scientist can then recognize whether our knowledge of the issue studied can be usefully advanced through its lens. As I described in chapter 1, social scientists must know whether they need a ruler or tape measure for the issue they study. In other words, should we use evolutionary theory to explain a social behavior? To answer this question, social scientists must decide whether they need to know the ultimate cause of a particular social behavior. To study an issue like warfare or ethnic conflict we must know an ultimate cause for a behavior for two reasons. First, the causes of both are grounded in human evolution. Second, because each has profound implications for human suffering, we must understand their causes as completely as possible, so we must study them from multiple levels of analysis and use various theoretical approaches. But for many of the issues social scientists address, we do not need an ultimate causal explanation. These issues are more usefully studied at different levels of analysis—ones that rely wholly on proximate causation—as do many social scientific theories.

The Darwinian Revolution and the Social Sciences

The union of the life sciences and the discipline of international relations—and social science generally—is just beginning. It will generate further insights as more scholars trained in the social sciences appreciate the value of evolutionary theory. But bringing Darwin into the social sciences will take time. There are three reasons why this is so. First, the objections I addressed in chapter 1—evolutionary theory is deterministic, reductionist, tautological, and Panglossian—still hinder the progress of creating a social science informed by human evolution. These are serious objections, but as I demonstrated in that chapter, they are misplaced criticisms of modern evolutionary theory. Thus, being invalid, they should not hinder the use of evolutionary theory in social science.

Second, given the gap between the social and life sciences, social scientists may be totally unaware of the contributions the life sciences may make. Indeed, to the extent that social scientists consider evolutionary theory, for example, they may associate it with the social Darwinist beliefs of Herbert Spencer and others who perverted Darwin's arguments. This perception is understandable but nonetheless unfortunate for several reasons, not the least of which being that it hinders consilience and greater knowledge about human behavior. The good news is that social Darwinist arguments have been thoroughly refuted by modern evolutionary theory. The bad news is that, due to the tragic legacy of social Darwinism, a general suspicion of evolutionary theory remains among many social scientists.

Perceptions aside, one reason for the ignorance of evolutionary biology in the social sciences is that its components, especially evolutionary theory, are seldom taught in undergraduate or graduate programs in the social sciences, except in psychology, where the growing sub-field of evolutionary psychology is becoming increasingly important. Thus, for example, a student of political science may go through an entire graduate program without ever encountering the ideas of Darwin or William Hamilton. This should change. Graduate programs should teach these ideas so that their students encounter them and apply them where appropriate in their research. If this is done, I am optimistic the benefits will be great not only for their disciplines, but also for them personally as intellectuals. Their disciplines will gain because new knowledge will be created. For intellectuals who are curious to know about life, how humans came to be what we are, an understanding of evolutionary theory is essential. Such understanding will aid the bridging of the gap between the life and social sciences. Additionally, if indeed this century is "the century of biology," then a general knowledge of evolutionary theory and the life sciences will be the coin of the realm—essential for comprehending the inevitable scientific discoveries and their wonderful (and it must be said, potentially horrific) implications for human life.

Third, individuals do not like to think of themselves as "survival machines" for their genes. Nonetheless, while it may be unpleasant, it is a fact at one level of understanding. But of course humans are more than survival machinery. Consequently, social scientists may be reluctant to incorporate evolutionary theory or other elements of evolutionary biology into their disciplines because human intelligence and capacity for culture have led them to believe that human behavior is now independent of evolutionary forces. In our transition from simple primates to modern humans, we crossed a fundamental boundary, placing humans in another category entirely, beyond the explanatory power of evolutionary theory. Put more forcefully: Humanity's uniqueness means that modern human behavior is incomprehensible in evolutionary terms.[6]

This view is eminently understandable. Human evolution is characterized by a remarkable expansion in intelligence, consciousness, and complex learning, as well as the ability to transmit culture. Nonetheless, as human culture becomes more sophisticated, evolutionary theory still obtains; humans continue to evolve through natural selection. We can think of culture as an important proximate means to the ultimate evolutionary ends (such as fitness in a specific environment) identified by Darwin, Hamilton, and other evolutionary theorists. Clearly, the proximate mechanisms that promote fitness may change over evolutionary time, and they are important, but they still depend on the previous adaptations of the species, genetic mutations, and other constraints, such as human physiology, emotions, or our inevitable death.

Having reflected on how evolutionary theory can help the social sciences advance our understanding of human behavior, we see clearly that they have not sufficiently confronted Darwin's argument. Indeed, scientific progress in the life sciences will force them to do so in this "century of biology." The continued advances of evolutionary biology—especially in ethology, genetics, and neuroscience—will reveal much about human behavior and make it difficult for social science to exclude their knowledge. Evolutionary psychologists have seen this, and encouragingly, some anthropologists recognize that their discipline must evolve to include, for example, the insights of genetic research into the human genome. As anthropologists Gísli Pálsson and Paul Rabinow argue: "the fracturing of anthropology . . . will be overcome only when social anthropologists learn more biology."[7] As social scientists recognize the value of evolutionary theory, other social science disciplines will likely create similar syntheses of knowledge. Indeed, Darwin's Revolution and the knowledge it generates require not only its penetration into social science disciplines but also the restructuring of the university's organization. As universities now stand, their academic structure largely falls into a college for the natural sciences and one for the social sciences. The stark distinction between the natural and the social sciences was appropriate in 1900, around the time when many American universities were created. But a century later, such a strong division no longer holds due to the growth of the life sciences.

This is a problem for the advancement of knowledge because the present structure of most universities does not promote interdisciplinary research—such as intellectual exchange between the life and social sciences—but rather the reverse, it hinders it. The structure of the university is a barrier and is hard to overcome. In addition, scholars in most social science disciplines are not encouraged to conduct social research informed by the life sciences, and at this time graduate students certainly are not. Moreover, publication opportunities are limited in mainstream social science journals, and departments and academic deans may be reluctant to hire or promote such scholars.

To combat these problems, universities need to reorganize. It would be better for intellectual exchange if a typical university consisted of a college of the natural sciences, a college of the social sciences, and one of the evolutionary sciences where biologists, geneticists, ecologists, environmental studies scholars, and those who use the life sciences to study social issues—such as the evolutionary psychologists, evolutionary anthropologists, and evolutionary political scientists—are able to conduct research, publish, train graduate students, and teach undergraduates. Ultimately, it is the advancement of the life sciences that will require such a reorganization of the university. The university must evolve to adapt to the changes and requirements of the century of biology, with its promise of profound innovations and developments for humans—some positive and some not—and even to prepare for the greater demands of the twenty-second century.

My final point is that the use of evolutionary biology in social science is neither a threat nor a cure-all. Social scientists need not worry that it will colonize their disciplines. Evolutionary biology provides greater understanding of human behavior to be sure. However, human behavior is so complex that no single purely social or evolutionary approach can monopolize its study. This point may be obvious, but it is crucial. The usefulness of evolutionary arguments depends in large measure on the question the scholar addresses. Evolutionary theory or human ecology is relevant for a discussion of the rationality of humans or the origin of war, but not for a discussion of nuclear proliferation or the new institutionalism. Evolutionary biology and an evolutionary social science that is being created are not panaceas for social science. Their great benefit for scholars is improved insights into human behavior through an understanding of evolution, fitness, adaptation, the workings of the brain and mind, and how resource constraints and other environmental pressures have shaped humans.

But the economists' painful rule that nothing is free applies here as well: any benefit comes at a cost. Social scientists who use components of evolutionary biology to advance social scientific knowledge must first master its complexities. This is made more difficult than it need be because very few programs in political science, economics, or sociology teach students its value. Second, social scientists must be sensitive to past abuses of evolutionary theory and alert to those who might now abuse these arguments. Third, these scholars must be aware of the appropriate place of evolutionary explanations in social science, considering which types of questions and issues it may usefully address. Recognizing these caveats, much can be accomplished. The union of evolution and social science has allowed scholars to advance our understanding of human behavior, and it will continue to do so.

As science advances, we will truly understand the effect of evolution on humans: how we are like other animals, how we evolved. In time, we will know in detail more about how we survived in many dangerous environments in our evolutionary past, how we suffered and were often prey. When we recognize that humans came in other forms, now extinct, and when we grasp that evolution still operates upon us, we will understand how much humans deserve to be exulted for having evolved in our unique way, as indeed all life does. When we recognize this, we may appreciate fully Sophocles's words in *Antigone:*

> Numberless are the world's wonders, but none
> More wonderful than man.[8]

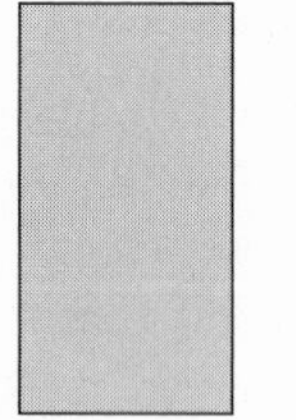

Notes

Introduction: Recognizing Darwin's Revolution

1. The debate over nature versus nurture has long been replaced in biology by the interaction principle that argues that the behavior of phenotypes derives from the interaction of genotype and the environment. See Robert A. Hinde, *Animal Behavior* (New York: McGraw-Hill, 1970).

2. I am indebted to evolutionary theorists Patrick Bateson and Richard Dawkins for this metaphor.

3. In this effort, the Human Genome Project (formally the International Human Genome Sequencing Consortium) faced competition from the Celera Genomics Corporation, a U.S. biotechnology company, which also sequenced part of the human genotype. One of the major goals of the Human Genome Project was to map the complete set of the genetic instructions carried by DNA in the twenty-three pairs of human chromosomes. In 2000, the Human Genome Project and Celera produced a "working draft" of the human genome sequence that was about 85 percent of the genome. Three years later, the Human Genome Project announced that most of the gaps were filled. While this work is largely completed, parts of the genome are still missing, although these parts are considered of minor importance. For a good introduction to this fascinating project, see Robert Cook-Deegan, *The Gene Wars: Science, Politics, and the Human Genome* (New York: Norton, 1995); Richard Lewontin, *It Ain't Necessarily So: The Dream of the Human Genome and Other Illusions* (New York: New York Review of Books, 2000), pp. 133–186; Matt Ridley, *Genome: The Autobiography of a Species in 23 Chapters* (New York: HarperCollins, 1999); and Edward Smith and Walter Sapp, eds., *Plain Talk about the Human Genome Project: A Tuskegee University Conference on Its Promise and Perils . . . and Matters of Race* (Tuskegee, Ala.: Tuskegee University, 1997).

4. For a solid, general analysis of the impact of genes on human behavior, see William R. Clark and Michael Grunstein, *Are We Hardwired? The Role of Genes in Human Behavior* (New York: Oxford University Press, 2000). John Avise's scholarship analyzes many of these illnesses in what he terms the "Chromosmal House of Horrors" present, or potentially so, in each human. See John C. Avise, *The Genetic*

Gods: Evolution and Belief in Human Affairs (Cambridge, Mass.: Harvard University Press, 1998), pp. 52–80.

5. For an introduction to the science of cloning; the development of the clone "Dolly"; the cloned rams "Cedric," "Cyril," "Cecil," and "Tuppence"; and its implications for humans, see Ian Wilmut, Keith Campbell, and Colin Tudge, *The Second Creation: Dolly and the Age of Biological Control* (New York: Farrar, Straus and Giroux, 2000). "Dolly" was euthanized in February 2003 after she developed a lung infection that was progressing and which would have caused her greater suffering.

6. An excellent account is Luigi Luca Cavalli-Sforza, Paolo Menozzi, and Alberto Piazza, *The History and Geography of Human Genes,* abr. ed. (Princeton, N.J.: Princeton University Press, 1994). Paul R. Ehrlich also provides a masterful and succinct account of human evolution and migration in his *Human Natures: Genes, Cultures, and the Human Prospect* (Washington, D.C.: Island Press, 2000), pp. 68–107. For a more detailed study, see Milford H. Wolpoff, *Paleoanthropology,* 2nd ed. (Boston: McGraw-Hill, 1999).

7. Skulls of approximately the same age have been found in the Middle East as well. Luigi Luca Cavalli-Sforza, *Genes, Peoples, and Languages* (New York: North Point Press, 2000), pp. 58–60. Also see Russell L. Ciochon and John G. Fleagle, *Primate Evolution and Human Origins* (New York: Aldine de Gruyter, 1987); Robert Foley, *Humans before Humanity: An Evolutionary Perspective* (Oxford: Blackwell, 1995), pp. 80–104; Jonathan Kingdon, *Self-Made Man: Human Evolution from Eden to Extinction?* (New York: John Wiley and Sons, 1993), pp. 13–66; and C.B. Stringer, "The Origin of Early Modern Humans: A Comparison of the European and Non-European Evidence," in Paul Mellars and Chris Stringer, eds., *The Human Revolution: Behavioural and Biological Perspectives on the Origins of Modern Humans* (Princeton, N.J.: Princeton University Press, 1989), pp. 232–244.

8. Cavalli-Sforza, *Genes, Peoples, and Languages,* pp. 35, 59; and Cavalli-Sforza et al., *The History and Geography of Human Genes,* pp. 58–68. For a thorough discussion of Neanderthals, see Christopher Stringer and Clive Gamble, *In Search of the Neanderthals: Solving the Puzzle of Human Origins* (New York: Thames and Hudson, 1993); and Ezra Zubrow, "The Demographic Modelling of Neanderthal Extinction," in Mellars and Stringer, eds., *The Human Revolution,* pp. 212–231. Neanderthals may have been alive as recently as 28,000 years ago.

9. Steven Pinker, *How the Mind Works* (New York: Norton, 1997), p. 21.

10. Michael S. Gazzaniga, *Nature's Mind: The Biological Roots of Thinking, Emotions, Sexuality, Language, and Intelligence* (New York: Basic Books, 1992).

11. Elaborating on the importance of such an understanding for sound social policies and principles is Antonio Damasio, *Looking for Spinoza: Joy, Sorrow, and the Feeling Brain* (New York: Harcourt, 2003).

12. Pinker, *How the Mind Works,* p. 21.

13. David M. Buss, *The Evolution of Desire: Strategies of Human Mating* (New York: Basic Books, 1994); and *The Dangerous Passion: Why Jealousy Is as Necessary as Love and Sex* (New York: Free Press, 2000).

14. Jerome H. Barkow, Leda Cosmides, and John Tooby, eds., *The Adapted Mind: Evolutionary Psychology and the Generation of Culture* (New York: Oxford University Press, 1992).

15. Robert A. Hinde, *Ethology: Its Nature and Relations with Other Sciences* (New York: Oxford University Press, 1982); Konrad Lorenz, *On Aggression,* trans. by Marjorie Kerr Wilson (New York: Harcourt, Brace and World, 1966); Niko Tinbergen, "Use and Misuse in Evolutionary Perspective," in W. Barlow, ed., *More Talk of Alexander* (London: Gollancz, 1978), pp. 218–236; and Irenäus Eibl-Eibesfeldt, *Human Ethology* (New York: Aldine de Gruyter, 1989), pp. 618–629.

16. Martin Daly and Margo Wilson, *Homicide* (New York: Aldine de Gruyter, 1988); Glenn Hausfater and Sarah Blaffer Hrdy, eds., *Infanticide: Comparative and Evolutionary Perspectives* (New York: Aldine, 1984); Randy Thornhill and Craig T. Palmer, *A Natural History of Rape: Biological Bases of Sexual Coercion* (Cambridge, Mass.: MIT Press, 2000); Wenda R. Trevathan, *Human Birth: An Evolutionary Perspective* (New York: Aldine, 1987); Jane B. Lancaster and Beatrix A. Hamburg, eds., *School-Age Pregnancy and Parenthood: Biosocial Dimensions* (New York: Aldine, 1986).

17. See Donald E. Brown, *Human Universals* (New York: McGraw-Hill, 1991).

18. Paul Ekman, "All Emotions Are Basic," in Paul Ekman and Richard J. Davidson, eds., *The Nature of Emotion: Fundamental Questions* (New York: Oxford University Press, 1994), pp. 15–19. Also see Alan J. Fridlund, *Human Facial Expression: An Evolutionary View* (New York: Academic Press, 1994).

19. See Edward O. Wilson, *Consilience: The Unity of Knowledge* (New York: Knopf, 1998).

20. Ernst Mayr, *Toward a New Philosophy of Biology: Observations of an Evolutionist* (Cambridge, Mass.: Harvard University Press, 1988), p. 193.

21. Mayr, *Toward a New Philosophy of Biology,* p. 194. For excellent discussions of the impact of Darwin's theory, see Peter J. Bowler, *Evolution: The History of an Idea,* rev. ed. (Berkeley: University of California Press, 1989); Daniel C. Dennett, *Darwin's Dangerous Idea: Evolution and the Meanings of Life* (New York: Simon and Schuster, 1995); Michael Ruse, *Taking Darwin Seriously: A Naturalistic Approach to Philosophy,* 2nd ed. (New York: Prometheus Books, 1998). Dennett's poignant work clearly discusses the implications of the Darwinian Revolution for the meaning and purpose of life.

22. Dennett, *Darwin's Dangerous Idea,* p. 21. For a similar reflection on Darwin's importance, see Richard Dawkins, "Darwin Triumphant: Darwinism as a Universal Truth," in Michael H. Robinson and Lionel Tiger, eds., *Man and Beast Revisited* (Washington, D.C.: Smithsonian Institution Press, 1991), pp. 23–39.

23. Ernst Mayr correctly argues that there "is no such thing as *the* Darwinian theory, since Darwin's evolutionary paradigm is highly composite." As I explore in more detail in the next chapter, Mayr argues that the Darwinian paradigm is five theories: "(1) evolution as such, (2) common descent, (3) gradualism, (4) multiplication of species, and (5) natural selection. Having provided this caveat, however, I will refer to Darwin's theory of evolution for reasons of simplicity. Mayr, *Toward a New Philosophy of Biology,* pp. 165, 198, emphasis original; and Mayr, *The Growth of Bio-*

logical Thought: Diversity, Evolution, and Inheritance (Cambridge, Mass.: Belknap Press of Harvard University Press, 1982), p. 506.

24. Wilson has made this argument since 1975; see more recently his *Consilience*, pp. 8–14, 197–228; and Roger D. Masters, "The Biological Nature of the State," *World Politics*, Vol. 35, No. 2 (January 1983), pp. 189–190. Also see Robin Fox, ed., *Biosocial Anthropology* (New York: Wiley and Sons, 1975); Albert Somit, "Human Nature as the Central Issue in Political Philosophy," in Elliott White, ed., *Sociobiology and Human Politics* (Lexington, Mass.: D.C. Heath, 1981), pp. 167–180; and Thomas C. Wiegele, ed., *Biology and the Social Sciences: An Emerging Revolution* (Boulder, Colo.: Westview Press, 1982). Richard Alexander claims that contemporary evolutionary theory provides "the first simple, general theory of human nature with a high likelihood of widespread acceptance." Richard D. Alexander, *Darwinism and Human Affairs* (Seattle: University of Washington Press, 1979), p. xii.

25. Masters, "The Biological Nature of the State," pp. 185–89. Also see Gary R. Johnson, "The Evolutionary Origins of Government and Politics," in Albert Somit and Steven A. Peterson, eds., *Research in Biopolitics*, Vol. 3 (Greenwich, Conn.: JAI Press, 1995), pp. 243–305.

26. Edward O. Wilson, *On Human Nature* (Cambridge, Mass.: Harvard University Press, 1978).

27. See Steven Pinker, *The Language Instinct: The New Science of Language and Mind* (New York: William Morrow, 1994); and Pinker, *How the Mind Works* (New York: Norton, 1997). Pinker and Paul Bloom advance the argument that language is the result of the evolutionary process. See Pinker and Bloom, "Natural Language and Natural Selection," in Barkow, Cosmides, and Tooby, eds., *The Adapted Mind: Evolutionary Psychology and the Generation of Culture*, pp. 451–493. Also see Robert Wright, *The Moral Animal: Evolutionary Psychology and Everyday Life* (New York: Vintage Books, 1994).

28. Steven Pinker, *The Blank Slate: The Modern Denial of Human Nature* (New York: Viking, 2002).

29. Ibid., p. 421.

30. Of course, these arguments and authors have been criticized by social scientists. In general, the criticism has been healthy because it has improved understanding. Many early critics saw the use of the life science approach to explain human social behavior as inappropriate because of the impact of culture or environment. Some of the critics, and only some, adopted a "hard" position. Environmental causes of human behavior were assumed to be the only relevant ones since humans were at birth the Lockean tabula rasa. See Kenneth Bock, *Human Nature and History: A Response to Sociobiology* (New York: Columbia University Press, 1980); Ashley Montagu, ed., *Sociobiology Examined* (New York: Oxford University Press, 1980); and Marshall D. Sahlins, *The Use and Abuse of Biology* (Ann Arbor: University of Michigan Press, 1976). A useful survey of this early debate is Arthur L. Caplan, ed., *The Sociobiology Debate* (New York: Harper and Row, 1978). The advances of evolutionary scholarship and the Human Genome Project are making such "hard" positions untenable because

they allow us to recognize that humans are not tabulae rasae at birth or thereafter; evolution affects humans throughout their lives.

Finally, an excellent analysis of the reaction to Darwin's thought in the United States is Ronald L. Numbers, *Darwinism Comes to America* (Cambridge, Mass.: Harvard University Press, 1998).

31. An excellent survey of how evolutionary theory can assist international relations is provided by Vincent S.E. Falger, "Biopolitics and the Study of International Relations," in Albert Somit and Steven A. Peterson, eds., *Research in Biopolitics,* Vol. 2 (Greenwich, Conn.: JAI Press, 1994), pp. 115–134; and Falger, "Evolution in International Relations?" in Steven A. Peterson and Albert Somit, eds., *Research in Biopolitics,* Vol. 8 (Amsterdam: Elsevier, 2001), pp. 91–118. An important concept I will not be able to address is the idea of the Evolutionary Stable Strategy (ESS) introduced by J. Maynard Smith and G.R. Price, "The Logic of Animal Conflict," *Nature,* Vol. 246 (November 2, 1973), pp. 15–18. This has helped inform Robert Axelrod and W.D. Hamilton's studying of the importance of tit-for-tat strategies for cooperation. See Axelrod and Hamilton, "The Evolution of Cooperation," *Science,* Vol. 211 (March 27, 1981), pp. 1390–1396; Axelrod, *The Evolution of Cooperation* (New York: Basic Books, 1984), chap. 3; and John Maynard Smith, *Evolution and the Theory of Games* (New York: Cambridge University Press, 1982). ESS is a behavior that results in the successful proliferation of genes so that they become almost universal and cannot be bettered by an alternative strategy. Put differently, the best strategy for any individual depends on what the majority of the population is doing. That is the best strategy to ensure survival of the phenotype, genes, or both. Once an ESS is established, it will stay because selection will penalize deviation.

32. Robert Jervis's use of psychological approaches to inform international relations is an ideal model of ways we can draw from another discipline to generate important insights in international relations. His work is essential for understanding decision making in crisis, deterrence theory, the origins of war through misperception, and alliance behavior. Robert Jervis, *The Logic of Images in International Relations* (Princeton, N.J.: Princeton University Press, 1970); Jervis, *Perception and Misperception in International Politics* (Princeton, N.J.: Princeton University Press, 1976); and Jervis, "Domino Beliefs and Strategic Behavior," in Robert Jervis and Jack Snyder, eds., *Dominoes and Bandwagons: Strategic Beliefs and Great Power Competition in the Eurasian Rimland* (New York: Oxford University Press, 1991), pp. 20–50.

The contribution of formal modeling, or rational choice, has also been significant for understanding voting behavior, deterrence theory, and collective action problems. See Anthony Downs, *An Economic Theory of Democracy* (New York: Harper and Row, 1957); Thomas C. Schelling, *Strategy of Conflict* (Cambridge, Mass.: Harvard University Press, 1960); and Mancur Olson, *The Logic of Collective Action: Public Goods and the Theory of Groups* (Cambridge, Mass.: Harvard University Press, 1965). For a debate concerning the continued relevance of rational choice in the subdiscipline of international security studies, see Stephen M. Walt, "Rigor or Rigor Mortis? Rational Choice and Security Studies," *International Security,* Vol. 23, No. 4 (Spring 1999), pp. 5–48; and the replies to Walt and his response in the special section "Formal Methods, Formal Complaints," *International Security,* Vol. 24, No. 2 (Fall 1999), pp. 56–130.

33. Ehrlich, *Human Natures,* p. 166.

34. Ibid.

35. I do not want to imply that evolutionary theory only influences social science while the reverse cannot be true. This is clearly not the case. It is worth remembering Darwin's claim that Thomas Malthus's study of population growth (*Essay on Population*) led to his famous epiphany on September 28, 1838. Darwin wrote he "read for amusement Malthus on *Population,* and being well prepared to appreciate the struggle for existence which everywhere goes on from long-continued observation of the habits of animals and plants, it at once struck me," he continues, "that under these circumstances favourable variations would tend to be preserved, and unfavorable ones be destroyed. The result of this would be the formation of new species. Here, then, I had at last got a theory by which to work." Charles Darwin, *The Autobiography of Charles Darwin, 1809–1882,* ed. by Nora Barlow (London: Collins, 1958), p. 120. Darwin's memory fails him in this passage of his autobiography, as he wrongly dates the event as occurring in October, rather than September 28, 1838, as Mayr's scholarship demonstrates. Malthus's work also influenced Alfred Russel Wallace, who also developed, independently of Darwin, the idea of natural selection. Mayr, *The Growth of Biological Thought,* pp. 494–495; and Michael Ruse, *Monad to Man: The Concept of Progress in Evolutionary Biology* (Cambridge, Mass.: Harvard University Press, 1996), pp. 191–201.

36. I will describe and contrast ultimate causation with proximate causation in chapter 1.

37. John J. Mearsheimer, *The Tragedy of Great Power Politics* (New York: Norton, 2001).

38. Jon Elster, *Nuts and Bolts for the Social Sciences* (New York: Cambridge University Press, 1989), p. 22.

39. Ibid., p. 28. But Elster acknowledges that doing "well for oneself" need not always be equated with egoism. See his discussion in chapter 6 of *Nuts and Bolts for the Social Sciences,* pp. 52–60.

40. George J. Stigler, "Smith's Travels on the Ship of State," in Andrew S. Skinner and Thomas Wilson, eds., *Essays on Adam Smith* (Oxford: Clarendon Press, 1975), p. 237. Also see Gary Becker, "The Economic Approach to Human Behavior," in Jon Elster, ed., *Rational Choice* (New York: New York University Press, 1986), pp. 108–122.

41. This is done perhaps most acutely and succinctly in Jane J. Mansbridge, "The Rise and Fall of Self-Interest in the Explanation of Political Life," in Mansbridge, ed., *Beyond Self-Interest* (Chicago: University of Chicago Press, 1990), pp. 3–22.

42. Roger D. Masters, *The Nature of Politics* (New Haven: Yale University Press, 1989); Eibl-Eibesfeldt, *Human Ethology*; and Wilson, *Consilience,* pp. 8–14, 197–228. Also see Richard D. Alexander, *The Biology of Moral Systems* (New York: Aldine de Gruyter, 1987); and Philip Kitcher, *Lives to Come: The Genetic Revolution and Human Possibilities* (New York: Simon and Schuster, 1996).

43. Émile Durkheim, *The Rules of Sociological Method,* trans. by Sarah A. Solovay and John H. Mueller (New York: Free Press, 1938), pp. 105–106.

44. Ibid., p. 106.

45. This argument has been carefully refined and perhaps reaches its apex in the work of such excellent anthropologists as Clifford Geertz and Marshall Sahlins. Clifford Geertz, *The Interpretation of Cultures* (New York: Basic Books, 1973); Geertz, *Local Knowledge: Further Essays in Interpretative Anthropology* (New York: Basic Books, 1983); and Marshall D. Sahlins, *How "Natives" Think: About Captain Cook, for Example* (Chicago: University of Chicago Press, 1995). Also see Carl N. Degler, *In Search of Human Nature: The Decline and Revival of Darwinism in American Social Thought* (New York: Oxford University Press, 1991), pp. 59–104.

46. For a presentation and elegant critique of the standard social science model, see John Tooby and Leda Cosmides, "The Psychological Foundations of Culture," in Barkow, Cosmides, and Tooby, eds., *The Adapted Mind,* pp. 19–136.

47. Making this point forcefully are Charles J. Lumsden and Edward O. Wilson, *Genes, Minds, and Culture: The Coevolutionary Process* (Cambridge, Mass.: Harvard University Press, 1981). The effect of culture cannot be minimized. In some societies, such as Pharaonic Egypt and the Incan Empire, its impact was so great that it overcame the incest taboo that both Edward Wilson and Claude Levi-Strauss have identified as universal for evolutionary and cultural reasons, respectively. Wilson, *On Human Nature,* p. 38; and Claude Levi-Strauss, *The Elementary Structures of Kinship,* rev. ed., trans. by James Harle Bell and John Richard von Sturmer (London: Eyre and Spottiswoode, 1969). Also see Brown, *Human Universals,* pp. 118–129. The fact that some tribes still practice cannibalism is another example of culture's ability to overcome the effects of evolution on human behavior.

48. Notably, advances in genetics now reveal that Durkheim's famous study of suicide, which he attributed in part to the differences in religious belief, is inadequate given what is now known about genetic proclivities to depression. See Lewis Wolpert, *Malignant Sadness: The Anatomy of Depression* (New York: Free Press, 1999); and Émile Durkheim, *Suicide: A Study in Sociology,* trans. by John A. Spaulding and George Simpson (New York: Free Press, 1951).

49. Richard Hofstadter, *Social Darwinism in American Thought* (Boston: Beacon Press, 1955).

50. For the impact of social Darwinian ideas on American immigration and the conception of American identity, see Desmond King, *Making Americans: Immigration, Race, and the Origins of the Diverse Democracy* (Cambridge, Mass.: Harvard University Press, 2000); and Rogers M. Smith, *Civic Ideas: Conflicting Visions of Citizenship in U.S. History* (New Haven: Yale University Press, 1997). On intelligence, see Degler, *In Search of Human Nature,* pp. 167–186.

A relatively large and growing literature describes eugenic policies in the United States, Britain, and other European countries. Nancy Gallagher documents the eugenics policies of the state of Vermont. See Nancy L. Gallagher, *Breeding Better Vermonters: The Eugenics Project in the Green Mountain State* (Hanover, N.H.: University

Press of New England, 1999). Randall Hansen and Desmond King explore eugenics policies in Britain and the United States in the interwar years in "Eugenic Ideas, Political Interests, and Policy Variance: Immigration and Sterilization Policy in Britain and the U.S.," *World Politics,* Vol. 53, No. 2 (January 2001), pp. 237–263. Stephen Jay Gould recounts how over seventy-five hundred people were sterilized in Virginia from 1924 to 1972 and the poignant story of Doris Buck, one victim of the sterilization program. See Stephen Jay Gould, *The Mismeasure of Man,* rev. and exp. (New York: Norton, 1996), pp. 365–366. Ridley reveals that over one hundred thousand people in the United States were sterilized for "feeble-mindedness" under more than thirty state laws passed between 1910 and 1935. Ridley, *Genome,* p. 290. Clark and Grunstein explain that similar programs were also found in Britain, Germany, and many Scandinavian countries. In 1999, Sweden compensated some sixty-three thousand people, mostly women, who had been sterilized by the government between 1936 and 1976 because they had been deemed to be "genetically inferior." Clark and Grunstein, *Are We Hardwired?,* pp. 279–294.

51. Stephen L. Chorover, *From Genesis to Genocide: The Meaning of Human Nature and Power of Behavior Control* (Cambridge, Mass.: MIT Press, 1979), pp. 77–109; Richard M. Lerner, *Final Solutions: Biology, Prejudice, and Genocide* (University Park, Pa.: Pennsylvania State University Press, 1992), pp. 21–49.

52. The animal rights movement began in 1975 with the publication of Peter Singer's *Animal Liberation.* The more recent 2002 edition contains his thoughts on the development of the movement since that time. See Singer, *Animal Liberation* (New York: HarperCollins, 2002).

53. Mary Midgley, *Beast and Man: The Roots of Human Nature* (Ithaca, N.Y.: Cornell University Press, 1978). Quoted in Degler, *In Search of Human Nature,* p. 329.

54. For an excellent overview of this scholarship, see William Irons and Lee Cronk, "Two Decades of a New Paradigm," in Lee Cronk, Napoleon Chagnon, William Irons, eds., *Adaptation and Human Behavior: An Anthropological Perspective* (New York: Aldine de Gruyter, 2000), pp. 3–26.

55. Brown, *Human Universals,* pp. 142–156. Also see Lee Cronk, *That Complex Whole: Culture and the Evolution of Human Behavior* (Boulder, Colo.: Westview Press, 1999).

56. Leda Cosmides and John Tooby, "From Evolution to Behavior: Evolutionary Psychology as the Missing Link," in John Dupré, ed., *The Latest on the Best: Essays on Evolution and Optimality* (Cambridge, Mass.: MIT Press, 1987), pp. 277–306; and Tooby and Cosmides, "Evolutionary Psychology and the Generation of Culture, Part I: Theoretical Considerations," *Ethology and Sociobiology,* Vol. 10, No. 1 (January 1989), pp. 29–49.

57. Francis Fukuyama, *The Great Disruption: Human Nature and the Reconstitution of Social Order* (New York: Free Press, 1999). Also see James Q. Wilson, *The Moral Sense* (New York: Free Press, 1993).

58. Sarah Blaffer Hrdy, *Mother Nature: A History of Mothers, Infants, and Natural Selection* (New York: Pantheon, 1999).

59. Arnold M. Ludwig, *King of the Mountain: The Nature of Political Leadership* (Lexington: University Press of Kentucky, 2002).

60. Stephen K. Sanderson, *The Evolution of Human Sociality: A Darwinian Conflict Perspective* (Lanham, Md.: Rowman and Littlefield Publishers, 2001).

61. Albert Somit and Steven A. Peterson, *Darwinism, Dominance, and Democracy: The Biological Bases of Authoritarianism* (Greenwich, Conn.: Praeger, 1997). Also see Laura L. Betzig, *Despotism and Differential Reproduction: A Darwinian View of History* (New York: Aldine, 1986). Evolutionary theory also provides insights into revolution: see Joseph Lopreato and F.P.A. Green, "The Evolutionary Foundations of Revolution," in J. van der Dennen and V. Falger, eds., *Sociobiology and Conflict: Evolutionary Perspectives on Competition, Cooperation, Violence and Warfare* (London: Chapman and Hall, 1990), pp. 107–122.

62. Notable efforts to develop a theory of human behavior sensitive to the interaction of evolutionary and cultural mechanisms are William H. Durham, "Toward a Coevolutionary Theory of Human Biology and Culture," in Napoleon A. Chagnon and William Irons, eds., *Evolutionary Biology and Human Social Behavior: An Anthropological Perspective* (North Scituate, Mass.: Duxbury Press, 1979), pp. 39–59; and Edward O. Wilson, "Biology and Anthropology: A Mutual Transformation?" in Chagnon and Irons, eds., *Evolutionary Biology and Human Social Behavior,* pp. 519–521.

63. Masters, *The Nature of Politics,* p. 1.

64. Lionel Tiger, *Men in Groups,* cited in Stephanie Gutmann, *The Kinder, Gentler Military: Can America's Gender-Neutral Fighting Force Still Win Wars?* (New York: Scribner, 2000), p. 244. Also see Tiger, *The Decline of Males* (New York: Golden Books, 1999), pp. 11–28, 231–266.

65. On the desirability of constructing verifiable scientific explanations, see Gary King, Robert O. Keohane, Sidney Verba, *Designing Social Inquiry: Scientific Inference in Qualitative Research* (Princeton, N.J.: Princeton University Press, 1994), p. 15. Stephen Van Evera argues that better theories have a broad explanatory range. See his *Guide to Methods for Students of Political Science* (Ithaca, N.Y.: Cornell University Press, 1997), pp. 17–18.

66. See for example Walker Connor, *Ethnonationalism: The Quest for Understanding* (Princeton, N.J.: Princeton University Press, 1994); Harold R. Isaacs, *Idols of the Tribe: Group Identity and Political Change* (New York: Harper and Row, 1975); Robert Kaplan, *Balkan Ghosts: A Journey through History* (New York: St. Martin's Press, 1993); and Anthony D. Smith, *Theories of Nationalism,* 2nd ed. (New York: Holmes and Meier, 1983).

67. For an extended investigation of the three major levels of analysis used in international relations, see Kenneth N. Waltz, *Man, the State, and War: A Theoretical Analysis* (New York: Columbia University Press, 1959), pp. 1–15; and J. David Singer, "The Level-of-Analysis Problem in International Relations," *World Politics,* Vol. 14, No. 1 (October 1961), pp. 77–92.

68. Formally, from a philosophical perspective, evolutionary theory is not strictly a first-level (or first image) or third-level theory, and evolutionary theorists do not use

this terminology or taxonomy when discussing the theory. The taxology of evolution by natural selection is different: the gene (or even molecule), individual, and perhaps group. To place the theory in terms easily identifiable for social scientists (although imperfect), evolution by natural selection may be thought of as a "loose" structural theory. The process of evolution is not a structural theory as is commonly defined in social science: while the interaction of unit and system is a structural element, occurrences at the level of the genotype are critically important, some theorists would say all important, and influenced by many factors described in chapter 1, such as mutation and chance. Consequently, what occurs at "the level of the unit," again to use a term familiar to social scientists, is highly significant. Reciprocally, evolutionary theory helps explain human behavior by explaining the human genotype and how it evolves, so some may prefer to classify it as falling under the first-level, or as a "loose" first-level theory.

69. Antonio Damasio and Albert Somit have been influential in promoting this concept. See Antonio R. Damasio, *Descartes' Error: Emotion, Reason, and the Human Brain* (New York: Grosset/Putnam, 1994); and Albert Somit, "Biopolitics," *British Journal of Political Science,* Vol. 2, Part 2 (April 1972), pp. 209–238. A good overview of the history of the use of evolutionary theory in political science is Roger D. Masters, "Political Science," in Mary Maxwell, ed., *The Sociobiological Imagination* (Albany: State University of New York, 1991), pp. 141–156. For a survey of this small but growing literature, see Albert Somit and Steven A. Peterson, *Biopolitics and Mainstream Political Science: A Master Bibliography* (DeKalb, Ill.: Program for Biosocial Research, 1990).

70. See William R. Thompson, "Evolving Toward an Evolutionary Perspective," in William R. Thompson, ed., *Evolutionary Interpretations of World Politics* (New York: Routledge, 2001), pp. 1–14, 2–4.

71. Kenneth N. Waltz, *Theory of International Politics* (Reading, Mass.: Addison-Wesley, 1979), pp. 74–77, 127–128.

1. Evolutionary Theory and Its Application to Social Science

1. Alfred Russel Wallace, "On the Tendency of Varieties to Depart Indefinitely from the Original Type," *Journal of the Proceedings of the Linnean Society, Zoology,* Vol. 3 (1858), pp. 53–62, 54. Quoted in Ernst Mayr, *The Growth of Biological Thought: Diversity, Evolution, and Inheritance* (Cambridge, Mass.: Belknap Press of Harvard University Press, 1982), p. 495. For a discussion of the similarities and differences in Darwin's and Wallace's thought, see Mayr, *The Growth of Biological Thought,* pp. 494–498; and Robert Wright, *The Moral Animal: Evolutionary Psychology and Everyday Life* (New York: Vintage, 1994), pp. 301–310.

2. Mayr demonstrates that only Darwin, among all the evolutionists of his day, understood all five of these important ideas and incorporated them into a paradigm. Mayr, *The Growth of Biological Thought,* pp. 505–510. Also see Stephen Jay Gould, *The Structure of Evolutionary Theory* (Cambridge, Mass.: Harvard University Press,

2002), pp. 93–169. Gertrude Himmelfarb skillfully analyzes the Victorian *Zeitgeist* and the Victorians' reaction to Darwin's ideas in *Darwin and the Darwinian Revolution* (Garden City, N.Y.: Doubleday, 1962).

3. Ernst Mayr, *Toward a New Philosophy of Biology: Observations of an Evolutionist* (Cambridge, Mass.: Harvard University Press, 1988), p. 198.

4. For a discussion of the beliefs at that time, see Stephen Jay Gould, *Time's Arrow, Time's Cycle: Myth and Metaphor in the Discovery of Geological Time* (Cambridge, Mass.: Harvard University Press, 1987), pp. 61–179.

5. Mayr, *Toward a New Philosophy of Biology,* p. 215. Emphasis omitted.

6. Charles Darwin, *On the Origin of Species by Means of Natural Selection, or the Preservation of Favoured Races in the Struggle for Life* (Cambridge, Mass.: Harvard University Press, 1964 [1859]), pp. 402–404.

7. Ibid., p. 488.

8. Ibid., pp. 488–489.

9. Mayr, *Toward a New Philosophy of Biology,* p. 203. Also see Ernst Mayr, *One Long Argument: Charles Darwin and the Genesis of Modern Evolutionary Thought* (Cambridge, Mass.: Harvard University Press, 1991), pp. 68–89.

10. Thomas H. Huxley, *Life and Letters of Thomas Henry Huxley, by His Son Leonard Huxley,* 2 Vols. (London: Macmillan, 1900), Vol. 2, p. 27. Quoted in Mayr, *Toward a New Philosophy of Biology,* p. 203.

11. Niles Eldredge and Stephen Jay Gould, "Punctuated Equilibria: An Alternative to Phyletic Gradualism," in Thomas J.M. Schopf, ed., *Models in Paleobiology* (San Francisco: Freeman Cooper, 1972), pp. 82–115; and Gould, "More Things in Heaven and Earth," in Hilary Rose and Steven Rose, eds., *Alas, Poor Darwin: Arguments against Evolutionary Psychology* (London: Jonathan Cape, 2000), pp. 85–105. Also see George G. Simpson, *Tempo and Mode in Evolution* (New York: Columbia University Press, 1944). For a critical response to Eldredge and Gould, see John Maynard Smith, *Did Darwin Get It Right? Essays on Games, Sex, and Evolution* (New York: Chapman and Hall, 1989). An excellent survey of this discussion is Albert Somit and Steven A. Peterson, eds., *The Dynamics of Evolution: The Punctuated Equilibrium Debate in the Natural and Social Sciences* (Ithaca, N.Y.: Cornell University Press, 1992).

12. Several authors have argued that the idea of punctuated equilibrium was well established in evolutionary theory before Eldredge and Gould. Richard Dawkins has been particularly forceful: "What needs to be said now, loud and clear, is the truth: that the theory of punctuated equilibrium lies firmly within the neo-Darwinian synthesis. It always did. It will take time to undo the damage wrought by the overblown rhetoric, but it will be undone." Richard Dawkins, *The Blind Watchmaker: Why the Evidence of Evolution Reveals a Universe without Design* (New York: Norton, 1986), pp. 223–252, 251. Also see Daniel C. Dennett, *Darwin's Dangerous Idea: Evolution and the Meanings of Life* (New York: Simon and Schuster, 1995), pp. 282–303; Steven Pinker, *How the Mind Works* (New York: Norton, 1997), pp. 133–134, 168–169; and Jonathan K. Waage and Patricia Adair Gowaty, "Myths of Genetic Determinism," in Patricia Adair Gowaty, ed., *Feminism and Evolutionary Biology: Boundaries, Intersec-*

tions, and Frontiers (London: Chapman and Hall, 1997), pp. 592–594. For Eldredge's reply, see Niles Eldredge, *Reinventing Darwin: The Great Debate at the High Table of Evolutionary Theory* (New York: John Wiley, 1995).

13. Mayr, *Toward a New Philosophy of Biology,* p. 205.

14. Mayr, *The Growth of Biological Thought,* pp. 403–408.

15. See, for example, Edward O. Wilson, *The Diversity of Life* (New York: Norton, 1999).

16. Darwin, *On the Origin of Species,* pp. 322ff.

17. Ibid., p. 61.

18. Ibid.

19. Mayr, *Toward a New Philosophy of Biology,* p. 216. Emphasis omitted.

20. More precisely, Theodosius Dobzhansky describes the genotype as the "sum total of genes of an individual or population," whereas the phenotype "is what is perceived by direct observation," the "biological system constructed by successive interactions of the individual's genotype with the environments in which the development takes place." Theodosius Dobzhansky, *Genetics and the Origin of Species,* 3rd ed., rev. (New York: Columbia University Press, 1961), p. 20; and Theodosius Dobzhansky, *Genetics of the Evolutionary Process* (New York: Columbia University Press, 1970), p. 36.

21. Evolutionary theorist Richard Dawkins describes a replicator as "anything in the universe of which copies are made." Richard Dawkins, "Replicators and Vehicles," King's College Sociobiology Study Group, ed., *Current Problems in Sociobiology* (Cambridge: Cambridge University Press, 1982), pp. 45–64, 46.

22. Robert N. Brandon, "The Levels of Selection: A Hierarchy of Interactors," in David L. Hull and Michael Ruse, eds., *The Philosophy of Biology* (New York: Oxford University Press, 1998), pp. 176–197, 178.

23. David L. Hull, "Interactors versus Vehicles," in H.C. Plotkin, ed., *The Role of Behavior in Evolution* (Cambridge, Mass.: MIT Press, 1988), pp. 19–50, 27. Also see David L. Hull, "Individuality and Selection," *Annual Review of Ecology and Systematics,* Vol. 11 (Palo Alto, Calif.: Annual Reviews, Inc., 1980), pp. 311–332.

24. Mayr, *Toward a New Philosophy of Biology,* p. 209.

25. Mendel did not use the term genes; the term was not used until 1909. Instead he referred to "particulate" transmission for hereditary factors. See John C. Avise, *Genetic Gods: Evolution and Belief in Human Affairs* (Cambridge, Mass.: Harvard University Press, 1998), p. 8.

26. Michel Morange, *The Misunderstood Gene,* trans. by Matthew Cobb (Cambridge, Mass.: Harvard University Press, 2001), p. 9.

27. Julian Huxley, *Evolution: The Modern Synthesis* (New York: Harper and Brothers, 1942).

28. Elliott Sober, *Philosophy of Biology* (Boulder, Colo.: Westview, 1993), p. 18. Huxley, *Evolution,* pp. 229–231.

29. Enrico Coen, *The Art of Genes: How Organisms Make Themselves* (New York: Oxford University Press, 1999), p. 27. Also see Austin L. Hughes, *Adaptive Evolution of Genes and Genomes* (New York: Oxford University Press, 1999), pp. 15–37.

30. Coen, *The Art of Genes,* p. 29.

31. Dobzhansky, *Genetics of the Evolutionary Process,* pp. 44–45. Emphasizing the role of chance in mutation is Jeffrey K. McKee, *The Riddled Chain: Chance, Coincidence, and Chaos in Human Evolution* (New Brunswick, N.J.: Rutgers University Press, 2000), pp. 135–139, 159–165.

32. Dobzhansky, *Genetics and the Origin of Species,* p. 156. Also see Ronald A. Fisher, *The Genetical Theory of Natural Selection,* 2nd ed., rev. (New York: Dover Publications, 1958). According to Hull, genetic drift is not a cause of natural selection because natural selection requires interactors and "an entity counts as an interactor only if it is functioning as one in the process in question. It is not enough that in past interactions it functioned as an interactor.... In instances of drift, there may be genes and organisms, but there are no interactors, only replicators." Hull, "Interactors versus Vehicles," p. 28.

33. Dobzhansky, *Genetics and the Origin of Species,* pp. 157–163. Also see Graham Bell, *Selection: The Mechanism of Evolution* (London: Chapman and Hall, 1997), p. 94; and Motoo Kimura, *The Neutral Theory of Molecular Evolution* (New York: Cambridge University Press, 1983).

34. Bell, *Selection,* p. 94.

35. Dobzhansky, *Genetics and the Origin of Species,* pp. 164–165.

36. Emphasizing the amalgamation of natural selection with chance are Stuart Kauffman, *At Home in the Universe: The Search for Laws of Self-Organization and Complexity* (New York: Oxford University Press, 1995); and McKee, *The Riddled Chain,* pp. 2–5, 197–199. Mayr notes that in 1859 many critics chided Darwin for including chance as an unscientific concept while ignoring what they saw as the hand of God in the design of creatures. Ernst Mayr, *This Is Biology: The Science of the Living World* (Cambridge, Mass.: Harvard University Press, 1997), p. 26.

37. Mayr, *The Growth of Biological Thought,* pp. 485–487.

38. Darwin, *On the Origin of Species,* pp. 87–90; and Darwin, *The Descent of Man, and Selection in Relation to Sex* (Princeton, N.J.: Princeton University Press, 1981 [1871], *passim.*

39. Darwin, *The Descent of Man,* Vol. 1, pp. 256–263, Vol. 2, pp. 355–384, quote from Vol. 1, p. 278. Also see Geoffrey F. Miller, "Sexual Selection for Cultural Displays," in Robin Dunbar, Chris Knight, and Camilla Power, eds, *The Evolution of Culture: An Interdisciplinary View* (Edinburgh: Edinburgh University Press, 1999), pp. 71–91; and Miller, *The Mating Mind: How Sexual Choice Shaped the Evolution of Human Nature* (New York: Doubleday, 2000), pp. 8–12, 33–67.

40. Darwin, *The Descent of Man,* Vol. 2, pp. 135–141, 154–182.

41. Bell, *Selection,* p. 584. Also see George C. Williams, *Sex and Evolution* (Princeton, N.J.: Princeton University Press, 1975), pp. 124–132.

42. Mayr, *The Growth of Biological Thought,* p. 597.

43. Mayr, *Toward a New Philosophy of Biology,* p. 140.

44. Philip Kitcher, *Vaulting Ambition: Sociobiology and the Quest for Human Nature* (Cambridge, Mass.: MIT Press, 1985), p. 42.

45. Sober, *Philosophy of Biology,* p. 9. Richard D. Alexander provides an extended and more detailed discussion in *Darwinism and Human Affairs* (Seattle: University of Washington Press, 1979), pp. 15–18.

46. Luigi Luca Cavalli-Sforza and Francesco Cavalli-Sforza, *The Great Human Diasporas: The History of Diversity and Evolution,* trans. by Sarah Thorne (Reading, Mass.: Addison-Wesley, 1995), pp. 92–104; Richard Dawkins, *The Extended Phenotype: The Long Reach of the Gene* (Oxford: Oxford University Press, 1999) pp. 38–54; Sober, *Philosophy of Biology,* pp. 18–19. Gene mutation may occur by either random changes in the chemical composition of extant genes or by alternations in the structure or number of chromosomes.

47. Stephen Jay Gould, *The Structure of Evolutionary Theory* (Cambridge, Mass.: Harvard University Press, 2002), p. 155.

48. In addition, what is ultimately important is relative, not absolute, fitness. That is, it is not only the number of offspring you produce, but the fact that you produce more than others. Elliott Sober and David Sloan Wilson, *Unto Others: The Evolution and Psychology of Unselfish Behavior* (Cambridge, Mass.: Harvard University Press, 1998), p. 23.

49. For excellent discussions of the complexities associated with fitness, see John Beatty, "Fitness: Theoretical Contexts," in Evelyn Fox Keller and Elisabeth A. Lloyd, eds., *Keywords in Evolutionary Biology* (Cambridge, Mass.: Harvard University Press, 1992), pp. 115–119; W.D. Hamilton, "Innate Social Aptitudes of Man: An Approach from Evolutionary Genetics," in Robin Fox, ed., *Biosocial Anthropology* (New York: Wiley and Sons, 1975), pp. 133–155; Susan K. Mills and John Beatty, "The Propensity Interpretation of Fitness," in Elliott Sober, ed., *Conceptual Issues in Evolutionary Biology* (Cambridge, Mass.: MIT Press, 1994), pp. 3–23; and Alexander Rosenberg, *The Structure of Biological Science* (New York: Cambridge University Press, 1985), pp. 154–164.

50. Of course, in this example there are genetic, phenotypic, and environmental limits to the speed of zebras. Furthermore, as Sober notes, the environment may be especially important, especially if offspring receive better nutrition. So purely environmental reasons might explain similarity between parental and offspring phenotypes. Sober, *Philosophy of Biology,* p. 11. Also see Ernst Mayr, *What Evolution Is* (New York: Basic Books, 2001), pp. 115–121.

51. As I explain in more detail below, evolutionary theory is concerned with ultimate causes of behavior rather than proximate causes.

52. John Maynard Smith, *The Theory of Evolution,* Canto ed. (New York: Cambridge University Press, 1993), p. 26.

53. Ibid.

54. Ibid., pp. 26–27.

55. Mayr, *What Evolution Is,* p. 150.

56. While it is critical to survive long enough to reproduce, human evolution allows our lifetimes to overlap with those of other generations, permitting (among other benefits) extensive parental and alloparental care. As Sarah Blaffer Hrdy describes it, "We need to keep in mind Mother Nature's cardinal rule for mothers: It's

not enough to produce offspring; to succeed through evolutionary time mothers must produce offspring who will survive and prosper." Hrdy, *Mother Nature: A History of Mothers, Infants, and Natural Selection* (New York: Pantheon Books, 1999), p. 64.

57. Dawkins describes this with his customary clarity: our traits arise through "cumulative evolution by nonrandom survival of random hereditary changes." Richard Dawkins, "Darwin Triumphant: Darwinism as a Universal Truth," in Michael H. Robinson and Lionel Tiger, eds., *Man and Beast Revisited* (Washington, D.C.: Smithsonian Institution Press, 1991), p. 24.

58. More precisely, human behavior is the result of the environment and genotype. The perspective begins with Darwin's description of natural selection. For excellent introductions to Darwin's original theory of evolution and how it has changed, see Peter J. Bowler, *Evolution: The History of an Idea,* rev. ed. (Berkeley: University of California Press, 1989); and Robert J. Richards, *Darwin and the Emergence of Evolutionary Theories of Mind and Behavior* (Chicago: University of Chicago Press, 1987). It still remains difficult for some people to accept this. As Stephen Gould notes, "The evolutionary unity of humans with all other organisms is the cardinal message of Darwin's revolution for nature's most arrogant species." Stephen J. Gould, *The Mismeasure of Man,* rev. and exp. (New York: Norton, 1996), p. 354.

59. Perhaps most compellingly making this point is Theodosius Dobzhansky, *Mankind Evolving: The Evolution of the Human Species* (New Haven: Yale University Press, 1962).

60. Edward O. Wilson, *Sociobiology: The New Synthesis* (Cambridge, Mass.: Harvard University Press, 1975), p. 144.

61. Steven Rose, "Escaping Evolutionary Psychology," in Hilary Rose and Steven Rose, eds., *Alas, Poor Darwin: Arguments against Evolutionary Psychology* (London: Jonathan Cape, 2000), pp. 247–265, 259.

62. Rose emphasizes this point in Ibid., p. 259.

63. Alexander Rosenberg, *Darwinism in Philosophy, Social Science and Policy* (New York: Cambridge University Press, 2000), p. 124.

64. Ibid., pp. 124–125.

65. Quoted in Bell, *Selection,* p. 25. Emphasis in original.

66. Ibid., p. 23.

67. Ibid.

68. Paul R. Ehrlich, *Human Natures: Genes, Cultures, and the Human Prospect* (Washington, D.C.: Island Press, 2000), p. 34.

69. Richard Dawkins, *The Selfish Gene,* new ed. (Oxford: Oxford University Press, 1989), p. 258.

70. A detailed analysis of how this could have occurred, building upon Dawkins's insights, is by Eörs Szathmáry, "The First Replicators," in Laurent Keller, ed., *Levels of Selection in Evolution* (Princeton, N.J.: Princeton University Press, 1999), pp. 31–52.

71. Dawkins, *The Selfish Gene,* p. 19.

72. Ibid., p. 20.

73. This argument is strongly articulated in Barry Schwartz, *The Battle for Human Nature: Science, Morality and Modern Life* (New York: Norton, 1986), pp. 96–97.

74. For a cogent critique of selfish gene theory, see Sober and Wilson, *Unto Others,* pp. 87–92.

75. George C. Williams, *Natural Selection: Domains, Levels, and Challenges* (New York: Oxford University Press, 1992), p. 15.

76. Ibid.

77. Ibid., p. 16.

78. Ibid.

79. Ibid.

80. Ibid.

81. Ibid.

82. Ibid.

83. W.D. Hamilton, "The Genetical Evolution of Social Behavior. I," *Journal of Theoretical Biology,* Vol. 7, No. 1 (July 1964), pp. 1–16; Hamilton, "The Genetical Evolution of Social Behavior. II," *Journal of Theoretical Biology,* Vol. 7, No. 1 (July 1964), pp. 17–52. John Maynard Smith terms the concept kin selection in his "Group Selection and Kin Selection," *Nature,* Vol. 201, No. 4924 (March 14, 1964), pp. 1145–1147. Also see Schwartz, *The Battle for Human Nature,* p. 97.

84. Roger D. Masters, "The Biological Nature of the State," *World Politics,* Vol. 35, No. 2 (January 1983), p. 165.

85. As David Barash explains, we "should be selected to show altruism toward others in direct proportion to how closely related they are to us genetically. . . . We should be willing to suffer greater risks in aiding individuals who are more closely related and should withhold aid to more distantly related individuals. . . . Similarly, we should require that a distant relative be in greater need . . . in order for us to render the same assistance that we would dispense to a closer relative." David P. Barash, *Sociobiology and Behavior* (New York: Elsevier, 1977), p. 309.

86. The motivation could be proximate and not ultimate.

87. Hamilton, "The Genetical Evolution of Social Behavior. I," pp. 15–16.

88. Mayr, *The Growth of Biological Thought,* p. 598.

89. Among vertebrates, this behavior has been found in colonies of naked mole rats in Kenya, Ethiopia, and Somalia. See John Alcock, *The Triumph of Sociobiology* (New York: Oxford University Press, 2001), pp. 97–99.

90. John Cartwright, *Evolution and Human Behavior: Darwinian Perspectives on Human Nature* (Cambridge, Mass.: MIT Press, 2000), p. 77; and Raghavendra Gadagkar, *Survival Strategies: Cooperation and Conflict in Animal Societies* (Cambridge, Mass.: Harvard University Press, 1997), pp. 95–104.

91. Cartwright, *Evolution and Human Behavior,* p. 77.

92. Bell, *Selection,* p. 524.

93. Group and individual selection are not necessarily incompatible; both may be operative in a population simultaneously. The concept of group selection was largely developed by Ronald Fisher and V.C. Wynne-Edwards. See Ronald A. Fisher, *The*

Genetical Theory of Natural Selection (Oxford: Clarendon Press, 1930); and V.C. Wynne-Edwards, *Animal Dispersion in Relation to Social Behaviour* (Edinburgh: Oliver and Boyd, 1962). Also see Darwin, *The Descent of Man,* Vol. 1, p. 166. J.B.S. Haldane is sometimes cited in support of group selection, but in fact he argued that group selection was possible but fragile and likely to be reversed by modest changes in conditions. See J.B.S. Haldane, *The Causes of Evolution* (London: Longmans Green, 1932).

94. See, for example, Dawkins, *The Selfish Gene,* pp. 7–11, 110, 297; Dawkins, "A Survival Machine," in J. Brockman, ed., *The Third Culture* (New York: Simon and Schuster, 1995), pp. 74–95; Alan Grafen, "Natural Selection, Kin Selection and Group Selection," in J. Krebs and N.B. Davies, eds., *Behavioural Ecology,* 2nd ed. (Oxford: Blackwell Scientific Publications, 1984), pp. 62–84; John Maynard Smith, "The Origin of Altruism," review of *Unto Others, Nature,* Vol. 393 (1998), pp. 639–640; Robert L. Trivers, "As They Would Do to You," Review of Elliott Sober and David Sloan Wilson, *Unto Others, The Skeptic,* Vol. 6, No. 4 (1998), pp. 81–83; and Trivers, "Think for Yourself," Trivers replies to Sober and Wilson, *The Skeptic,* Vol. 6, No. 4 (1998), pp. 86–87. George C. Williams is also critical, arguing that higher levels of selection are unnecessary. See Williams, *Adaptation and Natural Selection: A Critique of Some Current Evolutionary Thought* (Princeton, N.J.: Princeton University Press, 1966). Perhaps the strongest criticism of group selection may be found in H. Kern Reeve and Laurent Keller, "Levels of Selection: Burying the Units-of-Selection Debate and Unearthing the Crucial New Issues," in Laurent Keller, ed., *Levels of Selection in Evolution* (Princeton: Princeton University Press, 1999), pp. 3–14. An excellent collection of arguments over the proper unit of selection in evolution is Robert N. Brandon and Richard M. Burian, eds., *Genes, Organisms, Populations: Controversies over the Units of Selection* (Cambridge, Mass.: MIT Press, 1984).

95. David Sloan Wilson, *The Natural Selection of Populations and Communities* (Menlo Park, Calif.: Benjamin/Cummings, 1980), p. 45.

96. Ibid., pp. 76–78.

97. Classic studies are Robert L. Trivers, "The Evolution of Reciprocal Altruism," *The Quarterly Review of Biology,* Vol. 46, No. 1 (March 1971), pp. 35–57; and Wilson, *The Natural Selection of Populations and Communities.* For a critique of Trivers's conception of altruism, see Elliott Sober, "What Is Evolutionary Altruism?" in David L. Hull and Michael Ruse, eds., *The Philosophy of Biology* (New York: Oxford University Press, 1998), pp. 459–478.

98. Sober and Wilson, *Unto Others,* p. 10.

99. Darwin, *On the Origin of Species,* p. 201.

100. Sober and Wilson, *Unto Others,* p. 10. Also see Christopher Boehm, *Hierarchy in the Forest: The Evolution of Egalitarian Behavior* (Cambridge, Mass.: Harvard University Press, 1999); and Rose, "Escaping Evolutionary Psychology," p. 258.

101. Sober and Wilson, *Unto Others,* p. 26.

102. Ibid., p. 27. Also see Boehm, *Hierarchy in the Forest,* pp. 203–211.

103. Williams, *Natural Selection,* p. 46.

104. Bell, *Selection,* p. 530.

105. Ibid.

106. Gadagkar, *Survival Strategies,* pp. 26–27.

107. Ibid., p. 27. Emphasis in original.

108. Donald Symons, "An Evolutionary Approach: Can Darwin's View of Life Shed Light on Human Sexuality?" in James H. Geer and William T. O'Donohue, eds., *Theories of Human Sexuality* (New York: Plenum Press, 1987), p. 95.

109. See John Damuth and I. Lorraine Heisler, "Alternative Formulations of Multilevel Selection," *Biology and Philosophy,* Vol. 3 (1988), pp. 407–430; Len Nunney, "Group Selection, Altruism, and Structured-Deme Models," *The American Naturalist,* Vol. 126, No. 2 (August 1985), pp. 212–230; and Nunney, "Female-Biased Sex Ratios: Individual or Group Selection," *Evolution,* Vol. 39, No. 2 (March 1985), pp. 349–361.

110. The distinction between ultimate and proximate causation is commonly made in biology. See Mayr, *Toward a New Philosophy of Biology,* pp. 24–37. I base the distinction between universal and proximate causation upon the distinction that Mayr and Wesley Salmon make between fundamental and derivative theories. A universal cause is a fundamental, lawlike sentence. It is purely universal and adequately supported by empirical evidence. A proximate cause is derivable from fundamental theories. Wesley C. Salmon, *Four Decades of Scientific Explanation* (Minneapolis: University of Minnesota Press, 1989), pp. 18–20. Salmon's argument is based on the distinction Carl Hempel and Paul Oppenheim make between fundamental laws and derivative laws. Hempel and Oppenheim, "Studies in the Logic of Explanation," *Philosophy of Science,* Vol. 15 (1948), pp. 135–148. Reprinted in Carl G. Hempel, *Aspects of Scientific Explanation* (New York: Free Press, 1965), pp. 245–290, 267.

111. Niko Tinbergen and Paul Sherman both note that proximate causes may further be broken down into questions of individual ontogeny and physiological substrates, while ultimate causes may be considered in terms of evolutionary origins and current adaptive value. See N. Tinbergen, "On the Aims and Methods of Ethology," *Zeitschrift für Tierpsychologie,* Vol. 20 (1963), pp. 410–433; and Paul. W. Sherman, "The Levels of Analysis," *Animal Behaviour,* Vol. 36, No. 2 (April 1988), pp. 616–619.

112. Most explanations of behavior or events implicitly use ultimate and proximate causal analysis, although in most everyday explanations, proximate ones are sufficient. See Bruce Winterhalder and Eric Alden Smith, "Evolutionary Ecology and the Social Sciences," in Eric Alden Smith and Bruce Winterhalder, eds., *Evolutionary Ecology and Human Behavior* (New York: Aldine de Gruyter, 1992), pp. 3–23.

113. Edward O. Wilson, *Consilience: The Unity of Knowledge* (New York: Knopf, 1998), pp. 85–86. Emphasis in original. Also see Symons, "An Evolutionary Approach: Can Darwin's View of Life Shed Light on Human Sexuality?" p. 94.

114. Mayr, *This Is Biology,* p. 67.

115. Put differently, proximate causes explain the mechanisms behind a flock taking to the air: the physiological mechanisms that are activated by environmental cues. Why birds migrate is a question of ultimate causation: the migration confers

evolutionary advantage to the birds. By so doing, they leave more progeny. For a discussion of the types of questions ethologists should ask and how these questions should be pursued, see Tinbergen, "On the Aims and Methods of Ethology," pp. 410–433.

116. Symons, "An Evolutionary Approach," p. 94.

117. John Alcock, *The Triumph of Sociobiology* (New York: Oxford University Press, 2001), p. 15.

118. Ibid.

119. For a discussion of culture among chimpanzees, see Frans de Waal, *Good Natured: The Origins of Right and Wrong in Humans and Other Animals* (Cambridge, Mass.: Harvard University Press, 1996), pp. 178–180, 210.

120. Richard Dawkins credits Patrick Bateson as the originator of the cake metaphor. See Richard Dawkins, review of *Not in Our Genes, New Scientist* (January 24, 1985), pp. 59–60; and Dawkins, "In Defence of Selfish Genes," *Philosophy,* October 1981. Also see Patrick Bateson and Paul Martin, *Design for A Life: How Biology and Psychology Shape Human Behavior* (New York: Simon and Schuster, 2000), pp. 16, 220–221. Ian Stewart prefers the metaphor of a cookbook, in which genes are the recipes but the recipe is not the same as the meal. Ian Stewart, *Life's Other Secret: The New Mathematics of the Living World* (New York: John Wiley and Sons, 1998), p. x.

121. Randy Thornhill and Craig T. Palmer, *A Natural History of Rape: Biological Bases of Sexual Coercion* (Cambridge, Mass.: MIT Press, 2000), pp. 13–14.

122. Particularly interesting are those who study human behavior from the perspective of evolutionary theory in order to study how social behavior is shaped by natural selection. The locus classicus is Edward O. Wilson, *Sociobiology: The New Synthesis* (Cambridge, Mass.: Harvard University Press, 1975). Also see Edward O. Wilson, *On Human Nature* (Cambridge, Mass.: Harvard University Press, 1978); and Charles J. Lumsden and Edward O. Wilson, *Promethean Fire: Reflections on the Origin of Mind* (Cambridge, Mass.: Harvard University Press, 1983).

The two major types of criticisms of the application of evolutionary theory to humans come from the biological and cultural perspectives. The literature in both is extensive; for one of the best in the biological perspective, see Gould, *The Mismeasure of Man,* pp. 356–363; Kitcher, *Vaulting Ambition*; R.C. Lewontin, Steven Rose, and Leon J. Kamin, *Not in Our Genes: Biology, Ideology, and Human Nature* (New York: Pantheon Books, 1984); and Ashley Montagu, ed., *Sociobiology Examined* (New York: Oxford University Press, 1980). Criticisms from a cultural perspective include Kenneth Bock, *Human Nature and History: A Response to Sociobiology* (New York: Columbia University Press, 1980); and Marshall D. Sahlins, *The Use and Abuse of Biology* (Ann Arbor: University of Michigan Press, 1976). A useful survey is still Arthur L. Caplan, ed., *The Sociobiology Debate* (New York: Harper and Row, 1978). Also see Joshua S. Goldstein, "The Emperor's New Genes," *International Studies Quarterly,* Vol. 31, No. 1 (March 1987), pp. 33–43.

123. Leda Cosmides and John Tooby, "Cognitive Adaptations for Social Exchange," in Jerome H. Barkow, Leda Cosmides, and John Tooby, eds., *The Adapted Mind: Evolutionary Psychology and the Generation of Culture* (New York: Oxford Uni-

versity Press, 1992), pp. 163–228. Also see Pinker, *How the Mind Works,* pp. 336–337, 403–405.

124. Trivers developed the idea of reciprocal altruism in his article "The Evolution of Reciprocal Altruism."

125. Notable for their work at the nexus of biology and social science, in addition to the authors noted elsewhere, are Cavalli-Sforza and Cavalli-Sforza, *The Great Human Diasporas*; Peter A. Corning, "The Biological Bases of Behavior and Some Implications for Political Science," *World Politics,* Vol. 23, No. 3 (April 1971), pp. 321–370; Jared Diamond, *Guns, Germs, and Steel: The Fates of Human Societies* (New York: Norton, 1997); and Francis Fukuyama, *The Great Disruption: Human Nature and the Reconstitution of Social Order* (New York: Free Press, 1999).

126. Thornhill and Palmer, *A Natural History of Rape.*

127. See Sarah Blaffer Hrdy and Glenn Hausfater, "Comparative and Evolutionary Perspectives on Infanticide," in Glenn Hausfater and Sarah Blaffer Hrdy, eds., *Infanticide: Comparative and Evolutionary Perspectives* (New York: Aldine, 1984), pp. xiii–xxxv; Martin Daly and Margo Wilson, *Homicide* (New York: Aldine de Gruyter, 1988); and Frank J. Sulloway, *Born to Rebel: Birth Order, Family Dynamics, and Creative Lives* (New York: Pantheon, 1996).

128. Natural sciences use multiple levels of analysis as well. To take an example from evolutionary theory, the question of "Are there genes for trait X?" may be answered at three different levels: ontogeny, heritability, and adaptation. The ontogenetic question is always affirmative because, according to Donald Symons, genes assist in the development of every trait since traits emerge from the "interaction among genes, gene products, and myriad environmental phenomena." The heritability question is: "Is any of the population variance in trait X caused by genetic variance?" This level of analysis informs the work of population and behavioral geneticists. Finally, at the adaptationist level of analysis the issue is: "Was trait X per se designed by selection to serve some function; i.e. is it an adaptation?" Donald Symons, "On the Use and Misuse of Darwinism in the Study of Human Behavior," in Barkow, Cosmides, and Tooby, eds., *The Adapted Mind,* pp. 137–159, 140–141.

129. J. David Singer, "The Level-of-Analysis Problem in International Relations," *World Politics,* Vol. 14, No. 1 (October 1961), pp. 77–92; and Kenneth N. Waltz, *Theory of International Politics* (Reading, Mass.: Addison-Wesley, 1979), pp. 6–11.

130. Steven Rose first published a major attack on alleged determinism in his edited work, *Against Biological Determinism* (London: Allison and Busby, 1982).

131. Lewontin, Rose, and Kamin, *Not in Our Genes,* p. 6.

132. Ibid.

133. Ibid., p. 7.

134. Ibid.

135. Sober, *Philosophy of Biology,* p. 192. As Allison Jolly notes, improved nutrition has allowed modern humans to increase their body mass by 50 percent and their longevity by 100 percent over the past three hundred years; as a result, "we are just today returning to the height and size of the Cro-Magnons." Allison Jolly, *Lucy's*

Legacy: Sex and Intelligence in Human Evolution (Cambridge, Mass.: Harvard University Press, 1999), p. 218.

136. Steven Pinker, *The Blank Slate: The Modern Denial of Human Nature* (New York: Viking, 2002), p. 48.

137. John Maynard Smith, "Commentary" in Gowaty, ed., *Feminism and Evolutionary Biology,* pp. 522–526, 524. For an interesting overview of this controversy, see Ullica Segerstråle, *Defenders of the Truth: The Battle for Science in the Sociology Debate and Beyond* (New York: Oxford University Press, 2000), pp. 391–396.

138. Sewall Wright, "Comments," in Paul S. Moorhead and Martin M. Kaplan, eds., *Mathematical Challenges to the Neo-Darwinian Interpretation of Evolution* (Philadelphia: Wistar Institute Press, 1967), p. 117.

139. Sober, *Philosophy of Biology,* pp. 192–193.

140. Ibid., p. 193. Emphasis in original.

141. Pinker, *The Blank Slate,* p. 113. Emphasis in original.

142. Ibid. p. 113.

143. David Hume, *A Treatise of Human Nature,* ed. L.A. Selby-Bigge (Oxford: Clarendon Press, 1964 [1739]), Book III, Part I, Section 1.

144. George Edward Moore, *Principia Ethica* (Cambridge: Cambridge University Press, 1903), pp. 10–14.

145. Pinker, *How the Mind Works,* p. 46. Francis Fukuyama argues that the naturalistic fallacy is itself fallacious and that human rights may indeed be grounded in nature. See Francis Fukuyama, "Natural Rights and Human History," *The National Interest,* No. 64 (Summer 2001), pp. 19–30.

146. Lewontin, Rose, and Kamin, *Not in Our Genes,* pp. 5–7. Also see Steven Rose, "From Causations to Translations: A Dialectical Solution to a Reductionist Enigma," in Rose, ed., *Towards a Liberatory Biology* (London: Allison and Busby, 1982), pp. 10–25.

147. For excellent responses to the creationist argument, see Niles Eldredge, *The Triumph of Evolution and the Failure of Creationism* (New York: W.H. Freeman, 2000); and Michael Ruse, *Taking Darwin Seriously: A Naturalistic Approach to Philosophy,* 2nd ed. (New York: Prometheus Books, 1998), pp. 280–297.

148. Dawkins, *The Blind Watchmaker,* p. 13. Emphasis in original. For another excellent defense of this position, see George C. Williams, "A Defense of Reductionism in Evolutionary Biology," in R. Dawkins and M. Ridley, eds., *Oxford Surveys in Evolutionary Biology,* Vol. 2 (New York: Oxford University Press, 1985), pp. 1–27. Also see Segerstråle, *Defenders of the Truth,* pp. 284–291.

149. Dawkins, *The Blind Watchmaker,* p. 13.

150. Dennett, *Darwin's Dangerous Idea,* p. 81.

151. Ibid., p. 82.

152. Dawkins, *The Selfish Gene,* p. 331. Also see Dawkins, *The Extended Phenotype,* pp. 113–114.

153. Dawkins, *The Selfish Gene,* pp. 331–332.

154. Ibid., p. 332.

155. Rosenberg, *The Structure of Biological Science,* pp. 69–120.

156. Bobbi S. Low, *Why Sex Matters: A Darwinian Look at Human Behavior* (Princeton, N.J.: Princeton University Press, 2000), p. 34.

157. Sober, *Philosophy of Biology,* p. 70. Also see Rosenberg, *The Structure of Biological Science,* pp. 126–129.

158. Sober, *Philosophy of Biology,* p. 70.

159. Stephen Jay Gould and Richard C. Lewontin, "The Spandrels of San Marco and the Panglossian Paradigm: A Critique of the Adaptationist Programme," in Elliott Sober, ed., *Conceptual Issues in Evolutionary Biology,* 2nd ed. (Cambridge, Mass.: MIT Press, 1994), pp. 73–90. For a discussion of its influence, see Jack Selzer, "Introduction," in Jack Selzer, ed., *Understanding Scientific Prose* (Madison: University of Wisconsin Press, 1993), pp. 3–19.

160. Gould has elsewhere termed this hyper-selectionism or pan-adaptationism. Stephen Jay Gould, *The Panda's Thumb* (New York: Norton, 1980), pp. 56–58.

161. For their alternative, see Gould and Lewontin, "The Spandrels of San Marco and the Panglossian Paradigm," pp. 85–89.

162. Williams, *Natural Selection,* p. 78.

163. Rudyard Kipling, *Just So Stories* (Garden City, N.Y.: Doubleday, 1952). The argument is made in Stephen Jay Gould, "Sociobiology: The Art of Storytelling," *New Scientist,* Vol. 80, No. 1129 (November 16, 1978), pp. 530–533. Gould later develops this criticism in his "Sociobiology and the Theory of Natural Selection," in George W. Barlow and James Silverberg, eds., *Sociobiology: Beyond Nature/Nurture?* (Boulder, Colo.: Westview Press, 1980), pp. 257–269. Also see Lewontin, Rose, and Kamin, *Not in Our Genes,* pp. 258–264.

164. For perhaps the most forceful and pithy response to Gould and Lewontin, see David C. Queller, "The Spaniels of St. Marx and the Panglossian Paradox: A Critique of a Rhetorical Programme," *The Quarterly Review of Biology,* Vol. 70, No. 4 (December 1995), pp. 485–489. Queller argues that the "political-cultural context" of the "Spandrels" article was "the attempted intellectual lynching of a young science, sociobiology, which at its most uppity claimed to account for human nature in ways that were distasteful to many, not the least those with Marxist inclinations [Gould and Lewontin]." Ibid., p. 486.

165. Mayr, *Toward a New Philosophy of Biology,* p. 155. For a detailed response to Gould and Lewontin, see Dawkins, *The Extended Phenotype,* chap. 3. Also see Hughes, *Adaptive Evolution of Genes and Genomes,* pp. 3–14, 222–239. For a solid analysis of Gould and Lewontin that stresses the hidden debate with adaptationism within the text—their effort to cast adaptationists as foolish and advocates of evolutionary pluralism as wise—see Charles Bazerman, "Intertextual Self-Fashioning: Gould and Lewontin's Representations of Literature," in Selzer, ed., *Understanding Scientific Prose,* pp. 20–41.

166. Williams argues that evolutionary theorists are reluctant to call something an adaptation unless firm evidence suggests it is. "Adaptation is a special and onerous concept that should be used only where it is really necessary." Williams, *Adaptation*

and Natural Selection, p. 4. For other criteria, see Mayr, *Toward a New Philosophy of Biology,* pp. 148–159.

167. Williams, *Adaptation and Natural Selection,* p. 6.

168. Ibid.

169. Dennett, *Darwin's Dangerous Idea,* p. 247.

170. Williams, *Adaptation and Natural Selection,* pp. 9–10.

171. Dennett, *Darwin's Dangerous Idea,* pp. 247–248.

172. Darwin, *On the Origin of Species,* p. 201.

173. George G. Simpson, *The Meaning of Evolution* (New Haven: Yale University Press, 1949). Quoted in Dobzhansky, "On Some Fundamental Concepts of Darwinian Biology," in Theodosius Dobzhansky, Max K. Hecht, and William C. Steere, eds., *Evolutionary Biology,* Vol. 2 (New York: Appleton-Century-Crofts, 1968), p. 31.

174. Ibid.

175. Mayr, *Toward a New Philosophy of Biology,* pp. 153, 105.

176. Ibid., p. 106.

177. Evidence of the comet's impact some 250 million years ago is based on analysis of molecules (buckyballs) containing trapped gases that could only have come from space. Scientists postulate that such an impact probably also caused volcanic eruptions that further affected life on earth. See Kenneth Chang, "Scientists Find Signs of Meteor Crash That Led to Extinctions in Era before Dinosaurs," *New York Times,* February 23, 2001, <http://www.nytimes.com/2001/02/23/science/23EXTI.html/>; and Guy Gugliotta, "Comet Tied to Mass Extinction," *Washington Post,* February 23, 2001, p. A3. For a thorough analysis of the effect of the "big five" mass extinctions in the (1) Late Ordovician, (2) Late Devonian, (3) Late Permian, (4) Late Triassic, and (5) Late Cretaceous, see A. Hallam and P.B. Wignall, *Mass Extinctions and Their Aftermath* (New York: Oxford University Press, 1997). Also see Mayr, *What Evolution Is,* pp. 201–203.

178. Mayr, *Toward a New Philosophy of Biology,* p. 106.

179. François Jacob, "Evolution and Tinkering," *Science,* Vol. 196 (June 10, 1977), pp. 1161–1166, 1163. He masterfully develops this idea further in his *Of Flies, Mice, and Men,* trans. by Giselle Weiss (Cambridge, Mass.: Harvard University Press, 1998).

180. Dennett, *Darwin's Dangerous Idea,* pp. 267–282. For a detailed discussion of the architectural merit of a spandrel and its merit as a metaphor in evolutionary thought, see Robert Mark, "Architecture and Evolution," *The American Scientist,* Vol. 84, No. 4 (July–August 1996), pp. 383–389.

181. Williams, *Natural Selection,* p. 78.

182. Gould, *The Panda's Thumb,* pp. 57–58.

183. Williams, *Natural Selection,* p. 79.

184. George C. Williams, *The Pony Fish's Glow: And Other Clues to Plan and Purpose in Nature* (New York: Basic Books, 1997), p. 17.

185. Ibid.

186. Ibid., pp. 17–19.

187. Ibid., p. 19.

188. Ibid. Emphasis in original.
189. Ibid.
190. Mayr, *Toward a New Philosophy of Biology,* p. 151.
191. Ibid.
192. Ibid., p. 130.

2. Evolutionary Theory, Realism, and Rational Choice

1. While I focus on these theories, evolutionary theory may inform much more of social science and philosophy because it can account for human physiology, including emotions. Thus psychology, sociology, ethical philosophy, and epistemology may benefit from an understanding of how the brain (e.g. the hypothalamus and limbic system) works and how the human emotions—love, hate, fear, guilt, and compassion—are created.

2. Robert L. Trivers, "The Evolution of Reciprocal Altruism," *The Quarterly Review of Biology,* Vol. 46, No. 1 (March 1971), pp. 35–57. Also see Robert Trivers, *Social Evolution* (Menlo Park, Calif.: Benjamin/Cummings Publishing, 1985), pp. 386–394.

3. The principal works of the realist theory of international relations include Raymond Aron, *Peace and War: A Theory of International Relations,* trans. by Richard Howard and Annette Baker Fox (Garden City, N.Y.: Doubleday, 1966); E.H. Carr, *The Twenty Years' Crisis, 1919–1939: An Introduction to the Study of International Relations,* 2nd ed. (London: Macmillan, 1946); Robert Gilpin, *U.S. Power and the Multinational Corporation: The Political Economy of Foreign Direct Investment* (New York: Basic Books, 1975); Gilpin, *War and Change in World Politics* (Cambridge: Cambridge University Press, 1981); Gilpin, "The Richness of the Tradition of Political Realism," in Robert O. Keohane, ed., *Neorealism and Its Critics* (New York: Columbia University Press, 1986), pp. 301–321; George F. Kennan, *Realities of American Foreign Policy* (Princeton, N.J.: Princeton University Press, 1954); Friedrich Meinecke, *Machiavellism: The Doctrine of Raison d'Etat and Its Place in Modern History* (Boulder, Colo.: Westview, 1984) [originally published as *Die Idee der Staatsräson* (Munich: R. Oldenbourg Verlag, 1924)]; Hans J. Morgenthau, *Scientific Man vs. Power Politics* (Chicago: University of Chicago Press, 1946); Morgenthau, *Politics among Nations: The Struggle for Power and Peace,* 5th ed., rev. (New York: Knopf, 1978); and Martin Wight, *Power Politics* (London: Royal Institute of International Affairs, 1946). An overview of realist thought is Michael Joseph Smith, *Realist Thought from Weber to Kissinger* (Baton Rouge: Louisiana State University Press, 1986).

4. Reinhold Niebuhr, *The Nature and Destiny of Man: A Christian Interpretation,* 2 Vols. (New York: Charles Scribner's Sons, 1941, 1943), Vol. 1, p. 179.

5. Ibid., Vol. 1, pp. 178–179.

6. Reinhold Niebuhr, *Faith and History: A Comparison of Christian and Modern Views of History* (London: Nisbet, 1938), p. 9. See also Niebuhr, *The Children of Light and the Children of Darkness: A Vindication of Democracy and a Critique of Its Traditional Defence* (New York: Charles Scribner's Sons, 1944), p. 55.

7. Niebuhr, *The Nature and Destiny of Man,* Vol. 1, p. 189.

8. Reinhold Niebuhr, *Christianity and Power Politics* (New York: Charles Scribner's Sons, 1940) p. 156.

9. Ibid., pp. 156–157. As Niebuhr observes, "All of his intellectual and cultural pursuits . . . become infected with the sin of pride. Man's pride and will-to-power disturb the harmony of creation." Niebuhr, *The Nature and Destiny of Man,* Vol. 1, p. 179.

10. This is true even for a liberal democracy such as the United States. See Niebuhr, *The Children of Light and the Children of Darkness,* pp. 20–21, 183–186. Kenneth N. Waltz also makes this point in *Man, the State, and War: A Theoretical Analysis* (New York: Columbia University Press, 1959), pp. 18ff.

11. Niebuhr, *Christianity and Power Politics,* p. 26.

12. Ibid., p. 47.

13. Classical realism contains many assumptions that are not addressed here, such as: states desire survival; states are the key actors in international relations; and the nature of international relations is inherently one of conflict.

14. Demonstrating the brutality and prevalence of war in prehistoric times is Lawrence H. Keeley, *War before Civilization: The Myth of the Peaceful Savage* (New York: Oxford University Press, 1996). See also Carol R. Ember, "Myths about Hunter-Gatherers," *Ethnology,* Vol. 17, No. 4 (October 1978), pp. 439–448; and Harry H. Turney-High, *Primitive War: Its Practice and Concepts* (Columbia: University of South Carolina Press, 1949).

15. Thomas Hobbes, *Leviathan,* ed. C.B. Macpherson (Harmondsworth: Penguin, 1985 [1651]), chap. 11, p. 161.

16. Ibid., chap. 11, p. 161. See also Arnold Wolfers for a discussion of the human desire for power, *Discord and Collaboration: Essays in International Politics* (Baltimore, Md.: Johns Hopkins University Press, 1965), pp. 82–85.

17. Hobbes, *Leviathan,* chap. 13, p. 186.

18. Morgenthau, *Scientific Man vs. Power Politics,* p. 192.

19. Ibid., p. 193.

20. Morgenthau, *Politics among Nations,* p. 37.

21. Quoted in Morgenthau, *Scientific Man vs. Power Politics,* p. 193.

22. Here, I am concerned only with the minimal essential human traits necessary to construct the realist argument. I am not making claims about what individual realists do or should do. It is only important to my argument that realists consider these traits significant.

23. Richard D. Alexander makes precisely this point in *Darwinism and Human Affairs* (Seattle: University of Washington Press, 1979), Figure 4, p. 44. For a discussion of egoism, see Elliott Sober and David Sloan Wilson, *Unto Others: The Evolution and Psychology of Unselfish Behavior* (Cambridge, Mass.: Harvard University Press, 1998), pp. 2, 203–204, 224–227.

24. Morgenthau, *Politics among Nations,* pp. 4, 37.

25. For a discussion of the theoretical differences between realism and neorealism, see Kenneth N. Waltz, "Realist Thought and Neorealist Theory," *Journal of Interna-*

tional Affairs, Vol. 44, No. 1 (Spring/Summer 1990), pp. 21–37. Also valuable is Vincent S.E. Falger, "Human Nature in Modern International Relations: Part I. Theoretical Backgrounds," in Albert Somit and Steven A. Peterson, eds., *Research in Biopolitics,* Vol. 5: *Recent Explorations in Biology and Politics* (Greenwich, Conn.: JAI Press, 1997), pp. 155–175.

26. Kenneth N. Waltz, *Theory of International Politics* (Reading, Mass.: Addison-Wesley, 1979), pp. 114–116.

27. For a good introduction to structuralism, see Peter Caws, *Structuralism: The Art of the Intelligible* (London: Humanities Press International, 1988); and Richard Harlan, *Superstructuralism: The Philosophy of Structuralism and Post-Structuralism* (London: Routledge, 1987), pp. 11–32. For Waltz's discussion, see *Theory of International Politics,* pp. 73–74, 77–101.

28. Waltz, *Theory of International Politics,* pp. 161–193.

29. Mearsheimer also argues that because states have offensive military capabilities and can never be certain of the intentions of other states, they are always afraid of each other. John J. Mearsheimer, *The Tragedy of Great Power Politics* (New York: Norton, 2001), p. 3.

30. Mearsheimer, *The Tragedy of Great Power Politics,* pp. 338–344.

31. Ibid., pp. 1–8. For Waltz's argument, see *Theory of International Politics,* pp. 161–193.

32. Mearsheimer, *The Tragedy of Great Power Politics,* p. 21.

33. For an analysis of offensive realism and an alternative termed "defensive realism," see Sean M. Lynn-Jones, "Realism and America's Rise: A Review Essay," *International Security,* Vol. 23, No. 2 (Fall 1998), pp. 157–182. An exceptional study of realism, and in some respects the fountainhead of offensive realism, is Ashley Joachim Tellis, "The Drive to Domination: Towards a Pure Realist Theory of Politics," (Chicago: Ph.D. diss., University of Chicago, 1994).

34. The leading proponents of offensive realism are Christopher Layne and John Mearsheimer. Layne distinguishes between two types of offensive realism. In Type I, as a state's relative power increases, the scope of its external interests—and the definition of threats to those interests—expands. In Type II, hegemony is the optimal grand strategy for assuring the state's security. See Christopher Layne, *The Peace of Illusions: International Relations Theory and American Grand Strategy in the Post–Cold War Era* [tentative title] (Ithaca, N.Y.: Cornell University Press, forthcoming), chap. 2; Mearsheimer, "The False Promise of International Institutions," *International Security,* Vol. 19, No. 3 (Winter 1994/95), pp. 5–49; and Mearsheimer, *The Tragedy of Great Power Politics,* pp. 1–167.

35. Eric J. Labs, "Beyond Victory: Offensive Realism and the Expansion of War Aims," *Security Studies,* Vol. 6, No. 4 (Summer 1997), pp. 1–49, 12.

36. Fareed Zakaria proposes a variation of the offensive realist argument, which he terms "state-centered realism," that is significantly informed by neorealism. He submits that in creating a foreign policy a great power's intentions will be shaped by its capabilities, but he also recognizes that "state structure limits the availability of

national power." Zakaria thus combines unit-level and systemic causes in his theory. Fareed Zakaria, *From Wealth to Power: The Unusual Origins of America's World Role* (Princeton, N.J.: Princeton University Press, 1998), p. 9.

37. I do not want to suggest that my argument in this chapter is the limit of how evolutionary biology may assist realism. For example, Jennifer Sterling-Folker provides an exceptional analysis of evolutionary theory's usefulness for the ontology of realism in her "Realism and the Constructivist Challenge: Rejecting, Reconstructing, or Rereading," *International Studies Review,* Vol. 4, No. 1 (Spring 2002), pp. 73–97.

38. Charles Darwin, *On the Origin of Species by Means of Natural Selection, or the Preservation of Favoured Races in the Struggle for Life* (Cambridge, Mass.: Harvard University Press, 1964 [1859]), p. 402.

39. These are common standards but certainly not the only ones; other have been developed by Roy Bhaskar, Thomas Kuhn, and Imre Lakatos. I discuss the D-N model and Popper's falsification because they are commonly accepted. If one of these alternative standards were used it would not affect the result, so the use of Hempel's and Popper's models will be sufficient to demonstrate that evolutionary theory provides superior ultimate causation for realism than either evil or *animus dominandi.*

40. See Carl G. Hempel, *Philosophy of Natural Science* (Englewood Cliffs, N.J.: Prentice-Hall, 1966), pp. 49–54; and *Aspects of Scientific Explanation,* pp. 174, 232–233, 247–249.

41. Karl R. Popper, *The Logic of Scientific Discovery* (London: Hutchinson, 1959), pp. 40–42, 78–92, 419.

42. Elliott Sober, *Philosophy of Biology* (Boulder, Colo.: Westview, 1993), p. 48.

43. Karl Popper, "Intellectual Autobiography," in Paul Arthur Schilpp, ed., *The Philosophy of Karl Popper,* 2 Vols. (La Salle, Ill.: Open Court, 1974), Vol. 1, pp. 3–181, 134; and Popper, *Objective Knowledge,* rev. ed. (Oxford: Clarendon Press, 1979), pp. 256–280.

44. Popper's idea of falsifiability also has some significant problems that I cannot address here. See Sober, *Philosophy of Biology,* pp. 48–52.

45. Karl Popper, "Natural Selection and the Emergence of Mind," *Dialectica,* Vol. 32, Nos. 3–4 (1978), pp. 339–355, 344–345; and Popper, "Evolution," letter, *New Scientist,* Vol. 87, No. 1215 (August 21, 1980), p. 611.

46. Mary B. Williams, "Falsifiable Prediction of Evolutionary Theory," *Philosophy of Science,* Vol. 40, No. 4 (December 1973), pp. 518–537. Also see Susan K. Mills and John Beatty, "The Propensity Interpretation of Fitness," in Elliott Sober, ed., *Conceptual Issues in Evolutionary Biology* (Cambridge, Mass.: MIT Press, 1994), pp. 3–23, 9; M. Ruse, "Confirmation and Falsification of Theories in Evolution," *Scientia,* Vol. 104, Nos. 7–8 (1969), pp. 329–357; Sober, *Philosophy of Biology,* p. 47; and Mary B. Williams, "The Logical Status of the Theory of Natural Selection and Other Evolutionary Controversies," in Mario Bunge, ed., *The Methodological Unity of Science* (Dordrecht, Holland: Reidel, 1973), pp. 84–102.

47. Lamarckism is the belief that inheritable characteristics can be formed in direct response to the organism's needs in the environment. For example, giraffes have

long necks to reach high into the trees. Saltationism submits that evolutionary change can occur instantaneously though dramatic "macromutations" or evolutionary "jumps" or "leaps," such as reptiles evolving into mammals in one generation. Creationism is the belief that humans were created by God as explained in the Book of Genesis. Orthogenesis is a purported evolutionary force, trend, or momentum internal to the organism that allegedly drives features beyond adaptive advantage. The teeth of the saber-toothed tiger, horns of the Irish elk, and the size of certain dinosaurs have been cited as examples of orthogenesis. Ernst Mayr, *What Evolution Is* (New York: Basic Books, 2001), pp. 74–82.

48. Peter J. Bowler provides an excellent account of this struggle in *Evolution: The History of an Idea,* rev. ed. (Berkeley: University of California Press, 1989).

49. R.A. Fisher advances this argument in *The Genetical Theory of Natural Selection,* 2nd ed., rev. (New York: Dover, 1958). For a discussion demonstrating the testability of evolutionary theory, see Elliott Sober, "Six Sayings about Adaptationism," in David L. Hull and Michael Ruse, eds., *The Philosophy of Biology* (New York: Oxford University Press, 1998), pp. 80–82.

50. For an assessment of this threat, see *The Global Infectious Disease Threat and Its Implications for the United States,* National Intelligence Estimate NIE 99–17D, January 2000, available at <http://www.odci.gov/cia/publications/nie/report/nie99–17d.html>. An excellent and worrying account of the arms race between humans and these diseases is found in Stephen R. Palumbi, *The Evolution Explosion: How Humans Cause Rapid Evolutionary Change* (New York: Norton, 2001).

51. Gina Kolata, "Kill All the Bacteria!" *New York Times,* January 7, 2001, <http://www.nytimes.com/2001/01/07/weekinreview/07KOLA.html>.

52. Ernst Mayr, *This Is Biology: The Science of the Living World* (Cambridge, Mass.: Harvard University Press, 1997), p. 191.

53. Although, as I discuss below, inclusive fitness may modify this argument.

54. Richard Dawkins, *The Selfish Gene,* new ed. (Oxford: Oxford University Press, 1989), p. 258. Dawkins has been criticized for his argument that natural selection occurs at the level of the gene. For his convincing reply emphasizing a nested hierarchy of natural selection, see his *The Extended Phenotype: The Long Reach of the Gene,* rev. ed. (New York: Oxford University Press, 1999), pp. 112–114, 239.

55. J.B.S. Haldane, "Population Genetics," *New Biology,* Vol. 18 (1955) pp. 34–51, 44.

56. These are general categories of behavior and not the alpha and omega of ethological behavior. Christopher Boehm argues elegantly that humans are disposed to resent being dominated, and at least among humans, submissive individuals may only be biding their time until they see an opportunity to escape from their submissive status. See Christopher Boehm, *Hierarchy in the Forest: The Evolution of Egalitarian Behavior* (Cambridge, Mass.: Harvard University Press, 1999), pp. 237–238.

57. Dorothy L. Cheney and Robert M. Seyfarth, *How Monkeys See the World: Inside the Mind of Another Species* (Chicago: University of Chicago Press, 1990), p. 35; Alexander H. Harcourt and Frans B.M. de Waal, eds., *Coalitions and Alliances in Humans and Other Animals* (Oxford: Oxford University Press, 1992); de Waal, *Chimpan-*

zee Politics: Power and Sex among Apes, rev. ed. (Baltimore, Md.: Johns Hopkins University Press, 1998).

58. For an excellent general analysis of the genetic origins of aggression and its chemical mediators in humans, such as the hormone testosterone, its derivative dihydroxytestosterone (DHT), neurotransmitters such as serotonin, and some of the differences in behavior caused by these factors in men and women, see William R. Clark and Michael Grunstein, *Are We Hardwired? The Role of Genes in Human Behavior* (New York: Oxford University Press, 2000), pp. 159–175.

59. Richard Wrangham and Dale Peterson, *Demonic Males: Apes and the Origins of Human Violence* (Boston: Houghton Mifflin, 1996), p. 199.

60. Ibid. Also see Konrad Lorenz, *On Aggression,* trans. by Marjorie Kerr Wilson (New York: Harcourt, Brace and World, 1966); and Charles J. Lumsden and E.O. Wilson, *Promethean Fire: Reflections on the Origin of Mind* (Cambridge, Mass.: Harvard University Press, 1983), p. 23.

61. David P. Barash, *Sociobiology and Behavior* (New York: Elsevier, 1977), p. 237. In this respect, animal behavior is like deterrence and coercion in international relations. Animals, like states, signal their intentions in efforts to deter and coerce. As Waltz notes, "Force is least visible where power is most fully and most adequately present." *Theory of International Politics,* p. 185.

62. See Joseph Lopreato, *Human Nature and Biocultural Evolution* (Boston: Allen and Unwin, 1984), pp. 161–176; Donald L. McEachron and Darius Baer, "A Review of Selected Sociobiological Principles: Application to Hominid Evolution II. The Effects of Intergroup Conflict," *Journal of Social and Biological Structures,* Vol. 5, No. 2 (April 1982), pp. 121–139, 122–126; K.E. Moyer, "The Biological Basis of Dominance and Aggression," in Diane McGuinness, ed., *Dominance, Aggression and War* (New York: Paragon House, 1987), pp. 1–34; E.O. Wilson, *Sociobiology: The New Synthesis* (Cambridge, Mass.: Harvard University Press, 1975), p. 287; and Fred H. Willhoite Jr., "Primates and Political Authority," *American Political Science Review,* Vol. 70, No. 4 (December 1976), pp. 1110–1126, 1118–1120. Stanley Milgram also notes the importance of what he terms "dominance structures." He argues that the "potential for obedience is the prerequisite of . . . social organization." Milgram, *Obedience to Authority: An Experimental View* (New York: Harper and Row, 1974), p. 124. Many in his experiments defined themselves as "open to regulation by a person of higher status. In this condition the individual no longer views himself as responsible for his own actions but defines himself as an instrument for carrying out the wishes of others." Milgram, *Obedience to Authority,* p. 134.

63. In this respect, international relations resembles animal behavior. As an alpha male provides stability to the group, so too a hegemon in international relations, as many scholars recognize, may provide stability for lesser states in both the realm of international security and for international political economy. On the importance of resource harvesting for the development of dominance hierarchies, see James L. Boone, "Competition, Conflict, and the Development of Social Hierarchies," in Eric Alden

Smith and Bruce Winterhalder, eds., *Evolutionary Ecology and Human Behavior* (New York: Aldine de Gruyter, 1992), pp. 301–337.

64. An excellent overview of much of the literature is found in Bruce M. Knauft, "Violence and Sociality in Human Evolution," *Current Anthropology*, Vol. 32, No. 4 (August–October 1991), pp. 391–409.

65. Donald McEachron and Darius Baer discuss the origin of intelligence in humans and argue that conflict in early human history selected for intelligence, which led people to develop better weapons, which in turn selected for even greater intelligence. McEachron and Baer, "A Review of Selected Sociobiological Principles," pp. 121–139. Also making the argument that warfare contributed to the evolution of human intelligence is Roger Pitt, "Warfare and Hominid Brain Evolution," *Journal of Theoretical Biology*, Vol. 72, No. 3 (June 1978), pp. 551–575.

66. Denise Dellarosa Cummins, "Social Norms and Other Minds," in Cummins and Colin Allen, eds., *The Evolution of Mind* (New York: Oxford University Press, 1998), p. 30.

67. Ibid., p. 37, emphasis omitted. Also see Richard D. Alexander, *The Biology of Moral Systems* (New York: Aldine de Gruyter, 1987), pp. 114–117. R.A. Foley emphasizes the role of foraging for resources in the development of human intelligence. R.A. Foley, "Evolutionary Ecology of Fossil Hominids," in Smith and Winterhalder, eds., *Evolutionary Ecology and Human Behavior*, pp. 131–164.

68. E.W. Menzel, "A Group of Chimpanzees in a 1–Acre Field," in A.M Schrier and F. Stollnitz, eds., *Behavior of Nonhuman Primates* (New York: Academic Press, 1974), pp. 83–153. Quoted in Cummins, "Social Norms and Other Minds," p. 37.

69. Barash, *Sociobiology and Behavior*, pp. 239–240.

70. Cummins, "Social Norms and Other Minds," p. 30.

71. For example, one study showed subjects pictures of males with accompanying brief biographies that imparted information about each man's social status and character. The researchers found that subjects were far better at remembering low-status cheaters than high-status cheaters or noncheaters of any rank. Linda Mealey, Christopher Daood, and Michael Krage, "Enhanced Memory for Faces of Cheaters," *Ethology and Sociobiology*, Vol. 17, No. 2 (March 1996), pp. 119–128.

Another study revealed that increases in blood pressure associated with anger or frustration in social situations can be eliminated if individuals are given an opportunity to aggress against the person who caused their distress, but they usually do so only if the target is of lower status. If the target of aggression is of higher status, individuals typically do not aggress and blood pressure remains at the elevated level caused by frustration. Jack E. Hokanson, "The Effect of Frustration and Anxiety on Overt Aggression," *Journal of Abnormal and Social Psychology*, Vol. 62, No. 2 (March 1961), pp. 346–351; and Hokanson and Sanford Shetler, "The Effect of Overt Aggression on Physiological Arousal," *Journal of Abnormal and Social Psychology*, Vol. 63, No. 2 (September 196[illegible]), pp. 446–448.

72. Wilson, *Sociobiology*, p. 562. Emphasis in original. See also Lopreato, *Human*

Nature and Biocultural Evolution, pp. 177–186; and Albert Somit and Steven A. Peterson, *Darwinism, Dominance, and Democracy: The Biological Bases of Authoritarianism* (Greenwich, Conn.: Praeger, 1997), pp. 77–84.

73. Donald T. Campbell, "On the Genetics of Altruism and the Counter-Hedonic Components in Human Culture," *Journal of Social Issues,* Vol. 28, No. 3 (1972), pp. 21–37.

74. Alexander, *Darwinism and Human Affairs,* p. 64.

75. Campbell, "On the Genetics of Altruism and the Counter-Hedonic Components in Human Culture," pp. 21–37; Irenäus Eibl-Eibesfeldt, *Human Ethology* (New York: Aldine de Gruyter, 1989); Frank Kemp Salter, "Indoctrination as Institutionalized Persuasion," in Irenäus Eibl-Eibesfeldt and Frank Kemp Salter, eds., *Indoctrinability, Ideology, and Warfare: Evolutionary Perspectives* (New York: Berghahn Books, 1998), pp. 421–452; and Wilson, *Sociobiology,* p. 562. Roger D. Masters explores the biological foundation of political organization fully in his *The Nature of Politics* (New Haven: Yale University Press, 1989). Of course, evolution does not inform the specific form of the political organization, ideology, or religion.

76. See Robert Bigelow, *The Dawn Warriors: Man's Evolution toward Peace* (Boston: Little, Brown, 1969). Bigelow describes the relationship between external threat and cooperation. He argues that the need for defense against man's deadliest opponent, other humans, was instrumental for creating group cooperation, what he terms "cooperation-for-conflict," and also served to eliminate those groups who cooperated least well.

77. Masters, *The Nature of Politics.* See also Willhoite, "Primates and Political Authority," 1118–1123. Aspects of international relations may also be conceived of as a dominance hierarchy. Think of hegemony and challenges to it. The hegemon is the alpha male, and challengers are younger males seeking to supplant him.

78. There are simply too few studies of Bonobo behavior to reach solid conclusions. For an analysis of the Bonobo's hierarchy, see Frans B.M. de Waal, *Bonobo: The Forgotten Ape* (Berkeley: University of California Press, 1997).

79. Milgram acknowledges the importance of the evolutionary process for obedience: "we are born with a *potential* for obedience, which then interacts with the influence of society to produce the obedient man." The capacity for obedience, he argues, is like the capacity for language, in that both mental structures and a social milieu must be present. "In explaining the causes of obedience, we need to look both at the inborn structures and . . . social influences. . . . The proportion of influence exerted by each is a moot point. From the standpoint of evolutionary survival, all that matters is that we end up with organisms that can function in hierarchies." Milgram, *Obedience to Authority,* p. 125. Emphasis in original. For a detailed description of Milgram's experiments, see Arthur G. Miller, *The Obedience Experiments: A Case Study of Controversy in Social Science* (New York: Praeger, 1986).

80. Frank J. Sulloway, *Born to Rebel: Birth Order, Family Dynamics, and Creative Lives* (New York: Pantheon, 1996).

81. Although my focus is international relations, these motivations are present in domestic politics as well.

82. Process-tracing methodology is perhaps the best methodology to use to create such a focused empirical test. On process-tracing methodology, see Alexander L. George and Timothy J. McKeown, "Case Studies and Theories of Organizational Decision Making," in Robert F. Coulam and Richard A. Smith, eds., *Advances in Information Processing in Organizations: A Research Annual* (Greenwich, Conn.: JAI Press, 1985), Vol. 2, pp. 34–41; and Stephen Van Evera, *Guide to Methods for Students of Political Science* (Ithaca, N.Y.: Cornell University Press, 1997), pp. 64ff.

83. I recognize that rational choice theory is a diverse theory that has generated a vast literature. However, it is centered on core ideas, from which powerful analytical tools have been developed to explain significant political and economic behavior. My intent here, as with realism, is to focus on its core ideas rather than on the major debates within the theory, or specific areas of study, in order to explain how evolutionary theory can support it.

84. This definition of rational choice theory draws on Jon Elster, "Introduction" in Elster, ed., *Rational Choice* (New York: New York University Press, 1986), pp. 1–33, 1–2, 4, and 12–16; and James D. Morrow, *Game Theory for Political Scientists* (Princeton, N.J.: Princeton University Press, 1994), pp. 7–8. Also useful is Elster, "When Rationality Fails," in Karen Schweers Cook and Margaret Levi, eds., *The Limits of Rationality* (Chicago: University of Chicago Press, 1990), pp. 19–51.

85. Jon Elster, *Nuts and Bolts for the Social Sciences* (New York: Cambridge University Press, 1989), p. 22.

86. Elster, "Introduction," in Elster, ed., *Rational Choice,* p. 27.

87. Elster, *Nuts and Bolts for the Social Sciences,* p. 28. But Elster acknowledges that doing "well for oneself" need not always be equated with egoism. See his discussion in chapter 6 of Ibid., pp. 52–60.

88. As Kenneth Arrow writes, rational behavior is identified with "maximization of some sort." Kenneth J. Arrow, *Social Choice and Individual Values* (New Haven: Yale University Press, 1951), p. 14. Another question is whether individuals maximize utility or, as Herbert Simon argued, they "satisfice," that is, they limit themselves to what they believe to be sufficient or satisfactory. Herbert A. Simon, *Administrative Behavior: A Study of Decision-Making Processes in Administrative Organization,* 3rd ed. (New York: Free Press, 1976), pp. xxviii–xxxi and 38–41.

89. George Stigler and Gary S. Becker, "De Gustibus Non Est Disputandum," in Cook and Levi, eds., *The Limits of Rationality,* pp. 191–221.

90. Mancur Olson, *The Logic of Collective Action: Public Goods and the Theory of Groups* (Cambridge, Mass.: Harvard University Press, 1965).

91. Ibid., p. 2.

92. Russell Hardin, *Collective Action* (Baltimore, Md.: Johns Hopkins University Press, 1982), p. 9. For a similarly detailed discussion, see Jon Elster, *The Cement of Society: A Study of Social Order* (New York: Cambridge University Press, 1989).

93. Gary Becker, "The Economic Approach to Human Behavior," in Jon Elster, ed., *Rational Choice* (New York: New York University Press, 1986), pp. 108–122, 110.

94. For classic analyses of voting behavior, see James M. Buchanan and Gordon

Tullock, *The Calculus of Consent: Logical Foundations of Constitutional Democracy* (Ann Arbor: University of Michigan Press, 1962); and Anthony Downs, *An Economic Theory of Democracy* (New York: Harper and Row, 1957). A similarly important study for deterrence is Thomas C. Schelling, *Strategy of Conflict* (Cambridge, Mass.: Harvard University Press, 1960). On crisis bargaining, see James D. Fearon, "Domestic Political Audiences and the Escalation of International Disputes," *American Political Science Review,* Vol. 88, No. 3 (September 1994), pp. 577–592. Fearon also provides an insightful analysis of the spread of ethnic conflict, "Commitment Problems and the Spread of Ethnic Conflict," in David A. Lake and Donald Rothchild, eds., *The International Spread of Ethnic Conflict* (Princeton, N.J.: Princeton University Press, 1998), pp. 107–126. For insights into institutional behavior, see James D. Morrow, "Modeling International Institutions," *International Organization,* Vol. 48, No. 3 (Summer 1994), pp. 387–423. Robert Powell addresses relative gains in "Absolute and Relative Gains in International Relations," *American Political Science Review,* Vol. 85, No. 4 (December 1991), pp. 1303–1320.

95. The scholarship of Elster, Hardin, and Olson cited above provides the best analysis of these problems. Group conflict is especially well studied in Hardin, *One for All: The Logic of Group Conflict* (Princeton, N.J.: Princeton University Press, 1995). For a solid analysis of individual and group behavior see Thomas C. Schelling, *Micromotives and Macrobehavior* (New York: Norton, 1978).

96. Amartya K. Sen, "Rational Fools," in Jane Mansbridge, ed., *Beyond Self-Interest* (Chicago: University of Chicago Press, 1990), pp. 25–43, 30. Similar critiques are made by Robert H. Frank, *Passions within Reason: The Strategic Role of the Emotions* (New York: Norton, 1988); and Edmund S. Phelps, ed., *Altruism, Morality, and Economic Theory* (New York: Russell Sage, 1975).

97. Donald P. Green and Ian Shapiro, *The Pathology of Rational Choice: A Critique of Applications in Political Science* (New Haven: Yale University Press, 1994), p. 9.

98. Stephen M. Walt, "Rigor or Rigor Mortis? Rational Choice and Security Studies," *International Security,* Vol. 23, No. 4 (Spring 1999), pp. 5–48. See the replies of several users of rational choice as well as Walt's reply in the symposium: "Formal Methods, Formal Complaints: Debating the Role of Rational Choice in Security Studies," *International Security,* Vol. 24, No. 2 (Fall 1999), pp. 56–130.

99. Amos Tversky and Daniel Kahneman, "Judgment under Uncertainty: Heuristics and Biases," in Daniel Kahneman, Paul Slovic, and Amos Tversky, eds., *Judgment under Uncertainty: Heuristics and Biases* (New York: Cambridge University Press, 1982), pp. 3–20.

100. Geoffrey Brennan, "Comment: What Might Rationality Fail to Do?" in Cook and Levi, eds., *The Limits of Rationality,* pp. 51–59, 53.

101. Eric Alden Smith and Bruce Winterhalder, "Natural Selection and Decision-Making," in Smith and Winterhalder, eds., *Evolutionary Ecology and Human Behavior,* pp. 25–60; and Edward O. Wilson, *Consilience: The Unity of Knowledge* (New York: Knopf, 1998), pp. 212–222.

102. See Armen A. Alchian, "Uncertainty, Evolution, and Economic Theory,"

The Journal of Political Economy, Vol. 58, No. 3 (June 1950), pp. 211–221; and Milton Friedman, *Essays in Positive Economics* (Chicago: University of Chicago Press, 1953), p. 22.

103. Corning argues for a new discipline, "bioeconomics," that studies how organisms acquire and use resources to meet biological needs. Peter A. Corning, "Biopolitical Economy: A Trail-Guide for an Inevitable Discipline," in Albert Somit and Steven A. Peterson, eds., *Research in Biopolitics,* Vol. 5 (Greenwich, Conn.: JAI Press, 1997), pp. 247–277; Richard R. Nelson and Sidney G. Winter, *An Evolutionary Theory of Economic Change* (Cambridge, Mass.: Belknap Press of Harvard University Press, 1982).

104. Adam Smith, *The Theory of Moral Sentiments* (Indianapolis, Ind.: Liberty Fund, 1982 [1759]) p. 135. While Smith recognizes the importance of egoism, Ronald Coase makes clear that his view of human nature is informed by other sentiments, such as benevolence, but benevolence, as Coase writes, "put in its place." R.H. Coase, "Adam Smith's View of Man," *The Journal of Law and Economics,* Vol. 19, No. 3 (October 1976), pp. 529–546, 542.

105. Adam Smith, *An Inquiry into the Nature and Causes of the Wealth of Nations* (Chicago: University of Chicago Press, 1976 [1776]), p. 18.

106. Robert Boyd and Peter J. Richardson, "Sociobiology, Culture and Economic Theory," *Journal of Economic Behavior and Organization,* Vol. 1, No. 2 (June 1980), pp. 97–121, 101.

107. Patricia S. Churchland, *Neurophilosophy* (Cambridge, Mass.: Bradford/MIT Press, 1986), pp. 13–14.

108. Steven Pinker, *How the Mind Works* (New York: Norton, 1997), p. 191. On the importance of color, also see Roger N. Shepard, "The Perceptual Organization of Colors: An Adaptation to Regularities of the Terrestrial World?" in Jerome H. Barkow, Leda Cosmides, and John Tooby, eds., *The Adapted Mind: Evolutionary Psychology and the Generation of Culture* (New York: Oxford University Press, 1992), pp. 495–532.

109. Pinker, *How the Mind Works,* p. 191.

110. Ibid. Emphasis in original.

111. Ibid., p. 194.

112. Ibid.

113. Ibid. Also see John Tooby and Irven DeVore, "The Reconstruction of Hominid Evolution through Strategic Modeling," Warren G. Kinzey, ed., *The Evolution of Human Behavior: Primate Models* (Albany, N.Y.: SUNY Press, 1987), pp. 183–237.

114. Leda Cosmides and John Tooby, "Better Than Rational: Evolutionary Psychology and the Invisible Hand," *The American Economic Review,* Vol. 84, No. 2 (May 1994), pp. 327–332.

115. Note that none of these are uniquely human: dolphins (and perhaps whales and porpoises) are now thought to have language ability. See Curt Suplee, "Dolphins May Communicate Individually," *Washington Post,* August 25, 2000, p. A3. Discussing animal cognition more broadly is Donald R. Griffin, *Animal Minds* (Chicago: University of Chicago Press, 1992).

116. Natalie Angier, "For Monkeys, a Millipede a Day Keeps Mosquitoes Away," *New York Times,* December 5, 2000, <http://www.nytimes.com/2000/12/05/science/05MONK.html>.

117. Lee Alan Dugatkin, *The Imitation Factor: Evolution beyond the Gene* (New York: Free Press, 2000), pp. 170–172.

118. The scholarship of Philip Kellman and Elizabeth Spelke is particularly important on this issue. See Renée Baillargeon, Elizabeth S. Spelke, and Stanley Wasserman, "Object Permanence in Five-Month-Old Infants," *Cognition,* Vol. 20, No. 3 (April 1985), pp. 191–208; Philip J. Kellman, Henry Gleitman, and Elizabeth S. Spelke, "Object and Observer Motion in the Perception of Objects by Infants," *Journal of Experimental Psychology,* Vol. 13, No. 4 (November 1987), pp. 586–593; Philip J. Kellman and Elizabeth S. Spelke, "Perception of Partly Occluded Objects in Infancy," *Cognitive Psychology,* Vol. 15, No. 4 (October 1983), pp. 483–524; and Arlette Streri and Elizabeth S. Spelke, "Haptic Perception of Objects in Infancy," *Cognitive Psychology,* Vol. 20, No. 1 (January 1988), pp. 1–23. Also of value are Michael S. Gazzaniga, *Nature's Mind: The Biological Roots of Thinking, Emotions, Sexuality, Language, and Intelligence* (New York: Basic Books, 1992); Marc Hauser and Susan Carey, "Building a Cognitive Creature from a Set of Primitives: Evolutionary and Developmental Insights," in Cummins and Allen, eds., *Evolution of Mind,* pp. 51–106; and Carolyn A. Ristau, "Cognitive Ethology: The Minds of Children and Animals," in Cummins and Allen, eds., *Evolution of Mind,* pp. 127–161.

119. Cosmides and Tooby, "Better Than Rational," p. 330.

120. Leda Cosmides and John Tooby, "Cognitive Adaptations for Social Exchange," in Jerome H. Barkow, Leda Cosmides, and John Tooby, eds., *The Adapted Mind: Evolutionary Psychology and the Generation of Culture* (New York: Oxford University Press, 1992), pp. 163–228; and Cosmides and Tooby, "Better Than Rational," p. 330. Also see Leda Cosmides and John Tooby, "From Evolution to Behavior: Evolutionary Psychology as the Missing Link," in John Dupré, ed., *The Latest on the Best: Essays on Evolution and Optimality* (Cambridge, Mass.: MIT Press, 1987), pp. 277–306.

121. Cosmides and Tooby, "Cognitive Adaptations for Social Exchange," p. 166.

122. Ibid., p. 164.

123. Ibid.

124. For evidence of reciprocity in primates, see Frans B.M. de Waal and Lesleigh M. Luttrell, "Mechanisms of Social Reciprocity in Three Primate Species: Symmetrical Relationship Characteristics or Cognition?" *Ethology and Sociobiology,* Vol. 9, No. 2 (1988), pp. 101–118; and Frans de Waal, *Chimpanzee Politics: Power and Sex among Apes,* rev. ed. (Baltimore, Md.: Johns Hopkins University Press, 1998), pp. 205–207.

125. Cosmides and Tooby, "Cognitive Adaptations for Social Exchange," p. 164.

126. Ibid., p. 205. Also see Leda Cosmides, "The Logic of Social Exchange: Has Natural Selection Shaped How Humans Reason?" *Cognition,* Vol. 31, No. 3 (April 1989), pp. 187–276. Robin Dunbar discusses this issue in "Culture, Honesty and the Freerider Problem," in Robin Dunbar, Chris Knight, and Camilla Power, eds., *The*

Evolution of Culture: An Interdisciplinary View (Edinburgh: Edinburgh University Press, 1999), pp. 194–213.

127. Cosmides and Tooby, "Cognitive Adaptations for Social Exchange," p. 207; and Tooby and Cosmides, "Evolutionary Psychology and the Generation of Culture, Part II: A Computational Theory of Social Exchange," *Ethology and Sociobiology,* Vol. 10, No. 1 (January 1989), pp. 51–97.

128. Antonio Damasio's scholarship demonstrates how much rational thought depends upon a healthy body. See Antonio R. Damasio, *Descartes' Error: Emotion, Reason, and the Human Brain* (New York: Grosset/Putnam, 1994).

129. J. Hirshleifer, "Natural Economy versus Political Economy," *Journal of Social and Biological Structures,* Vol. 1 (1978), pp. 319–337, 321. Also see Hirshleifer, "On the Emotions as Guarantors of Threats and Promises," in Dupré, *The Latest on the Best,* pp. 307–326.

130. Frank, *Passions within Reason*; and Jack Hirshleifer, "On the Emotions as Guarantors of Threats and Promises," in Dupré, ed., *The Latest on the Best,* pp. 307–326.

131. Hirshleifer, "Natural Economy versus Political Economy," p. 335. Emphasis in original.

132. See Shin-Ho Chung and Richard Herrnstein, "Choice and Delay of Reinforcement," *Journal of the Experimental Analysis of Behavior,* Vol. 10, No. 1 (January 1967), pp. 67–74; Richard Herrnstein, "On the Law of Effect," *Journal of the Experimental Analysis of Behavior,* Vol. 13, No. 2 (March 1970), pp. 242–266. Also see W. Baum and H. Rachilin, "Choice as Time Allocation," *Journal of the Experimental Analysis of Behavior,* Vol. 12 (1969), pp. 453–467; Frank, *Passions within Reason,* pp. 78–80; Jay V. Slotnik, Catherine H. Kannenberg, David A. Eckerman, and Marcus B. Waller, "An Experimental Analysis of Impulsivity and Impulse Control in Humans," *Learning and Motivation,* Vol. 11, No. 1 (February 1980), pp. 61–77.

133. Frank, *Passions within Reason,* p. 80.

134. Ibid. Also see George Ainslie and R.J. Herrnstein, "Preference Reversal and Delayed Reinforcement," *Animal Learning and Behavior,* Vol. 9, No. 4 (1981), pp. 476–482.

135. Frank, *Passions within Reason,* p. 80.

136. Pierre L. van den Berghe, "Ethnicity and the Sociobiology Debate," in John Rex and David Mason, eds., *Theories of Race and Ethnic Relations* (New York: Cambridge University Press, 1986), pp. 246–263, 260.

137. Ibid., p. 261.

138. Donald E. Brown, *Human Universals* (New York: McGraw-Hill, 1991), pp. 48–49. For a thoughtful perspective on how widespread behavior must be before it is considered universal behavior, see John Tooby and Leda Cosmides, "The Innate Versus the Manifest: How Universal Does Universal Have to Be?" *Behavior and Brain Sciences,* Vol. 12, No. 1 (March 1989), pp. 36–37.

139. For a discussion of these problems, see Yukiko Koshiro, *Trans-Pacific Racisms and the U.S. Occupation of Japan* (New York: Columbia University Press, 1999);

and Katherine H.S. Moon, *Sex among Allies: Military Prostitution in U.S.-Korea Relations* (New York: Columbia University Press, 1997).

140. Friedman, *Essays in Positive Economics*, p. 15.

3. Evolutionary Theory and War

1. Samuel P. Huntington, *The Clash of Civilizations and the Remaking of World Order* (New York: Simon and Schuster, 1996); J.A. Hobson, *Imperialism: A Study* (London: George Allen and Unwin, 1938); Kenneth N. Waltz, *Theory of International Politics* (Reading, Mass.: Addison-Wesley, 1979).

2. Classic studies of the causes of war remain Geoffrey Blainey, *The Causes of War*, 3rd ed. (New York: Free Press, 1988); and Quincy Wright, *A Study of War*, 2nd ed. (Chicago: University of Chicago Press, 1965). As ambitious and thoughtful as these earlier works is Stephen Van Evera's *The Causes of War: Power and the Roots of Conflict* (Ithaca, N.Y.: Cornell University Press, 1999).

3. Following the convention of anthropologist Marshall Sahlins, I will use the term "tribal" to refer to "a body of people of common derivation and custom, in possession and control of their own extensive territory," not necessarily "united under sovereign governing authority." Marshall D. Sahlins, *Tribesmen* (Englewood Cliffs, N.J.: Prentice-Hall, 1968), pp. vii–viii.

4. Harry Holbert Turney-High, *Primitive War: Its Practice and Concepts* (Columbia, S.C.: University of South Carolina Press, 1949).

5. Lawrence H. Keeley, *War before Civilization: The Myth of the Peaceful Savage* (New York: Oxford University Press, 1996).

6. Sigmund Freud, *The Ego and the Id*, trans. by Joan Riviere (New York: Norton, 1962).

7. Sigmund Freud, "Why War?" in Melvin Small and J. David Singer, eds., *International War: An Anthology* (Homewood, Ill.: Dorsey Press, 1985), p. 162. Freud's idea is found among the reasons Niall Ferguson provides in his explanation of why soldiers fought in World War I. Ferguson, *The Pity of War: Explaining World War I* (New York: Basic Books, 1999), pp. 339–366.

8. Erich Fromm, *The Anatomy of Human Destructiveness* (London: Jonathan Cape, 1974). Fromm develops the idea of malignant aggression. This arises not only out of innate predispositions to hurt or humiliate or control others, but also from a joint operation of a social and emotional climate that directs individuals to seek to meet their existential needs for meaning in life not through personal relationships but through hatred. Erik Erikson notes that humans often believe that other groups must be "annihilated or kept 'in their places' by periodic warfare or conquest." Erik H. Erikson, *Gandhi's Truth: On the Origins of Militant Nonviolence* (New York: Norton, 1969), p. 432. For a thoughtful study of aggression from the perspective of evolutionary psychology, see David M. Buss and Todd K. Shackelford, "Human Aggression in Evolutionary Psychological Perspective," *Clinical Psychological Review*, Vol. 17, No. 6

(1997), pp. 605–619; and Norbert Elias, "On Transformations of Aggressiveness," *Theory and Society,* Vol. 5, No. 2 (March 1978), pp. 229–242.

9. Robert Jervis, *The Logic of Images in International Relations* (Princeton, N.J.: Princeton University Press, 1970); and Jervis, *Perception and Misperception in International Politics* (Princeton, N.J.: Princeton University Press, 1976). Building on some of Jervis's insights to explain the origins of the Cold War is Deborah Welch Larson, *The Origins of Containment: A Psychological Explanation* (Princeton, N.J.: Princeton University Press, 1985).

10. Yuen Foong Khong, *Analogies at War: Korea, Munich, Dien Bien Phu, and the Vietnam Decisions of 1965* (Princeton, N.J.: Princeton University Press, 1992).

11. Jack S. Levy, "Domestic Politics and War," in Robert I. Rotberg and Theodore K. Rabb, eds., *The Origin and Prevention of Major Wars* (New York: Cambridge University Press, 1989), pp. 79–99. Levy provides an excellent tour of the horizon of domestic and other causes of war in his "The Causes of War: A Review of Theories and Evidence," in Philip E. Tetlock, Jo L. Husbands, Robert Jervis, Paul C. Stern, and Charles Tilly, eds., *Behavior, Society and Nuclear War: Volume One* (New York: Oxford University Press, 1989), pp. 209–333.

12. Jack Snyder, *Myths of Empire: Domestic Politics and International Ambition* (Ithaca, N.Y.: Cornell University Press, 1991). In an earlier work, Snyder explained the origins of World War I by analyzing the offensive proclivities of the militaries of France, Germany, and Russia. Jack Snyder, *The Ideology of the Offensive: Military Decision Making and the Disasters of 1914* (Ithaca, N.Y.: Cornell University Press, 1984).

13. Stephen M. Walt, *Revolution and War* (Ithaca, N.Y.: Cornell University Press, 1996).

14. Kenneth N. Waltz, *Theory of International Politics* (Reading, Mass.: Addison-Wesley, 1979). Waltz applied his argument to the post–Cold War world in Kenneth N. Waltz, "Structural Realism after the Cold War," *International Security,* Vol. 25, No. 1 (Summer 2000), pp. 5–41.

15. John J. Mearsheimer, *The Tragedy of Great Power Politics* (New York: Norton, 2001).

16. Christopher Layne, *The Peace of Illusions: International Relations Theory and American Grand Strategy in the Post–Cold War Era* [tentative title] (Ithaca, N.Y.: Cornell University Press, forthcoming).

17. Van Evera, *The Causes of War.* Sean M. Lynn-Jones provides a comprehensive analysis of the offense-defense balance and its implications for international relations in his "Offense-Defense Theory and Its Critics," *Security Studies,* Vol. 4, No. 4 (Summer 1995), pp. 660–691. Robert Jervis describes how the offense-defense balance affects the security dilemma in his "Cooperation under the Security Dilemma," *World Politics,* Vol. 30, No. 2 (January 1978), pp. 167–214. For a critique of the central concept of the theory, the relative offensive or defensive nature of military technology, see Keir A. Lieber, "Grasping the Technological Peace: The Offense-Defense Balance and International Security," *International Security,* Vol. 25, No. 1 (Summer 2000), pp. 71–104.

18. Thomas C. Schelling, *Arms and Influence* (New Haven: Yale University Press, 1966), p. 19.

19. Paul R. Ehrlich, *Human Natures: Genes, Cultures, and the Human Prospect* (Washington, D.C.: Island Press, 2000), p. 213.

20. Konrad Lorenz, *On Aggression,* trans. by Marjorie Kerr Wilson (New York: Harcourt, Brace and World, 1966); and Robert Ardrey, *The Territorial Imperative: A Personal Inquiry into the Animal Origins of Property and Nations* (New York: Atheneum, 1966), pp. 289–305.

21. Edward O. Wilson, *On Human Nature* (Cambridge, Mass.: Harvard University Press, 1978), p. 119. While Wilson's point is accurate, no implication concerning the importance of human ethology, the biology of human behavior, should be drawn. The insights of human ethology can be profound and are best illustrated by Irenäus Eibl-Eibesfeldt, "Human Ethology: Concepts and Implications for the Sciences of Man," *The Behavioral and Brain Sciences,* Vol. 2 (1979), pp. 1–26.

22. Robert Bigelow, *The Dawn Warriors: Man's Evolution toward Peace* (Boston: Little, Brown, 1969); Peter A. Corning, "The Biological Bases of Behavior and Some Implications for Political Science," *World Politics,* Vol. 23, No. 3 (April 1971), pp. 321–370; Vincent S.E. Falger, "The Missing Link in International Relations Theory?" *Journal of Social and Biological Structures,* Vol. 14, No. 1 (1991), pp. 73–77; Azar Gat, "The Human Motivational Complex: Evolutionary Theory and the Causes of Hunter-Gatherer Fighting, Part I. Primary Somatic and Reproductive Causes," *Anthropological Quarterly,* Vol. 73, No. 1 (January 2000), pp. 20–34; and Gat, "The Human Motivational Complex: Evolutionary Theory and the Causes of Hunter-Gatherer Fighting, Part II. Proximate, Subordinate, and Derivative Causes," *Anthropological Quarterly,* Vol. 73, No. 2 (April 2000), pp. 74–88; R. Paul Shaw and Yuwa Wong, *Genetic Seeds of Warfare: Evolution, Nationalism, and Patriotism* (Boston: Unwin Hyman, 1989); John Strate, "The Role of War in the Evolution of Political Systems and the Functional Priority of Defense," *Humboldt Journal of Social Relations,* Vol. 12, No. 2 (Spring/Summer 1985), pp. 87–114. See also Albert Somit, "Humans, Chimps, and Bonobos: The Biological Bases of Aggression, War, and Peacemaking," *The Journal of Conflict Resolution,* Vol. 34, No. 3 (September 1990), pp. 553–582.

The work of Johan van der Dennen is centrally important to this topic. See Johan Matheus Gerradus van der Dennen, *The Origin of War: The Evolution of a Male-Coalitional Reproductive Strategy,* 2 Vols. (Groningen, Netherlands: Origin Press, 1995); van der Dennen, "The Politics of Peace in Preindustrial Societies: The Adaptive Rationale behind Corroboree and Calumet," in Albert Somit and Steven A. Peterson, eds., *Research in Biopolitics,* Vol. 6: *Sociobiology and Politics* (Stamford, Conn.: JAI Press, 1998), pp. 159–192; van der Dennen, "Human Evolution and the Origin of War: A Darwinian Heritage," in Johan M.G. van der Dennen, David Smillie, and Daniel R. Wilson, eds., *The Darwinian Heritage and Sociobiology* (Westport, Conn.: Praeger, 1999), pp. 163–185; and van der Dennen, "(Evolutionary) Theories of Warfare in Preindustrial (Foraging) Societies," *Neuroendocrinology Letters,* Vol. 23 (Supplement 4) (December 2002), pp. 55–65.

Also important in this matter—as in almost all other applications of evolutionary theory to human behavior—is the scholarship of Edward Wilson. See Edward O. Wilson, *Sociobiology: The New Synthesis* (Cambridge, Mass.: Harvard University Press, 1975), pp. 572–574.

23. This chapter builds upon Bradley A. Thayer, "Bringing in Darwin: Evolutionary Theory, Realism, and International Politics," *International Security,* Vol. 25, No. 2 (Fall 2000), pp. 124–151.

24. An excellent overview is Paul Crook, *Darwinism, War and History: The Debate over the Biology of War from the "Origin of Species" to the First World War* (New York: Cambridge University Press, 1994).

25. Walter Bagehot, *Physics and Politics: Or Thoughts on the Application of the Principles of "Natural Selection" and "Inheritance" to Political Society* (New York: D. Appleton, 1887), p. 45.

26. Daniel J. Hughes, ed., *Moltke on the Art of War: Selected Writings,* trans. by Daniel J. Hughes and Harry Bell (Novato, Calif.: Presidio, 1993), p. 22.

27. General Friedrich von Bernhardi, *Germany and the Next War,* trans. by Allen H. Powles (London: Longman, Green, and Co., 1912); and von Bernhardi, *Britain as Germany's Vassal,* trans. by J. Ellis Barker (London: Longman, Green and Co., 1914).

28. Von Bernhardi, *Britain as Germany's Vassal,* pp. 106–107.

29. Sir Arthur Keith, *The Place of Prejudice in Modern Civilization* (London: Williams and Norgate, Ltd., 1931).

30. Charles Darwin, *The Descent of Man, and Selection in Relation to Sex,* 2 Vols. (Princeton, N.J.: Princeton University Press, 1981 [1871]), Vol. 2, p. 405.

31. Group living is adaptive in many species for five reasons. First, it helps protect the group from predators due to the security of numbers within the flock, herd, school, or troop or due to cooperative group defense to deter or defend against predator attack. In some species this behavior may help the young survive to maturation. Second, group living improves the ability to drive off competitors for resources. Third, it increases efficiency in locating and acquiring food. Fourth, reproductive efficiency may increase. Fifth, group living also improves the ability of species to modify their environment. See Strate, "The Role of War in the Evolution of Political Systems and the Functional Priority of Defense," pp. 92–93.

32. W.D. Hamilton, "The Genetical Evolution of Social Behavior. I," *Journal of Theoretical Biology,* Vol. 7, No. 1 (July 1964), pp. 1–16; Hamilton, "The Genetical Evolution of Social Behavior. II," *Journal of Theoretical Biology,* Vol. 7, No. 1 (July 1964), pp. 17–52. John Maynard Smith terms this concept "kin selection" in his "Group Selection and Kin Selection," *Nature,* Vol. 201, No. 4924 (March 14, 1964), pp. 1145–1147.

33. For excellent studies of inclusive fitness as a cause of war, see R. Paul Shaw and Yuwa Wong, "Ethnic Mobilization and the Seeds of Warfare: An Evolutionary Perspective," *International Studies Quarterly,* Vol. 31, No. 1 (March 1987), pp. 5–31; and Shaw and Wong, *Genetic Seeds of Warfare.*

34. This constant competition for resources is so ubiquitous that anthropologist Barbara Price terms it the "law of living systems." Barbara J. Price, "Competition,

Production Intensification, and Ranked Society: Speculations from Evolutionary Theory," in R. Brian Ferguson, ed., *Warfare, Culture, and Environment* (Orlando, Fla.: Academic Press, 1984), p. 212.

35. Christopher Stringer and Clive Gamble, *In Search of the Neanderthals: Solving the Puzzle of Human Origins* (New York: Thames and Hudson, 1993), p. 44.

36. For an excellent analysis of extinctions in the Pleistocene, see Paul S. Martin and Richard G. Klein, *Quaternary Extinctions: A Prehistoric Revolution* (Tucson: University of Arizona Press, 1984).

37. Peter A. Corning, "Synergy Goes to War," paper prepared for the annual meeting of the Association for Politics and the Life Sciences, Charleston, S.C., October 2001, pp. 1–90, 46; and Elisabeth S. Vrba "On the Connections between Paleoclimate and Evolution," in Elisabeth S. Vrba, George H. Denton, Timothy C. Partridge, Lloyd H. Burckle, eds., *Paleoclimate and Evolution, with Emphasis on Human Origins* (New Haven: Yale University Press, 1995), pp. 24–45.

38. See George H. Denton, "Cenozoic Climate Change," in Timothy G. Bromage and Friedemann Schrenk, eds., *African Biogeography, Climate Change, and Human Evolution* (New York: Oxford University Press, 1999), pp. 94–114; and Brian Fagan, *The Little Ice Age: How Climate Made History 1300–1850* (New York: Basic Books, 2000). For detailed discussions of human migration and human evolution, see Jonathan Kingdon, *Self-Made Man: Human Evolution from Eden to Extinction?* (New York: John Wiley and Sons, 1993), pp. 67–123; and Milford H. Wolpoff, *Paleoanthropology,* 2nd ed. (Boston: McGraw-Hill, 1999).

39. Peter Andrews and Louise Humphrey, "African Miocene Environments and the Transition to Early Hominines," in Bromage and Schrenk, eds., *African Biogeography, Climate Change, and Human Evolution,* pp. 282–300.

40. Quoted in Corning, "Synergy Goes to War," p. 56.

41. John Tooby and Irven DeVore, "The Reconstruction of Hominid Behavior Evolution through Strategic Modeling," in Warren G. Kinzey, ed., *The Evolution of Human Behavior: Primate Models* (Albany: State University of New York Press, 1987), p. 203; Paul Mellars, "Technological Change across the Middle-Upper Palaeolithic Transition: Economic, Social and Cognitive Perspectives," in Paul Mellars and Chris Stringer, eds., *The Human Revolution: Behavioural and Biological Perspectives on the Origins of Modern Humans* (Princeton, N.J.: Princeton University Press, 1989), pp. 338–365; and Norman Owen-Smith, "Ecological Links between African Savanna Environments, Climate Change, and Early Hominid Evolution," in Bromage and Schrenk, eds., *African Biogeography, Climate Change, and Human Evolution,* pp. 138–149.

42. Bobbi S. Low "An Evolutionary Perspective on War," in William Zimmerman and Harold K. Jacobson, eds., *Behavior, Culture, and Conflict in World Politics* (Ann Arbor: University of Michigan Press, 1993), pp. 13–55, 24.

43. Of course, scarcity can be a relative concept as well as absolute, and a tolerance of scarcity might be greater depending on a group's prior circumstances and in comparison to other groups. See R. Brian Ferguson, "Introduction: Studying War," in Ferguson, ed., *Warfare, Culture, and Environment,* p. 38.

44. Also noting this are David M. Buss and Todd K. Shackelford, "Human Aggression in Evolutionary Psychological Perspective," *Clinical Psychology Review,* Vol. 17, No. 6 (1997), pp. 605–619.

45. Strate emphasizes the importance of defense from attack by conspecifics, other humans; he argues that it caused the growth of human societies. Strate, "The Role of War in the Evolution of Political Systems and the Functional Priority of Defense," pp. 95–110.

46. Wilson, *On Human Nature,* p. 112.

47. Bobbi S. Low, *Why Sex Matters: A Darwinian Look at Human Behavior* (Princeton, N.J.: Princeton University Press, 2000), p. 215.

48. Richard D. Alexander, *Darwinism and Human Affairs* (Seattle: University of Washington Press, 1979), p. 223.

49. See Corning, "Synergy Goes to War."

50. Ibid., p. 61.

51. Darwin, *The Descent of Man, and Selection in Relation to Sex,* Vol. 2, pp. 368–369.

52. Ibid., Vol. 1, p. 161.

53. William H. Durham, "Resource Competition and Human Aggression. Part I: A Review of Primitive War," *The Quarterly Review of Biology,* Vol. 51, No. 3 (September 1976), pp. 385–415; Durham, *Coevolution: Genes, Culture, and Human Diversity* (Stanford, Calif.: Stanford University Press, 1991), pp. 375–384; and Edward O. Wilson, "Competitive and Aggressive Behavior," in J.F. Eisenberg and Wilton S. Dillon, eds., *Man and Beast: Comparative Social Behavior* (Washington, D.C.: Smithsonian Institution Press, 1971), pp. 181–217.

54. Durham, "Resource Competition and Human Aggression. Part I"; and Carol R. Ember and Melvin Ember, "Resource Unpredictability, Mistrust, and War," *The Journal of Conflict Resolution,* Vol. 36, No. 2 (June 1992), pp. 242–262. In the jargon of behavioral ecology, this strategy is always an option for *K*-selected species, like humans. Ecologists often draw a distinction between species based on their reproductive strategies (or mechanisms of population regulation). A *K*-selected species invests heavily in only a few offspring. Each is heavily protected until self-sufficient. *K*-selection often arises in environments where resources are scarce. The *K* symbolizes a saturated carrying capacity of the environment (in sum, too few resources) and in these conditions heavy investment in a few offspring is logical. In contrast, an *r*-selected species proliferates rapidly to fill all ecological niches with offspring. The *r* symbolizes the intrinsic rate of increase in a population and how prolific such a species may be. Large numbers of offspring with little or no parental investment is the right strategy in conditions in which there are few predators or resources are abundant, for example.

55. Durham, "Resource Competition and Human Aggression. Part I," p. 411.

56. Although Durham does not discuss this, I assume that aggression is logical in the conditions he describes if it does not produce a decrease in the fitness of the group. If it did, if the relative power, size, strength, and armament were too great, then fleeing or bandwagoning would be rational.

57. Peter Meyer, "Warfare in Social Evolution: A Look at Its Evolutionary Un-

derpinnings," paper presented at the annual meeting of the Association for Politics and the Life Sciences, Charleston, S.C., October 2001, p. 6.

58. Andrew P. Vayda, "Expansion and Warfare among Swidden Agriculturalists," *American Anthropologist,* Vol. 63, No. 2 (April 1961), pp. 346–358.

59. Formally, swidden agriculture is a system of cultivation "always involving the impermanent agricultural use of plots produced by the cutting back and burning off of vegetative cover." Harold C. Conklin, "An Ethnoecological Approach to Shifting Agriculture," *Transactions of the New York Academy of Sciences,* 2nd series, Vol. 17 (New York: New York Academy of Sciences, 1954), pp. 133–142, 133.

60. This motivation for warfare is suggested by Kalervo Oberg, "Types of Social Structure among the Lowland Tribes of South and Central America," *American Anthropologist,* Vol. 57 (1955), pp. 472–487; and Rev. Gerard A. Zegwaard, "Headhunting Practices of the Asmat of Netherlands New Guinea," *American Anthropologist,* Vol. 61, No. 6 (December 1959), pp. 1020–1041.

61. Waltz, *Theory of International Politics,* pp. 74–77.

62. Durham, "Resource Competition and Human Aggression. Part I," p. 391.

63. Asen Balikci, "The Netsilik Eskimos: Adaptive Processes," in Richard B. Lee and Irven DeVore, eds., *Man the Hunter* (Chicago: Aldine, 1968), pp. 78–82, 80.

64. See Johan M.G. van der Dennen, "The Politics of Peace in Preindustrial Societies," in Albert Somit and Steven A. Peterson, eds., *Research in Biopolitics,* Vol. 6 (Stamford, Conn.: JAI Press, 1998), pp. 159–192, 170. Also see van der Dennen, "Primitive War and the Ethnological Inventory Project," in J. van der Dennen and V. Falger, eds., *Sociobiology and Conflict: Evolutionary Perspectives on Competition, Cooperation, Violence and Warfare* (London: Chapman and Hall, 1990), pp. 247–269, 264–269; and van der Dennen, *The Origin of War,* pp. 497–537.

65. Alexander, *Darwinism and Human Affairs,* p. 229.

66. Ibid.

67. Ibid. For a contrary argument, see Donald Tuzin, "The Spectre of Peace in Unlikely Places: Concept and Paradox in the Anthropology of Peace," in Thomas Gregor, ed., *A Natural History of Peace* (Nashville, Tenn.: Vanderbilt University Press, 1996), pp. 3–33; and Thomas Gregor and Clayton A. Robarchek, "Two Paths to Peace: Semai and Mehinaku Nonviolence," in Gregor, ed., *A Natural History of Peace,* pp. 159–188.

68. Alexander, *Darwinism and Human Affairs,* p. 229.

69. Group and individual selection are not necessarily incompatible; both may operate in a population simultaneously. The idea of group selection was largely developed by Ronald Fisher and V.C. Wynne-Edwards. See Ronald A. Fisher, *The Genetical Theory of Natural Selection* (Oxford: Clarendon Press, 1930); and V.C. Wynne-Edwards, *Animal Dispersion in Relation to Social Behaviour* (Edinburgh: Oliver and Boyd, 1962).

70. Classic studies are Robert L. Trivers, "The Evolution of Reciprocal Altruism," *The Quarterly Review of Biology,* Vol. 46, No. 1 (March 1971), pp. 35–57; and David Sloan Wilson, *The Natural Selection of Populations and Communities* (Menlo Park, Calif.: Benjamin/Cummings, 1980).

71. Wilson, *The Natural Selection of Populations and Communities,* p. 45.

72. As Elliott Sober and David Sloan Wilson submit, "altruism is maladaptive with respect to individual selection but adaptive with respect to group selection." Elliott Sober and David Sloan Wilson, *Unto Others: The Evolution and Psychology of Unselfish Behavior* (Cambridge, Mass.: Harvard University Press, 1998), p. 27.

73. The four criteria of group selection are drawn from Ibid., p. 26.

74. Also noting the relationship between group selection and warfare are Peter A. Corning, "An Evolutionary Paradigm for the Study of Human Aggression," in Martin A. Nettleship, R. Dalegivens, and Anderson Nettleship, eds., *War, Its Causes and Correlates* (The Hague: Mouton Publishers, 1975), pp. 359–387; Irenäus Eibl-Eibesfeldt, "Us and the Others," in Irenäus Eibl-Eibesfeldt and Frank Kemp Salter, eds., *Indoctrinability, Ideology, and Warfare: Evolutionary Perspectives* (New York: Berghahn Books, 1998), pp. 21–53, 34–36; Eibl-Eibesfeldt, "Warfare, Man's Indoctrinability and Group Selection," *Zeitschrift für Tierpsychologie,* Vol. 60, No. 3 (1982), pp. 177–198; and Joseph Soltis, Robert Boyd, and Peter J. Richerson, "Can Group-Functional Behaviors Evolve by Cultural Group Selection?" *Current Anthropology,* Vol. 36, No. 3 (June 1995), pp. 473–483.

75. Darwin, *The Descent of Man,* Vol. 1, p. 162.

76. Ibid., Vol. 1, pp. 162–163.

77. Wilson, *On Human Nature,* p. 112.

78. Ibid.

79. Wilson, *Sociobiology,* p. 573.

80. Ibid.

81. Ibid.

82. On this point see Robin Dunbar, "The Sociobiology of War," *Medicine and War,* Vol. 1 (1985), pp. 201–208, 204. Eibl-Eibesfeldt argues that individual selection alone cannot account for such traits as human indoctrinability and the inclination to polarize that evolved to promote conformity, and must be informed by group selection. See Eibl-Eibesfeldt, "Warfare, Man's Indoctrinability and Group Selection," pp. 177–198.

83. Turney-High argues that there are five areas of difference between what he calls "primitive" warfare and "civilized" warfare: (1) tactical operations; (2) definite command and control or clear military hierarchy to maintain military discipline; (3) the ability to conduct a campaign against the enemy beyond one battle; (4) a clarity of motive; and (5) adequate supply or logistics. Turney-High, *Primitive War,* p. 30. I note but reject these criteria for a distinction between primitive and modern warfare because all are found in primitive warfare. Criteria that are more important to establish the differences between these forms of warfare include (1) technology; (2) the modern economy, which supports technological development and enables a relatively small proportion of the population to participate in military operations while still generating necessary goods and services for civilians; (3) the bureaucratization or professionalization of the military; and (4) the size of modern militaries and their scope of operations on land, air, and sea, as well as their breadth over considerably larger distances.

84. Aubrey Cannon, "Conflict and Salmon on the Interior Plateau of British Columbia," in Brian Hayden, ed., *A Complex Cultu¦re of the British Columbia Plateau: Traditional* Stl'átl'imx *Resource Use* (Vancouver: UBC Press, 1992), pp. 506–524; and Ferguson, "Introduction," in Ferguson, ed., *Warfare, Culture, and Environment,* pp. 1–81.

85. Ferguson, "Introduction," in Ferguson, ed., *Warfare, Culture, and Environment,* p. 41.

86. *The Oxford Dictionary of Quotations,* 5th ed., ed. by Elizabeth Knowles (New York: Oxford University Press, 1999), p. 809.

87. Pitirim A. Sorokin, *Social and Cultural Dynamics,* 4 Vols. (New York: Bedminster Press, 1962), Vol. 3, pp. 351–360, 469–506.

88. Quoted in Wilson, *On Human Nature,* pp. 119–120. Anthropologists disagree somewhat on the proper name of the tribe. Napoleon Chagnon calls the tribe "Yanomamö," while Brian Ferguson prefers "Yanomami." Napoleon A. Chagnon, *Yanomamö,* 4th ed. (Fort Worth: Harcourt Brace Jovanovich, 1992); and R. Brian Ferguson, *Yanomami Warfare: A Political History* (Santa Fe, N.M.: School of American Research Press, 1995), pp. 65–66.

89. For a discussion of the powerful effect of the security dilemma in international relations, see Jervis, "Cooperation under the Security Dilemma," pp. 167–214. For a useful discussion of how to minimize some of its influence and promote greater cooperation in international relations, see Kenneth A. Oye, ed., *Cooperation under Anarchy* (Princeton, N.J.: Princeton University Press, 1986).

90. Marilyn Keyes Roper, "Evidence of Warfare in the Near East from 10,000–4,300 B.C.," in Nettleship et al., eds., *War, Its Causes and Correlates,* pp. 302–303.

91. Ibid., p. 304.

92. Kathleen Kenyon, "Earliest Jericho," *Antiquity,* Vol. 33 (1959), pp. 5–9; quoted in Roper, "Evidence of Warfare in the Near East from 10,000–4,300 B.C.," p. 305.

93. James Mellaart, *Earliest Civilizations of the Near East* (New York: McGraw-Hill, 1965); quoted in Roper, "Evidence of Warfare in the Near East from 10,000–4,300 B.C.," p. 307.

94. Roper, "Evidence of Warfare in the Near East from 10,000–4,300 B.C." p. 329.

95. Ibid.

96. Claudio Cioffi-Revilla, "Ancient Warfare: Origins and Systems," in Manus I. Midlarsky, ed., *Handbook of War Studies II* (Ann Arbor: University of Michigan Press, 2000), pp. 59–89, 82–83. Also see Cioffi-Revilla, "Origins and Evolution of War and Politics," *International Studies Quarterly,* Vol. 40, No. 1 (March 1996), pp. 1–22.

97. Ibid., p. 74.

98. Ibid., pp. 74–75.

99. Ibid., pp. 75, 83–84.

100. Keeley, *War before Civilization,* pp. 27–28. Also see Robert Knox Dentan, "Band-Level Eden: A Mystifying Chimera," *Cultural Anthropology,* Vol. 3, No. 3 (August 1988), pp. 276–284.

101. Quincy Wright, *A Study of War* (Chicago: University of Chicago Press, 1942), p. 556.

102. Keith F. Otterbein, *The Evolution of War: A Cross-Cultural Study* (New Haven: HRAF Press, 1970), pp. 66, 145–148. The 50 were drawn from the 628 societies listed in the *Ethnographic Atlas,* a compendium of all known societies, from the most industrialized to the least.

103. Carol R. Ember, "Myths about Hunter-Gatherers," *Ethnology,* Vol. 17, No. 4 (October 1978), pp. 439–448, 443–444. Ember and Melvin Ember submit in another cross-cultural study that 75 percent of tribes had engaged in warfare at least once every two years before they were pacified or amalgamated into larger societies. Carol R. Ember and Melvin Ember, *Cultural Anthropology,* 6th ed. (Englewood Cliffs, N.J.: Prentice Hall, 1990), p. 255.

104. Marc Howard Ross, "Political Decision Making and Conflict: Additional Cross-Cultural Codes and Scales," *Ethnology,* Vol. 22, No. 2 (April 1983), pp. 169–192.

105. Joseph G. Jorgensen, *Western Indians: Comparative Environments, Languages, and Cultures of 172 Western American Indian Tribes* (San Francisco: W.H. Freeman, 1980), pp. 240–247, 503–506, 509–515, 613–614.

106. Making this point forcefully is Irenäus Eibl-Eibesfeldt, "The Myth of the Aggression-Free Hunter and Gatherer Society," in Ralph L. Holloway, ed., *Primate Aggression, Territoriality, and Xenophobia: A Comparative Perspective* (New York: Academic Press, 1974), pp. 435–457.

107. Otterbein, *The Evolution of War,* pp. 145–148; and Ross, "Political Decision Making and Conflict."

108. Ember and Ember, "Resource Unpredictability, Mistrust, and War," p. 250.

109. Keeley, *War before Civilization,* p. 33.

110. For an excellent analysis of warfare for land, see Melvin Ember, "Statistical Evidence for an Ecological Explanation of Warfare," *American Anthropologist,* Vol. 84, No. 3 (June 1982), pp. 645–649. On warfare for adultery and wife-stealing, see William Tulio Divale, *Warfare in Primitive Societies: A Bibliography* (Santa Barbara, Calif.: American Bibliographic Center—Clio Press, 1973), pp. xxii–xxiii; and William Tulio Divale and Marvin Harris, "Population, Warfare, and the Male Supremacist Complex," *American Anthropologist,* Vol. 78, No. 3 (September 1976), pp. 521–538. For an analysis of warfare over the capture of women, see Douglas R. White, "Rethinking Polygyny: Co-wives, Codes, and Cultural Systems, *Current Anthropology,* Vol. 29, No. 4 (August–October 1988), pp. 529–558; and Douglas R. White and Michael L. Burton, "Causes of Polygyny: Ecology, Economy, Kinship, and Warfare," *American Anthropologist,* Vol. 90, No. 4 (August 1988), pp. 871–887.

111. Ember and Ember, "Resource Unpredictability, Mistrust, and War," pp. 242–262. Also see Carol R. Ember and Melvin Ember, "Warfare, Aggression, and Resource Problems: Cross-Cultural Codes," *Behavior Science Research,* Vol. 26, Nos. 1–4 (1992), pp. 169–226.

112. Ember and Ember, "Resource Unpredictability, Mistrust, and War," p. 250.

113. Otterbein, *The Evolution of War,* p. 66.

114. Ibid., p. 21.

115. Raoul Narroll, *Warfare, Peaceful Intercourse and Territorial Change: A Cross-*

Cultural Survey, unpublished manuscript, no date, quoted in Otterbein, *The Evolution of War,* p. 65.

116. Joseph H. Manson and Richard W. Wrangham, "Intergroup Aggression in Chimpanzees and Humans," *Current Anthropology,* Vol. 32, No. 4 (August–October 1991), pp. 369–390, 374–375.

117. Bronislaw Malinowski noted similar motivations for warfare in his "War and Weapons among the Natives of the Trobriand Islands," *Man,* Vol. 20, No. 5 (January 1920), pp. 10–12. Of course, there may be exceptions to this generalization for any specific tribe. For example, D.J.J. Brown argues that for the Polopa of Papua New Guinea, stealing women or pigs is a common motivation for war, but war over land is relatively rare because it is so abundant in their area. D.J.J. Brown, "The Structure of Polopa Feasting and Warfare," *Man* (New Series [N.S.]), Vol. 14, No. 4 (December 1979), pp. 712–733. Also see Paul Sillitoe, "Big Men and War in New Guinea," *Man* (N.S.), Vol. 13, No. 2 (June 1978), pp. 252–271.

118. Mervyn Meggitt, *Blood Is Their Argument: Warfare among the Mae Enga Tribesmen of the New Guinea Highlands* (Palo Alto, Calif.: Mayfield Publishing, 1977), pp. 12–14. He also emphasizes the importance of land as a cause of Mae warfare in M.J. Meggitt, *The Lineage System of the Mae-Enga of New Guinea* (New York: Barnes and Noble, 1965), pp. 68–71, 77–80, 226–227.

119. Meggitt, *Blood Is Their Argument,* Table One, p. 13.

120. Ibid., p. 70.

121. Ibid., p. 14.

122. Ibid.

123. Ibid., pp. 14–15.

124. Ibid., p. 178.

125. Ibid., p. 71.

126. Ibid., p. 32.

127. Ibid.

128. Ibid., pp. 34–36.

129. Ibid., pp. 181, 186.

130. Ibid., pp. 168–169.

131. Andrew P. Vayda, "Phases of the Process of War and Peace among the Marings of New Guinea," *Oceania,* Vol. 42, No. 1 (September 1971), pp. 1–24, 4.

132. Ibid., pp. 22–23. Vayda reviews and confirms his findings of the causes of Maring warfare in Andrew P. Vayda, "Explaining Why Marings Fought," *Journal of Anthropological Research,* Vol. 45, No. 2 (Summer 1989), pp. 159–177.

133. Vayda, "Expansion and Warfare among Swidden Agriculturalists," p. 350.

134. Ibid., p. 348.

135. Ibid., p. 354.

136. Ibid. Emphasis in original. Vayda's analysis is further developed in *War in Ecological Perspective: Persistence, Change, and Adaptive Processes in Three Oceanian Societies* (New York: Plenum Press, 1976), pp. 43–74. Also emphasizing the frequency of

warfare among hunter-gatherer tribes is Ember, "Myths about Hunter-Gatherers," pp. 439–448.

137. Vayda, "Expansion and Warfare among Swidden Agriculturalists," p. 348.

138. George E.B. Morren Jr., "Warfare on the Highland Fringe of New Guinea: The Case of the Mountain Ok," in Ferguson, ed., *Warfare, Culture, and Environment,* p. 173

139. Ibid. Lawrence Keeley also reports this motivation among Canadian Indian tribes: "during hard winters, the Chilcotin of British Columbia would attack small isolated hamlets or family camps of other tribes, kill all the inhabitants and live off their stored food." Keeley, *Warfare before Civilization,* p. 65.

140. Morren, "Warfare on the Highland Fringe of New Guinea," p. 179.

141. Ibid.

142. Ibid., p. 194.

143. Karl G. Heider, *The Dugum Dani: A Papuan Culture in the Highlands of West New Guinea,* Viking Fund Publications in Anthropology Number Forty-Nine (New York: Wenner-Gren Foundation for Anthropological Research, 1970), pp. 100–101. Heider provides an account of a specific war in his *Grand Valley Dani: Peaceful Warriors* (New York: Holt, Rinehart and Winston, 1979), pp. 90–92.

144. Paul Shankman, "Culture Contact, Cultural Ecology, and Dani Warfare," *Man* (N.S.), Vol. 26, No. 2 (June 1991), pp. 299–321.

145. Camilla H. Wedgwood, "Some Aspects of Warfare in Melanesia," *Oceania,* Vol. 1 (1930), p. 11.

146. Zegwaard, "Headhunting Practices of the Asmat of Netherlands New Guinea," p. 1036.

147. Ibid., p. 1040.

148. Ibid., p. 1041.

149. Leopold Pospisil, *The Kapauku Papuans and Their Law* (New Haven: Human Relations Area Files Press, 1971), p. 89.

150. Morren, "Warfare on the Highland Fringe of New Guinea," p. 186.

151. Vayda, "Expansion and Warfare among Swidden Agriculturalists," pp. 348–349. Also see A.P. Vayda, *Maori Warfare* (Wellington: The Polynesian Society, 1960), pp. 19–24.

152. R. Brian Ferguson, "A Reexamination of the Causes of Northwest Coast Warfare," in Ferguson, ed., *Warfare, Culture, and Environment,* pp. 267–328.

153. Ibid., p. 273.

154. Ibid.

155. Franz Boas, *Tsimshian Mythology* (New York: Johnson Reprint Co., 1970), pp. 355–378; quoted in Ferguson, "A Reexamination of the Causes of Northwest Coast Warfare," p. 274.

156. Ferguson, "A Reexamination of the Causes of Northwest Coast Warfare," p. 288.

157. Ibid., p. 294.

158. Ibid., p. 296.

159. Ibid., p. 300.

160. Edward S. Curtis, *The North American Indian, Volume IX: The Salish* (Cambridge, Mass.: The University Press, 1913), p. 74; quoted in Ferguson, "A Reexamination of the Causes of Northwest Coast Warfare," p. 300. Ferguson notes that slavery was widely practiced and a critical aspect of the economies of many of these tribes (Ibid., p. 305). According to William Christie MacLeod, "slavery on the northwest coast among the natives was of nearly as much economic importance to them as was slavery to the plantation regions of the United States before the Civil War." MacLeod, "Economic Aspects of Indigenous American Slavery," *American Anthropologist,* Vol. 30 (1928), pp. 632–650, 649.

161. Ferguson, "A Reexamination of the Causes of Northwest Coast Warfare," p. 304.

162. Ibid., p. 297.

163. Edwin T. Denig, *Indian Tribes of the Upper Missouri* (Washington, D.C.: Bureau of American Ethnology, 1930), p. 477; quoted in Thomas Biolsi, "Ecological and Cultural Factors in Plains Indian Warfare," in Ferguson, ed., *Warfare, Culture, and Environment,* p. 150.

164. George Bird Grinnell, *The Cheyenne Indians: Their History and Ways of Life,* 2 Vols. (Lincoln: University of Nebraska Press, 1972), Vol. 2, p. 3; quoted in Biolsi, "Ecological and Cultural Factors in Plains Indian Warfare," pp. 154–155.

165. Biolsi, "Ecological and Cultural Factors in Plains Indian Warfare," p. 155.

166. Ibid.

167. Ibid., pp. 155–156. Matt Ridley describes "the conventional wisdom" on Indians: that they "were at one with nature, respecting and forbearing towards it, magically attuned to it and resolute in practising careful management so as not to damage the stock of their game." Based in part on an analysis of the data of intertribal warfare and natural resource management, however, Ridley argues: "Archaeological sites throw doubt upon these comforting myths." Matt Ridley, *The Origins of Virtue: Human Instincts and the Evolution of Cooperation* (New York: Viking Penguin, 1997), p. 216.

Even if Indians' practices were not up to the standards of today's conservationists, one advocate of Indian rights urges continuing such an argument because "any evidence of ecologically unsound activities by indigenous and traditional peoples undermines their basic rights to land, resources, and cultural practice." D.A. Posey, "Do Amazonian Indians Conserve Their Environment or Not? And Who Are We to Know If They Do or Do Not?" *Abstracts, 92nd Annual Meeting of the American Anthropological Association* (Washington, D.C.: American Anthropological Association, 1993), p. 473; quoted in William T. Vickers, "From Opportunism to Nascent Conservation: The Case of the Siona-Secoya," *Human Nature,* Vol. 5, No. 4 (1994), pp. 307–337, 308.

168. Douglas B. Bamforth, "Indigenous People, Indigenous Violence: Precontact Warfare on the North American Great Plains," *Man* (N.S.), Vol. 29, No. 1 (March 1994), pp. 95–115, 98.

169. Ibid., pp. 105–108; Shepard Krech III, "Genocide in Tribal Society," *Nature,* Vol. 371, No. 6492 (September 1, 1994), pp. 14–15; and P. Willey, *Prehistoric Warfare on the Great Plains: Skeletal Analysis of the Crow Creek Massacre Victims* (New York: Garland Publishing, 1990), pp. 93–152.

170. Bamforth, "Indigenous People, Indigenous Violence," p. 109.

171. Ibid.

172. See Keith B. Richburg, "Case Closed on the 'Iceman' Mystery," *Washington Post,* July 26, 2001, p. A1; and Brenda Fowler, "After 5,300 Years, Mystery of Iceman's Death Is Solved," *New York Times,* August 7, 2001, at <http://www.nytimes.com/2001/08/07/science/social/07ICEM.html>.

173. Fowler, "After 5,300 Years, Mystery of Iceman's Death Is Solved."

174. Ibid.

175. James C. Chatters, *Ancient Encounters: Kennewick Man and the First Americans* (New York: Simon and Schuster, 2001), pp. 131–138, 160–162.

176. Ibid., p. 161.

177. Ibid., pp. 163–264.

178. William Balée, "The Ecology of Ancient Tupi Warfare," in Ferguson, ed., *Warfare, Culture, and Environment,* pp. 241–265.

179. Ibid., p. 253.

180. Donald W. Lathrap, "The 'Hunting' Economics of the Tropical Forest Zone of South America: An Attempt at Historical Perspective," in Lee and DeVore, eds., *Man the Hunter,* pp. 23–29.

181. Donald W. Lathrap, *The Upper Amazon* (New York: Praeger Publishers, 1970), p. 20.

182. Oberg, "Types of Social Structure among the Lowland Tribes of South and Central America," p. 473.

183. Chagnon, *Yanomamö,* p. 191.

184. Ibid.

185. Ibid.

186. Ibid. Chagnon's finding that *unokais* have more children has been disputed by Bruce Albert, "Yanomami 'Violence': Inclusive Fitness or Ethnographer's Representation," *Current Anthropology,* Vol. 30, No. 4 (August–October 1989), pp. 637–640; and Ferguson, *Yanomami Warfare,* pp. 359–362.

187. Chagnon, *Yanomamö*; and Napoleon Chagnon and Paul Bugos, "Kin Selection and Conflict: An Analysis of a Yanomamö Ax Fight," in Napoleon A. Chagnon and William Irons, eds., *Evolutionary Biology and Human Social Behavior: An Anthropological Perspective* (North Scituate, Mass.: Duxbury Press, 1979), pp. 213–238.

188. Ferguson, *Yanomami Warfare,* p. 52. Ferguson stresses that the numbers of battles waged to capture women are relatively low. In contrast to Chagnon and Ferguson, and almost all other experts on this tribe, J. Lizot argues that the Yanomamö wage war only to punish insults, avenge deaths, and fulfill kinship obligations. J. Lizot, "Population, Resources and Warfare among the Yanomami," *Man* (N.S.), Vol. 12, Nos. 3/4 (December 1977), pp. 497–517.

189. Ferguson, *Yanomami Warfare,* p. 34.

190. Ibid., pp. 45, 55. For a detailed study of the ways a Western presence increases both the occurrence and intensity of warfare among tribes, see R. Brian Ferguson and Neil L. Whitehead, "The Violent Edge of Empire," in Ferguson and Whitehead, eds., *War in the Tribal Zone: Expanding States and Indigenous Warfare* (Santa Fe, N.M.: School of American Research Press, 1992), pp. 1–30. Also see Jeffrey P. Blick, "Genocidal Warfare in Tribal Societies as a Result of European-Induced Culture Conflict," *Man* (N.S.), Vol. 23, No. 4 (December 1988), pp. 654–670.

191. Chagnon, *Yanomamö,* p. 16.

192. Ibid., pp. 16–17.

193. Ibid., pp. 17–19. Similar experiences are related by Ferguson, *Yanomami Warfare,* pp. 323–325, 333–334.

194. After a close study of the causes of Maasai wars and raiding practices, Alan Jacobs says their reputation is not supported. See Alan H. Jacobs, "Maasai Inter-Tribal Relations: Belligerent Herdsmen or Peaceable Pastoralists," in Katsuyoshi Fukui and David Turton, eds., *Warfare among East African Herders* (Osaka: National Museum of Ethnology, 1979), pp. 33–52.

195. Ibid., pp. 44–45.

196. Ibid., p. 46.

197. Elliot Fratkin, "A Comparison of the Role of Prophets in Samburu and Maasai Warfare," in Fukui and Turton, eds., *Warfare among East African Herders,* pp. 53–67, 55.

198. Ibid., p. 62.

199. Uri Almagor, "Raiders and Elders: A Confrontation of Generations among the Dassanetch," in Fukui and Turton, eds., *Warfare among East African Herders,* pp. 119–145, 123–124.

200. Ibid., p. 124.

201. Serge Tornay, "Armed Conflicts in the Lower Omo Valley, 1970–1976: An Analysis from within Nyangatom Society," in Fukui and Turton, eds., *Warfare among East African Herders,* pp. 97–117, 104.

202. Ibid., p. 105.

203. P.T.W. Baxter, "Boran Age-Sets and Warfare," in Fukui and Turton, eds., *Warfare among East African Herders,* pp. 69–95, 89.

204. Ibid., pp. 82, 88–89.

205. Ibid., p. 82.

206. Morimichi Tomikawa, "The Migrations and Inter-Tribal Relations of the Pastoral Datoga," in Fukui and Turton, eds., *Warfare among East African Herders,* pp. 15–31, 27.

207. Wilson, *Sociobiology,* p. 573. Wilson notes that aggression and genetic usurpation occurs among other primates as well.

208. Chagnon, *Yanomamö,* pp. 203–204.

209. Meggitt, *Blood Is Their Argument,* p. 82.

210. For an analysis of the disputes among the allies during and after World War

I, see Bradley A. Thayer, "Creating Stability in New World Orders" (Chicago: Ph.D. diss., University of Chicago, 1996), pp. 229–260. For a similar analysis during and in the immediate aftermath of World War II, see Mark A. Stoler, *Allies and Adversaries: The Joint Chiefs of Staff, the Grand Alliance, and U.S. Strategy in World War II* (Chapel Hill: University of North Carolina Press, 2000).

211. Keeley, *War before Civilization,* p. 65.

212. Ibid. Also see Turney-High, *Primitive War,* p. 124; for a detailed account of a Mae battle, see Meggitt, *Blood Is Their Argument,* pp. 85–88.

213. Azar Gat, "The Pattern of Fighting in Simple, Small-Scale, Prestate Societies," *Journal of Anthropological Research,* Vol. 55, No. 4 (Winter 1999), pp. 563–583.

214. Vayda, "Phases of the Process of War and Peace among the Marings of New Guinea," p. 11.

215. Ibid.

216. Wedgwood, "Some Aspects of Warfare in Melanesia," p. 13.

217. Turney-High, *Primitive War,* p. 129.

218. Ibid.

219. Ibid.

220. Morren, "Warfare on the Highland Fringe of New Guinea," p. 173.

221. Ibid., pp. 186–191.

222. Pospisil, *Kapauku Papuans and Their Law,* p. 92.

223. Ibid.

224. Meggitt, *Blood Is Their Argument,* p. 74.

225. Ibid., pp. 50–51.

226. Ibid., pp. 51–52.

227. Chagnon, *Yanomamö,* p. 189.

228. Ibid.

229. Ibid. Also see Ferguson, *Yanomami Warfare,* pp. 46–49.

230. Chagnon, *Yanomamö,* p. 198.

231. Ibid.

232. Ibid.

233. Ibid.

234. Ibid.

235. Keeley, *War before Civilization,* p. 66. Leslie Sponsel argues in contrast to Keeley that human history is "relatively free" of systematic evidence of violence and that "nonviolence and peace were likely the norm throughout most of human prehistory." Leslie E. Sponsel, "The Natural History of Peace: The Positive View of Human Nature and Its Potential," in Gregor, ed., *A Natural History of Peace,* pp. 95–125, 103. Emphasis omitted. Sponsel may be right that the norm was peace in the sense that peace existed for a given community most of the time, as it does for the United States today. However, the paucity of evidence does not permit a definite conclusion. Furthermore, Sponsel's argument does not detract from Keeley's argument, given the tremendous costs of warfare for a society that is attacked and must defend itself. Nor does this mean that aggression would not be worth the potential risks. Thus Sponsel

may be correct that the norm is peace. But this does not speak to the causes of aggression. After all, large predators like lions or crocodiles spend most of their time resting, but that is not the behavior that most concerns the zebra or buffalo.

236. Keeley, *War before Civilization,* p. 66.

237. Ibid., pp. 66–67.

238. Ibid., p. 67.

239. This point is demonstrated forcefully in Margaret Rodman and Matthew Cooper, eds., *The Pacification of Melanesia* (Ann Arbor: University of Michigan Press, 1979).

240. Sterling Robbins, *Auyana: Those Who Held onto Home* (Seattle: University of Washington Press, 1982), p. 189.

241. Keeley, *War before Civilization,* pp. 59–60.

242. Meyer, "Warfare in Social Evolution," p. 6.

243. Wedgwood, "Some Aspects of Warfare in Melanesia," p. 17. Also see Turney-High, *Primitive War,* pp. 58–59.

244. Wedgwood, "Some Aspects of Warfare in Melanesia," p. 17.

245. Meggitt, *Blood Is Their Argument,* p. 93.

246. This point is acknowledged by most students of tribal warfare. See, for example, Carol R. Ember, Melvin Ember, and Bruce Russett, "Peace between Participatory Polities: A Cross-Cultural Test of the 'Democracies Rarely Fight Each Other' Hypothesis," *World Politics,* Vol. 44, No. 4 (July 1992), pp. 573–599, 581; Gat, "The Pattern of Fighting in Simple, Small-Scale, Prestate Societies," pp. 574–577; and Meggitt, *Blood Is Their Argument,* p. 201.

247. Keeley, *War before Civilization,* p. 65.

248. Ibid.

249. Vayda, "Phases of the Process of War and Peace among the Maring of New Guinea," p. 13.

250. Meggitt, *Blood Is Their Argument,* p. 110.

251. He has casualty rates for seventeen disputes over land; an average of 4.6 men died in each. Seventeen other conflicts were motivated by other causes, with an average casualty rate of 2.7. Ibid., p. 109.

252. Ibid., pp. 100–105.

253. Heider, *Grand Valley Dani,* p. 106.

254. Meggitt, *Blood Is Their Argument,* p. 111.

255. Ibid.

256. Pospisil, *Kapauku Papuans and Their Law,* p. 89.

257. Tornay, "Armed Conflicts in the Lower Omo Valley, 1970–1976," p. 111.

258. Chagnon, *Yanomamö,* p. 205. Emphasis in original.

259. Robbins, *Auyana,* p. 211.

260. Ibid., p. 211.

261. Allan R. Millett and Peter Maslowski, *For the Common Defense: A Military History of the United States of America,* rev. and exp. (New York: Free Press, 1994), p. 653.

262. Ibid.

263. Robbins, *Auyana,* p. 212.

264. Carl von Clausewitz, *On War,* ed. and trans. by Michael Howard and Peter Paret (Princeton, N.J.: Princeton University Press, 1976), p. 259.

265. Clifton B. Kroeber and Bernard L. Fontana, *Massacre on the Gila: An Account of the Last Major Battle between American Indians, with Reflections on the Origin of War* (Tucson: University of Arizona Press, 1986).

266. Keeley, *War before Civilization,* p. 67.

267. Ibid., pp. 67–68.

268. Ferguson, "A Reexamination of the Causes of Northwest Coast Warfare," p. 285.

269. Wedgwood, "Some Aspects of Warfare in Melanesia," p. 14.

270. Morren, "Warfare on the Highland Fringe of New Guinea," p. 183.

271. Heider, *Grand Valley Dani,* pp. 103–106.

272. Vayda, "Phases of the Process of War and Peace among the Marings of New Guinea," p. 12.

273. Ibid., pp. 12–13.

274. Pospisil, *Kapauku Papuans and Their Law,* p. 91.

275. Chagnon, *Yanomamö,* p. 189.

276. Ibid., pp. 189–190, also see p. 3. Ferguson mentions attacks under the ruse of a feast; see for example, his *Yanomami Warfare,* pp. 236–238, 251.

277. George R. Milner, Eve Anderson, and Virginia G. Smith, "Warfare in Late Prehistoric West-Central Illinois," *American Antiquity,* Vol. 56, No. 4 (1991), pp. 581–603, 589.

278. Ibid., p. 590.

279. Meggitt, *Blood Is Their Argument,* p. 81.

280. Fisher is quoted in Arthur J. Marder, *The Anatomy of British Seapower: A History of British Naval Policy in the Pre-Dreadnought Era, 1880–1905* (New York: Knopf, 1940), p. 473. For an analysis of the still-central role of geography for the grand strategy of the United States during the Cold War, see Michael C. Desch, "The Keys That Lock up the World: Identifying American Interests in the Periphery," *International Security,* Vol. 14, No. 1 (Summer 1989), pp. 86–121.

281. Sir Halford J. Mackinder, *Democratic Ideals and Reality* (New York: Henry Holt and Co., 1919), pp. ix, 73–114.

282. William H. Durham, *Scarcity and Survival in Central America: Ecological Origins of the Soccer War* (Stanford, Calif.: Stanford University Press, 1979).

283. Ehrlich, *Human Natures,* p. 277.

284. National Intelligence Council, *Global Trends 2015: A Dialogue about the Future with Nongovernment Experts,* <http://www.odci.gov/cia/reports/globaltrends2015/index.html>.

285. See Thomas F. Homer-Dixon, "On the Threshold: Environmental Changes as Causes of Acute Conflict," *International Security,* Vol. 16, No. 2 (Fall 1991), pp. 76–116; and "Environmental Scarcities and Violent Conflict: Evidence from Cases," *International Security,* Vol. 19, No. 1 (Summer 1994), pp. 5–40. These points are also

developed in his *Environment, Scarcity, and Violence* (Princeton, N.J.: Princeton University Press, 1999). Michael T. Klare, "The New Geography of Conflict," *Foreign Affairs,* Vol. 80, No. 3 (May/June 2001), pp. 49–61; and Daniel H. Deudney and Richard A. Matthew, eds., *Contested Grounds: Security and Conflict in the New Environmental Politics* (Albany: State University of New York Press, 1999). Also see Nazli Choucri and Robert C. North, *Nations in Conflict: National Growth and International Violence* (San Francisco: W.H. Freeman, 1975); and Arthur H. Westing, ed., *Global Resources and International Conflict* (Oxford: Oxford University Press, 1986).

286. Of course, local histories and social relations may influence the potency or degree to which environmental scarcity is a cause of conflict. Emphasizing this point are Nancy Lee Peluso and Michael Watts, "Violent Environments," in Nancy Lee Peluso and Michael Watts, eds., *Violent Environments* (Ithaca, N.Y.: Cornell University Press, 2001), pp. 3–38.

287. National Intelligence Council, *Global Trends 2015,* <http://www.odci.gov/cia/reports/globaltrends2015/index.html>.

288. On the problem of conflict over water resources in these areas, see Michael T. Klare, *Resource Wars: The New Landscape of Global Conflict* (New York: Metropolitan Books, 2001), pp. 138–189.

289. National Intelligence Council, *Global Trends 2015,* <http://www.odci.gov/cia/reports/globaltrends2015/index.html>.

290. Other significant sources of oil and gas in East Asia are the fields of the Tarim basin of western China and the Yadana gas field in the Andaman Sea.

291. Wilson, *On Human Nature,* p. 114.

292. For a description of the importance of the phalanx for ancient warfare, see Victor Davis Hanson, *The Western Way of Warfare: Infantry Battle in Classical Greece* (New York: Oxford University Press, 1989), pp. 9–39. For an analysis of information warfare, see Bradley A. Thayer, "The Political Effects of Information Warfare: Why New Military Capabilities Cause Old Political Dangers," *Security Studies,* Vol. 10, No. 1 (Autumn 2000), pp. 46–91.

293. Milner, Anderson, and Smith, "Warfare in Late Prehistoric West-Central Illinois," p. 590. Kroeber and Fontana document the rituals involved and the elaborate care the Maricopa, Mohaves, Quechan, and Pima tribes gave the scalps of their enemies in *Massacre on the Gila,* pp. 89–100.

294. Turney-High, *Primitive War,* pp. 158–159.

4. Implications of an Evolutionary Understanding of War

1. Geoffrey Parker, *The Military Revolution: Military Innovation and the Rise of the West, 1500–1800* (New York: Cambridge University Press, 1988); Michael Roberts, "The Military Revolution, 1560–1660," in Clifford J. Rogers, ed., *The Military Revolution Debate: Readings on the Military Transformation of Early Modern Europe* (Boulder, Colo.: Westview Press, 1995), pp. 13–35; Charles Tilly, "Reflections on the

History of European State-Making," in Charles Tilly, ed., *The Formation of National States in Western Europe* (Princeton, N.J.: Princeton University Press, 1975), pp. 3–83.

2. Robert Bigelow, *The Dawn Warriors: Man's Evolution toward Peace* (Boston: Little, Brown, 1969), pp. 5–19. Also see Roger Pitt, "Warfare and Hominid Brain Evolution," *The Journal of Theoretical Biology,* Vol. 72, No. 3 (June 1978), pp. 551–575.

3. Bigelow, *The Dawn Warriors,* pp. 5–6.

4. This evolution perhaps would be furthered if we assume mild polygamy among early humans so that the women were spared, incorporated into the victorious groups, and then propagated their genes. Ibid., pp. 6, 106–116.

5. Pitt, "Warfare and Hominid Brain Evolution," p. 564.

6. Ibid., p. 571.

7. Darius Baer and Donald L. McEachron, "A Review of Selected Sociobiological Principles: Application to Hominid Evolution I. The Development of Group Structure," *Journal of Social and Biological Structures,* Vol. 5, No. 1 (January 1982), pp. 69–90, 82. Lionel Tiger and Robin Fox also emphasize the importance of tools. They suggest that war began when tools were perfected into weapons. See Lionel Tiger and Robin Fox, *The Imperial Animal* (New York: Holt, Reinhart, and Winston, 1971).

8. Baer and McEachron, "A Review of Selected Sociobiological Principles: Application to Hominid Evolution I," p. 82.

9. Donald L. McEachron and Darius Baer, "A Review of Selected Sociobiological Principles: Application to Hominid Evolution II. The Effects of Intergroup Conflict," *Journal of Social and Biological Structures,* Vol. 5, No. 2 (April 1982), pp. 121–139, 130–131.

10. Ibid., p. 131.

11. Ibid.

12. W.D. Hamilton, "Innate Social Aptitudes of Man: An Approach from Evolutionary Genetics," in Robin Fox, ed., *Biosocial Anthropology* (New York: Wiley and Sons, 1975), pp. 133–155, 143.

13. Richard W. Byrne and Andrew Whiten, "Machiavellian Intelligence," in Andrew Whiten and Richard W. Byrne, eds., *Machiavellian Intelligence II: Extensions and Evaluations* (New York: Cambridge University Press, 1997), pp. 1–23. Also see Robin Dunbar, Chris Knight, and Camilla Power, eds., *The Evolution of Culture: An Interdisciplinary View* (Edinburgh: Edinburgh University Press, 1999); and Michael R. Rose, "The Mental Arms Race Amplifier," *Human Ecology,* Vol. 8, No. 3 (1980), pp. 285–293.

14. Richard D. Alexander, *Darwinism and Human Affairs* (Seattle: University of Washington Press, 1979), pp. 222–223.

15. Jared Diamond, *Guns, Germs, and Steel: The Fates of Human Societies* (New York: Norton, 1997), p. 289. A related argument is advanced in Andrew Bard Schmookler, *The Parable of the Tribes: The Problem of Power in Social Evolution* (Berkeley: University of California Press, 1984), pp. 74–92.

16. Diamond, *Guns, Germs, and Steel,* p. 289. Also see Robert Bigelow, "The Role of Competition and Cooperation in Human Evolution," in Martin A. Nettleship et al., eds., *War, Its Causes and Correlates* (The Hague: Mouton Publishers, 1975), pp. 235–258; and Lawrence H. Keeley, *War before Civilization: The Myth of the Peaceful Savage* (New York: Oxford University Press, 1996).

17. Diamond, *Guns, Germs, and Steel,* pp. 291–292; and Robert L. Carneiro, "A Theory of the Origin of the State," *Science,* August 21, 1970, pp. 734–737.

18. Diamond, *Guns, Germs, and Steel,* pp. 289–290.

19. For classic considerations, see Carneiro, "A Theory of the Origin of the State," and Franz Oppenheimer, *The State,* trans. by John Gitterman (New York: Free Life Editions, 1975).

20. Tilly, "Reflections on the History of European State-Making," p. 42. Also see Bruce D. Porter, *War and the Rise of the State: The Military Foundations of Modern Politics* (New York: Free Press, 1994); and Martin van Creveld, *The Rise and Decline of the State* (New York: Cambridge University Press, 1999).

21. Kenneth N. Waltz, *Theory of International Politics* (Reading, Mass.: Addison-Wesley, 1979), pp. 74–77, 127–128.

22. Carneiro, "A Theory of the Origin of the State," pp. 734–737.

23. Quoted in *The Oxford Dictionary of Thematic Quotations,* ed. by Susan Ratcliffe (New York: Oxford University Press, 2000), pp. 187–188.

24. Erich Fromm, *The Anatomy of Human Destructiveness* (London: Jonathan Cape, 1974).

25. The fact that warfare is not unique to humans may be difficult for some to accept; as Matt Ridley submits, humans tend to romanticize wildlife, emphasizing a perceived benevolence and overlooking viciousness. He views this perspective as an aspect of a "condescending sentimentalism" that causes humans to ignore unpleasant aspects of natural life. "Ground squirrels routinely eat baby ground squirrels; mallard drakes routinely drown ducks during gang rape; parasitic wasps routinely eat their victims alive from inside; chimpanzees—our nearest relatives—routinely pursue gang warfare." However, Ridley notes, these are not the images seen on television. As "supposedly objective television programmes about nature repeatedly demonstrate, human beings just do not want to know these facts. They bowdlerize nature, desperately play up the slimmest of clues to animal virtue (dolphins saving drowning people, elephants mourning their dead), and clutch at straws suggesting that humankind somehow caused aberrant cruelty." Matt Ridley, *The Origins of Virtue: Human Instincts and the Evolution of Cooperation* (New York: Viking, 1996), p. 215.

26. Magnus Enquist and Olof Leimar, "The Evolution of Fatal Fighting," *Animal Behaviour,* Vol. 39, No. 1 (January 1990), pp. 1–9.

27. L. David Mech et al., *The Wolves of Denali* (Minneapolis: University of Minnesota Press, 1998).

28. Bobbi S. Low, *Why Sex Matters: A Darwinian Look at Human Behavior* (Princeton, N.J.: Princeton University Press, 2000), p. 221.

29. H. [Hans] Kruuk, *The Spotted Hyena: A Study of Predation and Social Behavior* (Chicago: University of Chicago Press, 1972).

30. See Charles Packer, "The Ecology of Sociality in Felids," in Daniel I. Rubenstein and Richard W. Wrangham, eds., *Ecological Aspects of Social Evolution: Birds and Mammals* (Princeton, N.J.: Princeton University Press, 1986), pp. 429–451.

31. A.H. Harcourt and Frans B.M. de Waal, "Cooperation in Conflict: From Ants to Anthropoids," in Alexander H. Harcourt and Frans B.M. de Waal, eds., *Coalitions and Alliances in Humans and Other Animals* (New York: Oxford University Press, 1992), pp. 493–510, 494. For evidence of such behavior in dolphins, see Richard C. Conner, Rachel A. Smolker, and Andrew F. Richards, "Dolphin Alliances and Coalitions," in Harcourt and de Waal, eds., *Coalitions and Alliances in Humans and Other Animals,* pp. 415–443.

32. Bernard Chapais, "Alliances as a Means of Competition in Primates: Evolutionary, Developmental, and Cognitive Aspects," *Yearbook of Physical Anthropology,* Vol. 38 (New York: Wiley-Liss, 1995), pp. 115–136.

33. Edward O. Wilson, *On Human Nature* (Cambridge, Mass.: Harvard University Press, 1978), pp. 103–104. For a discussion of infanticide among primates, including humans, see Michael P. Ghiglieri, *The Dark Side of Man: Tracing the Origins of Male Violence* (Reading, Mass.: Perseus Books, 1999), pp. 129–138; Glenn Hausfater and Sarah Blaffer Hrdy, eds., *Infanticide: Comparative and Evolutionary Perspectives* (New York: Aldine, 1984); and J. Paul Scott, "The Biological Basis of Warfare," in J. Martin Ramirez, Robert A. Hinde, and Jo Groebel, eds., *Essays on Violence* (Sevilla: Publicaciones de la Universidad de Sevilla, 1987), pp. 21–43.

34. Ridley, *The Origins of Virtue,* p. 215.

35. Wilson, *On Human Nature,* p. 104.

36. Edward O. Wilson, *Sociobiology: The New Synthesis* (Cambridge, Mass.: Harvard University Press, 1975), p. 247.

37. I draw this definition from Dorothy L. Cheney, "Interactions and Relationships between Groups," in Barbara B. Smuts, Dorothy L. Cheney, Robert M. Seyfarth, Richard W. Wrangham, and Thomas T. Struhsaker, eds., *Primate Societies* (Chicago: University of Chicago Press, 1987), pp. 267–281. Also see N. Tinbergen, "On War and Peace in Animals and Man," *Science,* Vol. 160 (June 28, 1968), pp. 1411–1418.

38. Harcourt and de Waal, "Cooperation in Conflict," p. 494.

39. Ibid.

40. Ibid.

41. Harcourt argues that warfare is found among gorillas. See A.H. Harcourt, "Strategies of Emigration and Transfer by Primates, with Particular Reference to Gorillas," *Zeitschrift für Tierpsychologie,* Vol. 48 (1978), pp. 401–420.

42. On warfare among insect societies, see E.O. Wilson, *The Insect Societies* (Cambridge, Mass.: Belknap Press of Harvard University Press, 1971), pp. 447–452; among chimpanzees, see Jane Goodall, *The Chimpanzees of Gombe: Patterns of Behavior* (Cambridge, Mass.: Harvard University Press, 1986), pp. 530–534. For an excellent analy-

sis, see Johan Matheus Gerradus van der Dennen, *The Origin of War: The Evolution of a Male-Coalitional Reproductive Strategy,* 2 Vols. (Groningen, Netherlands: Origin Press, 1995), pp. 143–214.

43. Bert Hölldobler and Edward O. Wilson, *Journey to the Ants: A Story of Scientific Exploration* (Cambridge, Mass.: Harvard University Press, 1994), p. 59.

44. Ibid.

45. Wilson, *Sociobiology,* p. 50; and *The Insect Societies,* p. 451; and Hölldobler and Wilson, *Journey to the Ants,* pp. 59–73. Also see Eldridge S. Adams, "Boundary Disputes in the Territorial Ant *Azteca trigona:* Effects of Asymmetries in Colony Size," *Animal Behaviour,* Vol. 39, No. 2 (February 1990), pp. 321–328.

46. Wilson, *The Insect Societies,* p. 451.

47. E.S. Brown, "Immature Nutfall of Coconuts in the Solomon Islands: II, Changes in Ant Populations, and Their Relation to Vegetation," *Bulletin of Entomological Research,* Vol. 50, No. 3 (1959), pp. 523–558, 534.

48. Ibid., p. 534.

49. Ibid.

50. Henry C. McCook, *The Honey Ants of the Garden of the Gods, and Occident Ants of the American Plains* (Philadelphia: J.B. Lippincott, 1882), pp. 152–160. Also see Laurent Keller and H. Kern Reeve, "Dynamics of Conflicts within Insect Societies," in Keller, ed., *Levels of Selection in Evolution* (Princeton, N.J.: Princeton University Press, 1999), pp. 153–175.

51. Wilson, *Sociobiology,* p. 50.

52. Ibid.

53. Hölldobler and Wilson, *Journey to the Ants,* p. 63.

54. Bert Hölldobler, "Foraging and Spatiotemporal Territories in the Honey Ant *Myrmecocystus mimicus* Wheeler (Hymenoptera: Formicidae)," *Behavioral Ecology and Sociobiology,* Vol. 9 (1981), pp. 301–314, 309.

55. Ibid., p. 309.

56. Pitt, "Warfare and Hominid Brain Evolution," p. 570.

57. Chimpanzees also show tenderness, extraordinarily affectionate bonds among family members that often last throughout life, and male protection of females and young. Excellent ethological research by Goodall and others has significantly advanced human understanding of *Homo sapiens'* closest relative.

58. Jane Goodall, *Through a Window: My Thirty Years with the Chimpanzees of Gombe* (Boston: Houghton Mifflin, 1990), pp. 75–84, 98–111; and Goodall, *The Chimpanzees of Gombe,* pp. 313–356, 503–534.

Goodall's findings have been confirmed for chimpanzees and other primates. See Ghiglieri, *The Dark Side of Man,* pp. 165–166; Junichiro Itani, "Intraspecific Killing among Non-Human Primates," *Journal of Social and Biological Structures,* Vol. 5, No. 4 (1982), pp. 361–368; Frans de Waal, *Peacemaking among Primates* (Cambridge, Mass.: Harvard University Press, 1989), pp. 61–78; Bruce M. Knauft, "Violence and Sociality in Human Evolution," *Current Anthropology,* Vol. 32, No. 4 (August–October 1991), pp. 391–409, 403–407; and Richard Wrangham and Dale Peterson, *Demonic Males:*

Apes and the Origins of Human Violence (Boston: Houghton Mifflin, 1996), pp. 49–62, 127–152.

59. Goodall, *The Chimpanzees of Gombe,* p. 530.

60. Goodall implies this hypothesis in ibid., and Wrangham develops it fully in "Is Military Incompetence Adaptive?" *Evolution and Human Behavior,* Vol. 20, No. 1 (January 1999), pp. 1–17; Wrangham and Peterson, *Demonic Males*; and Joseph H. Manson and Wrangham, "Intergroup Aggression in Chimpanzees and Humans," *Current Anthropology,* Vol. 32, No. 4 (August–October 1991), pp. 369–390.

Wrangham's argument builds on the arguments of Alexander and van der Dennen. Alexander suggests that intergroup aggression in chimpanzees and humans arose due to the combination of male cooperation, strong territoriality, and transfer of females between social groups. Van der Dennen also stresses cooperative group living and hunting, high intelligence, and large and overlapping ranges. See Richard D. Alexander, "Evolution of the Human Psyche," in Paul Mellars and Chris Stringer, eds., *The Human Revolution: Behavioural and Biological Perspectives on the Origins of Modern Humans* (Princeton, N.J.: Princeton University Press, 1989), pp. 455–513; and van der Dennen, *The Origin of War,* Vol. 1, pp. 180–187.

61. Manson and Wrangham, "Intergroup Aggression in Chimpanzees and Humans," p. 370; and Wrangham, "Is Military Incompetence Adaptive?" p. 5.

62. Richard W. Wrangham, "Evolution of Coalitionary Killing," in *Yearbook of Physical Anthropology,* Vol. 42 (New York: Wiley-Liss, 1999), pp. 1–30, 11–12.

63. Ibid., p. 15.

64. Ibid.; and Wrangham, "Is Military Incompetence Adaptive?" p. 5.

65. Also on this point, see Low, *Why Sex Matters,* pp. 221–223.

66. Ibid., p. 222.

67. Toshisada Nishida, "The Social Structure of Chimpanzees of the Mahale Mountains," in David A. Hamburg and Elizabeth R. McCown, eds., *The Great Apes* (Menlo Park, Calif.: Cummings, 1979), pp. 73–121; Christophe Boesch and Hedwige Boesch-Achermann, *The Chimpanzees of the Taï Forest: Behavioural Ecology and Evolution* (New York: Oxford University Press, 2000).

68. On chimpanzee hunting group behavior, see Christophe Boesch and Hedwige Boesch, "Hunting Behavior of Wild Chimpanzees in the Taï National Park," *American Journal of Physical Anthropology,* Vol. 78, No. 4 (April 1989), pp. 547–573.

69. Goodall, *The Chimpanzees of Gombe,* p. 490.

70. Ibid.

71. Ibid.

72. Ibid. Boesch and Boesch-Achermann also note these behaviors in *The Chimpanzees of the Taï Forest,* p. 137.

73. Goodall, *The Chimpanzees of Gombe,* p. 491.

74. Boesch and Boesch-Achermann, *The Chimpanzees of the Taï Forest,* p. 137.

75. Ibid.

76. Ibid.

77. Wrangham, "Evolution of Coalitionary Killing," p. 7. Wrangham also reports

on frequent border avoidance as documented by observers and by higher prey densities in these areas. See Wrangham, "Evolution of Coalitionary Killing," and Craig B. Stanford, *Chimpanzee and Red Colobus: The Ecology of Predator and Prey* (Cambridge, Mass.: Harvard University Press, 1998).

78. Boesch and Boesch-Achermann, *The Chimpanzees of the Taï Forest,* p. 136.

79. Wrangham, "Evolution of Coalitionary Killing," p. 7.

80. Boesch and Boesch-Achermann, *The Chimpanzees of the Taï Forest,* p. 136.

81. Ibid., p. 144.

82. Ibid., pp. 137, 141–143.

83. Wrangham, "Evolution of Coalitionary Killing," p. 8.

84. In addition to the struggle between the Kasakela and Kahama communities described below, Goodall documents three such encounters between the Kasakela males and males from a northern group, three encounters between Kahama males and the Kalande males to their south, and at least one between Kasakela males and "Eastern" males. Goodall, *The Chimpanzees of Gombe,* pp. 492–493.

85. Ibid., p. 529. Emphasis in original.

86. Ibid., pp. 488–532.

87. Ibid., p. 504.

88. Ibid.

89. Hugh disappeared around 1973, when the two communities were formed, and Goodall suggests that because he was not old he may have fallen victim to intercommunity aggression. Godi was attacked in January 1974 and died shortly afterward. Dé was attacked and severely injured in February of the same year, and was never seen again after April 1974, despite intensive searching. Goliath was attacked in February 1975 and never seen again despite intensive searching. Charlie was attacked, although this was not observed, and his body found in May 1977. Sniff, the last male, was attacked in November 1977 and died shortly afterward. What happened to Willy Wally is not reported, although he failed to answer Sniff's calls in mid-1977, suggesting that he had died. Madam Bee was attacked several times by the Kasakela males, the final time fatally, in September 1975. Ibid., pp. 61, 506–514.

90. Goodall notes that when adult males encounter such females, aggression is rare and it is usually mild, unlike the twenty-five encounters with anestrous mothers and infants the Kasakela males had. Of these encounters, 76 percent were aggressive, and all but one of the fifteen actual attacks on the older females were extremely severe. Ibid., pp. 493–502.

91. Wrangham, "Evolution of Coalitionary Killing," p. 11.

92. See Nishida, "The Social Structure of Chimpanzees of the Mahale Mountains," pp. 73–121; and Nishida, Mariko Hiraiwa-Hasegawa, Toshikazu Hasegawa, and Yukio Takahata, "Group Extinction and Female Transfer in Wild Chimpanzees in the Mahale National Park, Tanzania," *Zeitschrift für Tierpsychologie,* Vol. 67 (1985), pp. 284–301.

93. Nishida et al., "Group Extinction and Female Transfer in Wild Chimpanzees in the Mahale National Park, Tanzania," p. 288. Also see Toshisada Nishida, Hiroyuki Takasaki, and Yukio Takahata, "Demography and Reproductive Profiles," in Toshisada

Nishida, ed., *The Chimpanzees of the Mahale Mountains* (Tokyo: University of Tokyo Press, 1990), pp. 63–97, 86.

94. Nishida et al., "Group Extinction and Female Transfer in Wild Chimpanzees in the Mahale National Park, Tanzania," p. 297.

95. Wrangham, "Evolution of Coalitionary Killing," pp. 8–9.

96. Nishida et al., "Group Extinction and Female Transfer in Wild Chimpanzees in the Mahale National Park, Tanzania," pp. 289–292.

97. Wrangham, "Evolution of Coalitionary Killing," p. 9. Also see Ghiglieri, *The Dark Side of Man*, pp. 174–175.

98. Wrangham, "Evolution of Coalitionary Killing," pp. 10–11. Also see Boesch and Boesch-Achermann, *The Chimpanzees of the Taï Forest*, pp. 153–157.

99. Wrangham, "Evolution of Coalitionary Killing," p. 12.

100. Goodall, *The Chimpanzees of Gombe*, p. 527. Emphasis in original.

101. Ibid.

102. Boesch and Boesch-Achermann, *The Chimpanzees of the Taï Forest*, Table 7.9, p. 139.

103. Ibid., Table 7.9, p. 139.

104. Ibid., pp. 144–145.

105. Wrangham, "Evolution of Coalitionary Killing," p. 12.

106. Keith F. Otterbein, *The Evolution of War: A Cross-Cultural Study*, 2nd ed. (New Haven: Human Relations Area Files Press, 1985), p. xxii; and Keith F. Otterbein, "The Origins of War," *Critical Review*, Vol. 11, No. 2 (Spring 1997), pp. 251–277, 253.

107. Otterbein, *The Evolution of War*, p. xxii.

108. Wrangham, "Evolution of Coalitionary Killing," p. 3; and Wrangham, "Is Military Incompetence Adaptive?" p. 6.

109. Goodall, *The Chimpanzees of Gombe*, p. 528. Emphasis in original.

110. Ibid.

111. Ibid.

112. Ibid., p. 534.

113. Ibid., p. 532.

114. Ibid.

115. This quote is taken from the "Note of 10 July 1827" concerning the revision of *On War*. See Carl von Clausewitz, *On War*, ed. and trans. by Michael Howard and Peter Paret (Princeton, N.J.: Princeton University Press, 1976), p. 69. Emphasis omitted. The same point is made in *On War* on pp. 88 and 605. Margaret Mead has provided the classic anthropological definition of war as intergroup conflict that is organized and socially sanctioned. Mead, "Alternatives to War," in Milton Fried, Marvin Harris, and Robert Murphy, eds., *War* (Garden City, N.Y.: Natural History Press, 1968), pp. 215–228.

116. Von Clausewitz, *On War*, p. 605.

117. Bernard Brodie expertly demonstrates this point in his *War and Politics* (New York: Macmillan, 1973).

118. Most famously, S.L.A. Marshall argued that in World War II on average only 15 percent, and at most 25 percent, of American infantrymen fired their weapons in combat. Marshall, *Men against Fire: The Problem of Battle Command in Future War* (New York: William Morrow and Co., 1947), p. 54. Also see Barbara Ehrenreich, *Blood Rites: Origins and History of the Passions of War* (New York: Metropolitan, 1997); Dave Grossman, *On Killing: The Psychological Cost of Learning to Kill in War and Society* (Boston: Little, Brown, 1995); Robert A. Hinde, "Aggression and War: Individuals, Groups, and States," in Philip E. Tetlock, Jo. L. Husbands, Robert Jervis, Paul C. Stern, and Charles Tilly, eds., *Behavior, Society, and International Conflict,* Vol. 3 (New York: Oxford University Press, 1993), pp. 8–70; John Keegan, *The Face of Battle: A Study of Agincourt, Waterloo, and the Somme* (New York: Viking Press, 1976); and Anthony Kellett, *Combat Motivation: The Behavior of Soldiers in Battle* (Boston: Kluwer, 1982). On cowardice in battle, see William Ian Miller, *The Mystery of Courage* (Cambridge, Mass.: Harvard University Press, 2000).

119. Quoted in Richard Holmes, *Acts of War: The Behavior of Men in Battle* (New York: Free Press, 1986), p. 31.

120. Quoted in Paul Fussell, *Wartime: Understanding and Behavior in the Second World War* (New York: Oxford University Press, 1989), p. 290.

121. William James, *Memories and Studies* (New York: Longmans Green, 1911), p. 282.

122. Ibid.

123. Winston S. Churchill, *The Story of the Malakand Field Force: An Episode of Frontier War* (New York: Norton, 1990 [1898]), p. 9.

124. J. Glenn Gray, *The Warriors: Reflections on Men in Battle* (New York: Harcourt, Brace and Company, 1959), p. 44.

125. Ibid.

126. Ibid.

127. James Jones, *The Thin Red Line* (New York: Charles Scribner's Sons, 1962), p. 200.

128. Ibid.

129. Jack Belden, *Still Time to Die* (New York: Harper and Brothers, 1943), p. 254.

130. Harry Brown, *A Walk in the Sun* (New York: Knopf, 1944), p. 116.

131. Ibid.

132. Gray, *The Warriors,* p. 33.

133. Ibid.

134. Ibid., pp. 33, 36.

135. Ibid., pp. 51–52.

136. Ibid., p. 51.

137. Ibid., p. 40.

138. Ibid.

139. Ibid.

140. Robert Ardrey also notes this in his *The Territorial Imperative: A Personal Inquiry into the Animal Origins of Property and Nations* (New York: Atheneum, 1966), pp. 334–

336. *New York Times* journalist Chris Hedges makes the same argument based on his extensive experiences covering wars in Central America, the Middle East, and the Balkans. Chris Hedges, *War Is a Force That Gives Us Meaning* (New York: PublicAffairs, 2002).

141. Robert F. Murphy, "Intergroup Hostility and Social Cohesion," *American Anthropologist,* Vol. 59, No. 6 (December 1957), p. 1034.

142. Camilla H. Wedgwood, "Some Aspects of Warfare in Melanesia," *Oceania,* Vol. 1 (1930), pp. 5–33, 32.

143. David M. Kennedy provides an excellent account of how the U.S. government and civil society worked together to mobilize resources in his *Over Here: The First World War and American Society* (New York: Oxford University Press, 1980). Bartholomew H. Sparrow provides an excellent similar discussion concerning World War II. See his *From the Outside In: World War II and the American State* (Princeton, N.J.: Princeton University Press, 1996). On social imperialism, see Lewis A. Coser, *The Function of Social Conflict* (Glencoe, Ill.: Free Press, 1956); and Richard N. Rosecrance, *Action and Reaction in World Politics* (Boston: Little, Brown, 1963).

144. E.B. Sledge, *With the Old Breed at Peleliu and Okinawa* (New York: Oxford University Press, 1990), p. 266.

145. Ibid., p. 315.

146. Ibid.

147. Guy Sajer, *The Forgotten Soldier* (New York: Harper and Row, 1971), p. 113. Another exceptional German memoir of conflict on the Eastern Front in World War II, particularly during the final years of the conflict, is Gottlob Herbert Bidermann, *In Deadly Combat: A German Soldier's Memoir of the Eastern Front,* trans. by Derek S. Zumbro (Lawrence: University Press of Kansas, 2000).

148. William Manchester, *Goodbye Darkness: A Memoir of the Pacific War* (New York: Little, Brown, 1980), p. 391.

149. Ibid.

150. Quoted in Marvin Jensen, *Strike Swiftly!: The 70th Tank Battalion—From North Africa to Germany* (Novato, Calif.: Presidio Press, 1997), p. 326.

151. David H. Hackworth and Julie Sherman, *About Face* (New York: Simon and Schuster, 1989), p. 111.

152. Ibid.

153. Ibid., p. 112. Emphasis in original.

154. Lt. Gen. Harold G. Moore, USA (Ret.) and Joseph L. Galloway, *We Were Soldiers Once . . . and Young: Ia Drang: The Battle That Changed the War in Vietnam* (New York: Random House, 1992), p. xviii.

155. William Broyles Jr., "Why Men Love War," *Esquire* (November 1984), p. 55.

156. Ibid.

157. Ibid.

158. Ibid., pp. 56, 58.

159. Ibid., p. 58.

160. Philip Caputo, *A Rumor of War* (New York: Holt, Rinehart and Winston, 1977), p. 81.

161. Joanna Bourke, *An Intimate History of Killing: Face-to-Face Killing in Twentieth-Century Warfare* (New York: Basic Books, 1999), pp. 1–32.

162. Al Santoli, ed., *Everything We Had: An Oral History of the Vietnam War by Thirty-Three American Soldiers Who Fought It* (New York: Random House, 1981), pp. 98–99; quoted in Bourke, *An Intimate History of Killing,* p. 20.

163. Ibid.

164. Murphy, "Intergroup Hostility and Social Cohesion," p. 1028.

165. Sledge, *With the Old Breed at Peleliu and Okinawa,* p. 34.

166. Ibid.

167. Ibid.

168. Ibid.

169. Ibid. Gerald F. Linderman provides a detailed description of the feelings of American soldiers, sailors, and Marines toward the Japanese in *The World within War: America's Combat Experience in World War II* (New York: Free Press, 1997), pp. 143–184. Equally illuminating and poignant are the interviews Patrick O'Donnell conducted with veterans of the Pacific War. See Patrick K. O'Donnell, *Into the Rising Sun: In Their Own Words, World War II's Pacific Veterans Reveal the Heart of Combat* (New York: Free Press, 2002).

170. Eric Bergerud, *Touched with Fire: The Land War in the South Pacific* (New York: Viking Penguin, 1996), pp. 406–407.

171. Ibid., p. 407.

172. Quoted in ibid., pp. 411–412. Sledge, *With the Old Breed at Peleliu and Okinawa,* pp. 120, 123, 152.

173. Quoted in Bergerud, *Touched with Fire,* p. 412.

174. Bourke, *An Intimate History of Killing,* p. 150.

175. Caputo, *A Rumor of War,* p. 231.

176. Ibid.

177. Quoted in Bourke, *An Intimate History of Killing,* p. 170.

178. Robert Graves, *Goodbye to All That* (London: Cassell and Co., 1929), p. 163.

179. Ibid.

180. Ibid., p. 164. Another exceptional memoir of the life of a British infantryman in World War I is W.H.A. Groom, *Poor Bloody Infantry: A Memoir of the First World War* (London: William Kimber, 1976).

181. While the Malmédy massacre has rightfully captivated American interest and the British remain concerned about the May 1940 massacre of ninety-nine prisoners of the Royal Norfolk Regiment by men of the SS *Totenkopf* Division, the war on the Eastern Front was much worse for both German and Soviet soldiers captured by the enemy. About 57 percent (3,300,000) of the total of 5,700,000 Soviet prisoners captured by the Germans in World War II perished. See Christian Streit, "The Fate of the Soviet Prisoners of War," in Michael Berenbaum, ed., *A Mosaic of Victims: Non-Jews Persecuted and Murdered by the Nazis* (New York: New York University Press, 1990), pp. 142–149.

182. The order is quoted in Stephen E. Ambrose, *Citizen Soldiers: The U.S. Army from the Normandy Beaches to the Bulge to the Surrender of Germany June 7,*

1944–May 7, 1945 (New York: Simon and Schuster, 1997), p. 354. Linderman found that killing SS soldiers was a relatively common event. See Linderman, *The World within War,* p. 120.

183. Belton Y. Cooper, *Death Traps: The Survival of an American Armored Division in World War II* (Novato, Calif.: Presidio, 1998), p. 172.

184. Ibid.

185. Quoted in Jensen, *Strike Swiftly!,* pp. 269–270.

186. Sajer, *The Forgotten Solder,* p. 186.

187. Holmes, *Acts of War,* p. 384.

188. Ibid.

189. An excellent account of Japanese atrocities against prisoners of war is Yuki Tanaka, *Hidden Horrors: Japanese War Crimes in World War II* (Boulder, Colo.: Westview Press, 1996). For a detailed account of biological weapons experiments conducted on prisoners of war, as well as mostly Chinese and Russian civilians, see Sheldon H. Harries, *Factories of Death: Japanese Biological Warfare, 1932–45, and the American Cover-Up* (New York: Routledge, 1994).

190. Tanaka, *Hidden Horrors,* p. 2.

191. Ibid.

192. Quoted in Jonathan Lewis and Ben Steele, *Hell in the Pacific: From Pearl Harbor to Hiroshima and Beyond* (London: Channel Four Books, 2001), p. 81.

193. Bourke, *An Intimate History of Killing,* p. 171.

194. Quoted in Bourke, *An Intimate History of Killing,* p. 170.

195. Ambrose, *Citizen Soldiers,* pp. 352–353.

196. Quoted in Bergerud, *Touched with Fire,* p. 413.

197. Quoted in Ibid.

198. Quoted in Ibid.

199. Quoted in Ibid.

200. Quoted in Ibid., pp. 413–414.

201. Quoted in Ibid., p. 414.

202. George MacDonald Fraser, *Quartered Safe Out Here: A Recollection of the War in Burma* (London: Harvill, 1992), p. 191. Emphasis in original.

203. Ibid.

204. Ibid.

205. Broyles, "Why Men Love War," p. 62. Also see Joshua S. Goldstein, *War and Gender: How Gender Shapes the War System and Vice Versa* (Cambridge: Cambridge University Press, 2001), pp. 333–356.

206. Quoted in Holmes, *Acts of War,* p. 57.

207. Mark Bowden, *Black Hawk Down: A Story of Modern War* (New York: Atlantic Monthly Press, 1999), p. 254.

208. Gray, *The Warriors,* p. 61.

209. Ibid., pp. 61–62.

210. Also noting this are David M. Buss and Todd K. Shackelford. "Human Aggression in Evolutionary Psychological Perspective," *Clinical Psychological Review,*

Vol. 17, No. 6 (1997), pp. 605–619, 612–613; and Paul R. Ehrlich, *Human Natures: Genes, Cultures, and the Human Prospect* (Washington, D.C.: Island Press, 2000), pp. 175–178.

211. David B. Adams, "Why Are There So Few Women Warriors," *Behavior Science Research,* Vol. 18, No. 3 (1983), pp. 196–212.

212. Harry Holbert Turney-High, *Primitive War: Its Practice and Concepts* (Columbia: University of South Carolina Press, 1949), p. 154.

213. Goldstein, *War and Gender,* pp. 60–64.

214. For an analysis of female Soviet soldiers, see Susanne Conze and Beate Fieseler, "Soviet Women as Comrades-in-Arms: A Blind Spot in the History of the War," in Robert W. Thurston and Bernd Bonwetsch, eds., *The People's War: Responses to World War II in the Soviet Union* (Urbana: University of Illinois Press, 2000), pp. 211–234. Also see Goldstein, *War and Gender,* pp. 64–70. The most detailed account in English of Soviet women aerial combatants in World War II is found in Reina Pennington, *Wings, Women, and War: Soviet Airwomen in World War II Combat* (Lawrence: University Press of Kansas, 2002).

215. Holmes, *Acts of War,* p. 102.

216. For examples of women warriors, see Bourke, *An Intimate History of Killing,* pp. 294–333.

217. For a discussion of Jewish women fighting against the Germans in World War II see Rich Cohen, *The Avengers: A Jewish War Story* (New York: Knopf, 2000). Also see Robert B. Edgerton, *Warrior Women: The Amazons of Dahomey and the Nature of War* (Boulder, Colo.: Westview Press, 2000). Other examples are found in Martin van Creveld's study, *Men, Women and War: Do Women Belong in the Front Line?* (London: Cassell and Co., 2001). Van Creveld argues that the greater presence of women in Western militaries is the result of the decline of these militaries due to the lack of great power wars caused by the development of nuclear weapons.

218. John Tooby and Leda Cosmides, "The Evolution of War and Its Cognitive Foundations," *Proceedings of the Institute of Evolutionary Studies,* Vol. 88 (1988), pp. 1–15.

219. The most fecund women appears to have been a nineteenth-century Russian woman who had sixty-nine children, many of whom were triplets. Ehrlich, *Human Natures,* p. 190; and Jared Diamond, *The Third Chimpanzee: The Evolution and Future of the Human Animal* (New York: HarperCollins Publishers, 1992), p. 88.

220. Low, *Why Sex Matters,* p. 215.

221. Adams, "Why Are There So Few Women Warriors," p. 210.

222. Low, *Why Sex Matters,* p. 217.

223. I focus on the use of biological weapons against humans, although biological weapons may also be directed against the animals upon which a population depends for food. Near the end of World War II in Europe, the German government was prepared to use the foot-and-mouth virus against British cattle. Roger Boyes, "Nazis Planned to Use Virus against Britain," *The Times,* March 12, 2001, <http://www.thetimes.co.uk/article/0,,2–97518,00.html>. An exceptional overview of the history of biological weapons is Erhard Geissler and John Ellis van Courtland Moon,

eds., *Biological and Toxin Weapons: Research, Development and Use from the Middle Ages to 1945* (Oxford: Oxford University Press, 1999).

224. See Arno Karlen, *Man and Microbes: Disease and Plagues in History and Modern Times* (New York: Simon and Schuster, 1995), pp. 79–91.

225. George W. Christopher, Theodore J. Cieslak, Julie A. Pavlin, and Edward M. Eitzen Jr., "Biological Warfare: A Historical Perspective," in Joshua Lederberg, ed., *Biological Weapons: Limiting the Threat* (Cambridge, Mass.: MIT Press, 1999), pp. 17–35, 18. The three major plague outbreaks in history are estimated to have killed 200 million people. The first plague pandemic swept the Western world beginning in A.D. 541, and the third started in China in 1894. By 1900, it had spread worldwide, killing 12.5 million in India alone.

226. Christopher Wills, *Yellow Fever, Black Goddess: The Coevolution of People and Plagues* (Reading, Mass.: Addison Wesley, 1996), p. 62.

227. Norman F. Cantor, *In the Wake of the Plague: The Black Death and the World It Made* (New York: The Free Press, 2001), p. 7. Cantor argues that a simultaneous outbreak of anthrax also contributed to the high mortality. See also Wills, *Yellow Fever, Black Goddess,* pp. 53–89.

228. Christopher et al., "Biological Warfare," pp. 18–19. Also see Noble David Cook, *Born to Die: Disease and New World Conquest, 1492–1650* (New York: Cambridge University Press, 1998), p. 214; and E. Wagner Stearn and Allen E. Stearn, *The Effect of Smallpox on the Destiny of the Amerindian* (Boston: Bruce Humphries, Inc., 1945), pp. 44–45.

229. Donald R. Hopkins, *Princes and Peasants: Smallpox in History* (Chicago: University of Chicago Press, 1983), pp. 258–259; and Jonathan B. Tucker, *Scourge: The Once and Future Threat of Smallpox* (New York: Atlantic Monthly Press, 2001), p. 21. See also Elizabeth Anne Fenn, *Pox Americana: The Great Smallpox Epidemic of 1775–82* (New York: Hill and Wang, 2001), pp. 44–103.

230. Hopkins, *Princes and Peasants,* pp. 260–261; and Tucker, *Scourge,* p. 22. Hopkins explains that variolation or inoculation is done by inserting "pus or powdered scabs containing smallpox virus from a previous patient into the skin of a susceptible person to cause" the infection deliberately. A person infected in this way could still transmit the disease. With vaccination, the weakened smallpox virus or a similar virus is inserted into the skin. Hopkins, *Princes and Peasants,* p. 7.

231. The scope of the Japanese program is well addressed by Harries, *Factories of Death.*

232. Ken Alibek did much to publicize this information. He had been Deputy Director of the Soviet/Russian program. See Ken Alibek with Stephen Handelman, *Biohazard: The Chilling True Story of the Largest Covert Biological Weapons Program in the World—Told from Inside by the Man Who Ran It* (New York: Delta, 1999); and Tom Mangold and Jeff Goldberg, *Plague Wars: A True Story of Biological Warfare* (New York: St. Martin's Press, 1999). On the Soviet program, with particular insight into the 1979 Sverdlovsk anthrax outbreak, see Jeanne Guillemin, *Anthrax: The Investigation of a Deadly Outbreak* (Berkeley: University of California Press, 1999). For an ac-

count of the offensive U.S. biological weapons program before President Richard Nixon ordered it terminated in 1969, see Ed Regis, *The Biology of Doom: The History of America's Secret Germ Warfare Project* (New York: Henry Holt and Co., 1999).

233. See Richard A. Falkenrath, Robert D. Newman, and Bradley A. Thayer, *America's Achilles' Heel: Nuclear, Biological, and Chemical Terrorism and Covert Attack* (Cambridge, Mass.: MIT Press, 1998); and Bruce Hoffman, *Inside Terrorism* (New York: Columbia University Press, 1998).

234. The anthrax attacks, their victims, and the aftermath are well detailed in Marilyn W. Thompson, *The Killer Strain: Anthrax and a Government Exposed* (New York: HarperCollins, 2003).

235. Particularly insightful is Hoffman, *Inside Terrorism,* pp. 87–130.

236. Both incidents are explained in Karl Vick, "Plea Bargain Rejected in Bubonic Plague Case," *Washington Post,* April 3, 1996, p. A8; for an analysis of the Larry Harris incidents, see Jessica Eve Stern, "Larry Wayne Harris (1998)," in Jonathan B. Tucker, ed., *Toxic Terror: Assessing Terrorist Use of Chemical and Biological Weapons* (Cambridge, Mass.: MIT Press, 2000), pp. 227–246.

237. David E. Kaplan, "Aum Shinrikyo (1995)," in Tucker, ed., *Toxic Terror,* pp. 207–226.

238. Richard Danzig and Pamela B. Berkowsky, "Why Should We Be Concerned about Biological Warfare," in Lederberg, ed., *Biological Weapons,* pp. 9–14.

239. Kathleen M. Vogel, "Pathogen Proliferation: Threats from the Former Soviet Bioweapons Complex," *Politics and the Life Sciences,* Vol. 19, No. 1 (March 2000), pp. 3–16.

240. Rick Weiss, "Germ Tests Point Away From Iraq," *Washington Post,* October 30, 2001, p. A9.

241. Jonathan B. Tucker, "Introduction," in Tucker, ed., *Toxic Terror,* pp. 1–14, 8.

242. Adrian V.S. Hill and Arno G. Motulsky, "Genetic Variation and Human Disease: The Role of Natural Selection," in Stephen C. Stearns, ed., *Evolution in Health and Disease* (New York: Oxford University Press, 1999), pp. 50–61; and Joshua Lederberg, "Viruses and Humankind: Intracellular Symbiosis and Evolutionary Competition," in Stephen S. Morse, ed., *Emerging Viruses* (New York: Oxford University Press, 1993), pp. 3–9. Tucker notes that the Soviet Union selected "India-1967" for weaponization because it killed about 30 percent of those infected, it was very stable as an aerosol, and it maintained its potency for long periods of time. Tucker, *Scourge,* p. 141. See also Frank Ryan, *Virus X: Tracking the New Killer Plagues* (Boston: Little, Brown, 1997).

243. On interspecies transfer of viruses, see Frank Fenner, "Human Monkeypox, A Newly Discovered Human Virus Disease," in Morse, ed., *Emerging Viruses,* pp. 176–183; and Colin R. Parish, "Canine Parvovirus 2: A Probable Example of Interspecies Transfer," in Morse, ed., *Emerging Viruses,* pp. 194–202. On climatic change and disease, see Thomas E. Lovejoy, "Global Change and Epidemiology: Nasty Synergies," in Morse, ed., *Emerging Viruses,* pp. 261–268.

244. William H. McNeill, "Patterns of Disease Emergence in History," in Morse, ed., *Emerging Viruses,* pp. 29–36, 36.

245. David Brown, "On the Trail of the 1918 Influenza Epidemic," *Washington Post,* February 27, 2001, p. A3. Also see Alfred W. Crosby, *America's Forgotten Pandemic: The Influenza of 1918* (New York: Cambridge University Press, 1989), pp. 20–21, 37. The deadlier fall strain first appeared in the same week of August 1918 in Brest, Boston, and Freetown. In September 1918, it is estimated that 3 percent of the entire indigenous population of Sierra Leone died. Crosby, *America's Forgotten Pandemic,* p. 38.

246. Tucker, *Scourge,* p. 159.

247. Rick Weiss, "Clarifying the Facts and Risks of Anthrax," *Washington Post,* October 18, 2001, p. A12. Also see Judith Miller, Stephen Engelberg, and William Broad, *Germs: Biological Weapons and America's Secret War* (New York: Simon and Schuster, 2001), pp. 66–97, 134–137.

248. Tucker, *Scourge,* p. 204. Also see Miller, Engelberg, and Broad, *Germs,* p. 42.

249. Alfred W. Crosby, *Ecological Imperialism: The Biological Expansion of Europe, 900–1900* (New York: Cambridge University Press, 1986); William H. McNeill, *Plagues and Peoples* (Garden City, N.Y.: Anchor Press/Doubleday, 1976); and Diamond, *Guns, Germs, and Steel.*

250. Diamond, *The Third Chimpanzee,* p. 187.

251. Ibid. Also see William H. McNeill, *The Global Condition: Conquerors, Catastrophes, and Community* (Princeton, N.J.: Princeton University Press, 1992), pp. 83–84.

252. Diamond, *The Third Chimpanzee,* p. 187.

253. Noting the influence of geography on economic development is Ricardo Hausmann, "Prisoners of Geography," *Foreign Policy,* No. 122 (January/February 2001), pp. 44–53; and David S. Landes, *The Wealth and Poverty of Nations: Why Some Are So Rich and Some So Poor* (New York: Norton, 1999).

254. Diamond, *Guns, Germs, and Steel,* p. 207; Crosby, *Ecological Imperialism,* p. 31; and McNeill, *The Global Condition,* p. 84.

255. McNeill, *The Global Condition,* p. 42.

256. Ibid.

257. The total population of the Americas is estimated to have been as low as 8.4 million or as high as 112 million at the conclusion of the fifteenth century.

258. Alfred W. Crosby, *Germs, Seeds, and Animals: Studies in Ecological History* (Armonk, N.Y.: M.E. Sharpe, 1994), p. 21.

259. Russell Thornton, *American Indian Holocaust and Survival: A Population History Since 1492* (Norman: University of Oklahoma Press, 1987), p. 25.

260. David E. Stannard, *American Holocaust: The Conquest of the New World* (New York: Oxford University Press, 1992), p. 268.

261. William Denevan, "Native American Populations in 1492: Recent Research and a Revised Hemispheric Estimate," in William M. Denevan, ed., *The Native Population of the Americas in 1492,* 2nd ed. (Madison: University of Wisconsin Press, 1992), p. xxviii.

262. Thornton, *American Indian Holocaust and Survival,* pp. 36–37.

263. Steven T. Katz, *The Holocaust in Historical Context,* 3 Vols. (New York: Oxford University Press, 1994), Vol. 1, p. 88.

264. Ibid.

265. Ibid., Vol. 1, pp. 88–89.

266. Ibid., Vol. 1, p. 89.

267. Ibid.

268. Thornton, *American Indian Holocaust and Survival,* p. 42.

269. Ibid., p. 43.

270. Katz, *The Holocaust in Historical Context,* Vol. 1, pp. 89–90.

271. Denevan, "Native American Populations in 1492," p. xxix; also see Cook, *Born to Die,* p. 5.

272. Katz, *The Holocaust in Historical Context,* Vol. 1, p. 91.

273. For an account of the cruelties of the Europeans and, later, Americans in the New World, see Richard Drinnon, *Facing West: The Metaphysics of Indian-Hating and Empire-Building* (Minneapolis: University of Minnesota Press, 1980); and Stannard, *American Holocaust.*

274. Alfred W. Crosby, *The Columbian Exchange: Biological and Cultural Consequences of 1492* (Westport, Conn.: Greenwood Press, 1972), p. 38.

275. Ibid., p. 39.

276. Cook, *Born to Die,* p. 206.

277. Crosby, *Germs, Seeds, and Animals,* p. 101.

278. Thornton, *American Indian Holocaust and Survival,* pp. 39–40.

279. Cook, *Born to Die,* p. 58.

280. Ibid., p. 207.

281. Crosby, *Germs, Seeds, and Animals,* p. xi.

282. Hopkins, *Princes and Peasants,* pp. 1–3.

283. Ibid., p. 4.

284. Ibid.

285. Crosby, *Germs, Seeds, and Animals,* p. 58; and Hopkins, *Princes and Peasants,* p. 204.

286. Crosby, *The Columbian Exchange,* p. 38; and Cook, *Born to Die,* p. 86.

287. Of all vulnerable populations, the Maori of New Zealand seem to have been the most fortunate. Europeans came to New Zealand after the spread of vaccination in Europe and so were poor carriers. Two-thirds of the Maori population had been vaccinated before smallpox came to New Zealand in the 1850s. Crosby, *Germs, Seeds, and Animals,* p. 37.

288. Henry F. Dobyns, *Their Number Become Thinned* (Knoxville: University of Tennessee Press, 1983), p. 14.

289. An exceptional account of the campaign is Victor Davis Hanson, *Carnage and Culture: Landmark Battles in the Rise of Western Power* (New York: Doubleday, 2001), pp. 170–232.

290. Dobyns, *Their Number Become Thinned,* p. 11. Also see Abel A. Alves, *Bru-*

tality and Benevolence: Human Ethology, Culture, and the Birth of Mexico (Westport, Conn.: Greenwood Press, 1996).

291. Dobyns, *Their Number Become Thinned,* p. 12; and Hopkins, *Princes and Peasants,* pp. 208–213.

292. Quoted in Crosby, *The Columbian Exchange,* p. 41.

293. Thornton, *American Indian Holocaust and Survival,* p. 65.

294. Ibid., p. 71.

295. Ibid., p. 74; Cook, *Born to Die,* p. 195.

296. Thornton, *American Indian Holocaust and Survival,* pp. 79–80.

297. Stearn and Stearn, *The Effect of Smallpox on the Destiny of the Amerindian,* p. 37.

298. R. Brian Ferguson, "A Reexamination of the Causes of Pacific Northwest Warfare," in R. Brian Ferguson, ed., *Warfare, Culture, and Environment* (Orlando, Fla.: Academic Press, 1984), pp. 267–328, 273.

299. Ibid., p. 273.

300. Dobyns, *Their Number Become Thinned,* pp. 15–23.

301. Ibid.

302. Thornton, *American Indian Holocaust and Survival,* pp. 94–95.

303. R. Brian Ferguson, *Yanomami Warfare: A Political History* (Santa Fe, N.Mex.: School of American Research Press, 1995), p. 22.

304. John Saffirio and Raymond Hames, "The Forest and the Highway," in *The Impact of Contact: Two Yanomamö Case Studies* (Bennington, Vt.: Working Papers on South American Indians, No. 6, 1983), pp. 1–52, 5; quoted in Ferguson, *Yanomami Warfare,* p. 26.

305. Crosby, *Germs, Seeds, and Animals,* p. 100.

306. Alan H. Jacobs, "Maasai Inter-Tribal Relations: Belligerent Herdsmen or Peaceable Pastoralists," in Katsuyoshi Fukui and David Turton, eds., *Warfare among East African Herders* (Osaka: National Museum of Ethnology, 1979), pp. 33–52, 46–47.

307. Diamond, *Guns, Germs, and Steel,* p. 212.

308. Ibid., pp. 212–213.

309. Philip D. Curtin, *Disease and Empire: The Health of European Troops in the Conquest of Africa* (New York: Cambridge University Press, 1998), p. 5.

310. Curtin, *Disease and Empire,* p. 5.

311. Philip D. Curtin, *Death by Migration: Europe's Encounter with the Tropical World in the Nineteenth Century* (New York: Cambridge University Press, 1989), p. 6.

312. Curtin, *Disease and Empire,* p. 5.

313. Ibid.

314. Ibid., pp. 49–52.

315. This was also true of isolated European settlements, such as those in Iceland. Smallpox reached that island only in 1241 or 1306, and returned after a long absence in 1707, when about eighteen thousand people, or one-third of the entire population, died. Crosby, *Ecological Imperialism,* p. 52.

316. Curtin, *Disease and Empire,* p. 113. During the Spanish-American War, U.S. troops were greatly affected by typhoid, even before moving overseas. U.S. soldiers experienced a typhoid morbidity of 192 per thousand, and 14.63 deaths per thousand, which is about eight times the morbidity and three times the mortality the British experienced at the same time in India. Curtin, *Disease and Empire,* p. 124.

317. Arthur Conan Doyle quoted in Curtin, *Disease and Empire,* p. 212.

318. Ibid., p. 219.

319. For example, the deaths from disease for British troops in garrison in Britain averaged only 4.31 per thousand from 1886–1894.

320. Disease was also transmitted from one continent to another continent. For example, the yellow fever spread from Africa to the New World in 1647, and leprosy from Asia to Polynesia in the late eighteenth century.

321. Sheldon Watts, *Epidemics and History: Disease, Power and Imperialism* (New Haven: Yale University Press, 1997), pp. 122–130.

322. Raymond Pearson, *The Longman Companion to European Nationalism 1789–1920* (London: Longman, 1994), p. 237.

323. Watts, *Epidemics and History,* pp. 174–175.

324. Holbrooke is quoted in Christopher S. Wren, "Era Waning, Holbrooke Takes Stock," *New York Times,* January 14, 2001, <http://www.nytimes.com/2001/01/14/world14HOLB.html>.

325. Samuel R. Berger, "A Foreign Policy for the Global Age," Remarks at the Intercultural Center, Georgetown University, Washington, D.C., October 19, 2000, released by the Office of the Press Secretary, The White House, <http://www.pub.whitehouse.gov/uri-res/I2R?urn:pdi://oma.eop.gov.us/2000/10/20/14.text.1>, p. 6. Also see National Intelligence Council, *Global Trends 2015: A Dialogue about the Future with Nongovernment Experts,* <http://www.odci.gov/cia/reports/globaltrends2015/index.html>.

326. Berger, "A Foreign Policy for the Global Age," p. 6. Indeed, this is not a concern isolated to Africa or the less developed world. Western epidemiologists share a great concern that Russia will become a "disease pump" with drug-resistant strains of encephalitis, hepatitis, syphilis, and tuberculosis, all of which are increasing there, spreading to the West. Abigail Zuger, "Infectious Diseases Rising Again in Russia," *New York Times,* December 5, 2000, <http://www.nytimes.com/2000/12/05/science/05INFE.html>.

327. Nicholas Eberstadt, "The Future of AIDS," *Foreign Affairs,* Vol. 81, No. 6 (November/December 2002), pp. 22–45, estimates from p. 35.

328. Thomas Homer-Dixon, *The Ingenuity Gap* (New York: Knopf, 2000), p. 34.

329. For a description of the objectives of the global fund, see George W. Bush, "Remarks by the President during Announcement of Proposals for Global Fund to Fight HIV/AIDS, Malaria and Tuberculosis," The White House, May 11, 2001, <http://www.whitehouse.gov/news/releases/2001/05/20010511–1.html>.

330. Wills suspects that the Madras plague is actually a different disease. Wills, *Yellow Fever, Black Goddess,* pp. 90–102.

5. Evolutionary Theory and Ethnic Conflict

1. An exceptional account of ethnic cleansing in Bosnia-Herzegovina is provided by Steven L. Burg and Paul S. Shoup, *The War in Bosnia-Herzegovina: Ethnic Conflict and International Intervention* (Armonk, N.Y.: M.E. Sharpe, 1999). Critical scholarship for understanding the causes and conduct of the genocide in Rwanda are Michael Barnett, *Eyewitness to a Genocide: The United Nations and Rwanda* (Ithaca, N.Y.: Cornell University Press, 2002); Bill Berkeley, *The Graves Are Not Yet Full: Race, Tribe and Power in the Heart of Africa* (New York: Basic Books, 2001); Alison Des Forges, *Leave None to Tell the Story: Genocide in Rwanda* (New York: Human Rights Watch, 1999); Alain Destexhe, *Rwanda and Genocide in the Twentieth Century* (New York: New York University Press, 1995); Mahmood Mamdani, *When Victims Become Killers: Colonialism, Nativism, and the Genocide in Rwanda* (Princeton, N.J.: Princeton University Press, 2001); Gérard Prunier, *The Rwandan Crisis: History of a Genocide* (New York: Columbia University Press, 1997); and Christopher C. Taylor, *Sacrifice as Terror: The Rwandan Genocide of 1994* (New York: Berg, 1999). Alan Kuperman argues that even the best possible intervention—a large force at the beginning of the genocide—could have saved at best half of the ultimate victims. See Alan J. Kuperman, "Rwanda in Retrospect," *Foreign Affairs,* Vol. 79, No. 1 (January/February 2000), pp. 94–118.

2. An excellent overview of ethnic conflict in the post–Cold War world is Tatu Vanhanen, *Ethnic Conflicts Explained by Ethnic Nepotism,* published as *Research in Biopolitics,* Vol. 7 (Stamford, Conn.: JAI Press, 1999).

3. See Michael Mandelbaum, "A Perfect Failure," *Foreign Affairs,* Vol. 78, No. 5 (September/October 1999), pp. 2–8. For an insightful analysis of the causes and conduct of U.S. intervention in Kosovo, see Christopher Layne, "Miscalculations and Blunders Lead to War," in Ted Galen Carpenter, ed., *NATO's Empty Victory: A Postmortem on the Balkan War* (Washington, D.C.: Cato Institute, 2000), pp. 11–20. Also see Ivo H. Daalder and Michael E. O'Hanlon, *Winning Ugly: NATO's War to Save Kosovo* (Washington, D.C.: Brookings Institution Press, 2000). Placing the Kosovo intervention in the context of Clinton's foreign policy achievements is Stephen M. Walt, "Two Cheers for Clinton's Foreign Policy," *Foreign Affairs,* Vol. 79, No. 2 (March/April 2000), pp. 63–79.

4. "Clinton: U.S. Is Fighting for the World's Future," *New York Times,* April 17, 1999, p. A6.

5. Scholars who have noted the evolutionary origins of these behaviors or the breadth of ethnocentrism among human populations include Marilynn B. Brewer and Donald T. Campbell, *Ethnocentrism and Intergroup Attitudes* (New York: Wiley, 1976); Brewer, "The Role of Ethnocentrism in Intergroup Conflict," in William G. Austin and Stephen Worchel, eds., *The Social Psychology of Intergroup Relations* (Monterey, Calif.: Brooks/Cole, 1979), pp. 71–84; Irenäus Eibl-Eibesfeldt, "Warfare, Man's Indoctrinability and Group Selection," *Zeitschrift für Tierpsychologie,* Vol. 60, No. 3 (1982), pp. 177–198; Robert A. Levine and Donald T. Campbell, *Ethnocentrism: Theories of Conflict, Ethnic Attitudes and Group Behavior* (New York: Wiley, 1972);

Pierre van den Berghe, *The Ethnic Phenomenon* (New York: Elsevier, 1981), pp. 15–36; Johan M.G. van der Dennen, "Ethnocentrism and In-Group/Out-Group Differentiation," in Vernon Reynolds, Vincent S.E. Falger, and Ian Vine, eds., *The Sociobiology of Ethnocentrism: Evolutionary Dimensions of Xenophobia, Discrimination, Racism and Nationalism* (London: Croom Helm, 1987), pp. 1–47, 7–8, 17–20; and A. Michael Warnecke, Roger D. Masters, and Guido Kempter, "The Roots of Nationalism: Nonverbal Behavior and Xenophobia," *Ethology and Sociobiology,* Vol. 13, No. 4 (1992), pp. 267–282.

6. Useful studies delineating circumstances in which ethnic conflict might occur are Thomas S. Szayna, ed., *Identifying Potential Ethnic Conflict: Application of a Process Model* (Santa Monica, Calif.: Rand Corp., 2000); and Ashley J. Tellis, Thomas S. Szayna, and James A. Winnefeld, *Anticipating Ethnic Conflict* (Santa Monica, Calif.: Rand Corp., 1997).

7. The modernists are also termed "instrumentalists." I use the term modernist because it more accurately captures the difference between the two paradigms by emphasizing that national identity is the product of modern forces, such as the French Revolution and economic modernization. Some scholars have also suggested a third paradigm, constructivism, for understanding the causes of nationalism and national identity. Its central argument is that, for all nations, national identity is created by social interaction and is consequently malleable. Benedict Anderson emphasized the malleability of national identity when he first advanced this argument in 1983. He argues that nations are "imagined communities," the product of a nationalist intelligentsia who create and re-create the nation through the media, art, and education, among other sources. See Benedict Anderson, *Imagined Communities: Reflections on the Origins and Spread of Nationalism* (London: Verso, 1983).

However, for my present purposes I consider constructivism to be part of the modernist paradigm because of its emphasis upon identity as a modern social phenomenon and the ability of elites to rely upon mass press and other media to manipulate identity. For other explicitly constructivist examinations of ethnic identity and conflict, see Rogers Brubaker, "National Minorities, Nationalizing States, and External National Homelands in the New Europe," *Daedalus* (Spring 1995), pp. 107–132; and Crawford Young, ed., *The Rising Tide of Cultural Pluralism: The Nation-State at Bay?* (Madison: University of Wisconsin Press, 1993).

8. Michael Brown has identified four underlying causes of ethnic conflict, which I will consider below. He argues that structural, political, economic, or cultural factors may be triggered by elite or mass movements, or by internal or external politics. Michael E. Brown, "The Causes of Internal Conflict: An Overview," in Brown, Owen R. Coté Jr., Sean M. Lynn-Jones, and Steven E. Miller, eds., *Nationalism and Ethnic Conflict* (Cambridge, Mass.: MIT Press, 1997), pp. 3–25.

9. Walker Connor, *Ethnonationalism: The Quest for Understanding* (Princeton, N.J.: Princeton University Press, 1994), p. 91.

10. Ibid., p. xi; Anthony D. Smith, *National Identity* (Reno: University of Nevada Press, 1991), pp. 19–28; and Smith, *Myths and Memories of the Nation* (New York:

Oxford University Press, 1999), pp. 12–17. In a later work, Smith defines nation as a "named human population occupying a historic territory or homeland and sharing common myths and memories; a mass, public culture; a single economy; and common rights and duties for all members." Smith, *The Nation in History: Historiographical Debates about Ethnicity and Nationalism* (Hanover, N.H.: University Press of New England, 2000), p. 3.

11. Ernest Renan, "What Is a Nation?" in Homi Bhabha, ed., *Nation and Narration* (New York: Routledge, 1990), pp. 8–22, 19.

12. Scholars representative of this paradigm are Connor, *Ethnonationalism*; Harold R. Isaacs, *Idols of the Tribe: Group Identity and Political Change* (New York: Harper and Row, 1975); Robert Kaplan, *Balkan Ghosts: A Journey through History* (New York: St. Martin's Press, 1993); and Anthony D. Smith, *Theories of Nationalism,* 2nd ed. (New York: Holmes and Meier, 1983).

13. Yael Tamir, *Liberal Nationalism* (Princeton, N.J.: Princeton University Press, 1993), p. 66.

14. Donald L. Horowitz, *Ethnic Groups in Conflict* (Berkeley: University of California Press, 1985), chaps. 1–2; and Anthony D. Smith, *The Ethnic Origins of Nations* (Oxford: Blackwell, 1986), chap. 2.

15. Connor, *Ethnonationalism,* p. 202.

16. Hugh Seton-Watson, *Nations and States* (London: Methuen, 1977), p. 7.

17. Tamir, *Liberal Nationalism,* p. 66.

18. David Miller, *On Nationality* (New York: Oxford University Press, 1995), p. 22. For Miller nation also embodies historical continuity, is an active identity, refers to a particular geographical place, and contains a national character or common public culture. Miller, *On Nationality,* pp. 22–27.

19. Max Weber, "Politics as a Vocation," in H.H. Gerth and C. Wright Mills, eds., *From Max Weber: Essays in Sociology* (New York: Oxford University Press, 1946), pp. 77–128, 78. Emphasis in original.

20. Seton-Watson, *Nations and States,* p. 1.

21. Connor, *Ethnonationalism,* p. 96. In the case of France at the time of the French Revolution Connor notes the French nation also contained many other nations, such as the Alsatians, Basques, Bretons, Catalans, Corsicans, Flemings, and Occitanians, and others. Connor, *Ethnonationalism,* p. 95. For an analysis of the process of forging the French nation, see Eugen Weber, *Peasants into Frenchmen: The Modernization of Rural France, 1870–1914* (Stanford, Calif.: Stanford University Press, 1976).

22. Data are drawn from the Central Intelligence Agency's *World Factbook 2000* available at <http://www.odci.gov/cia/publications/factbook/index.html>. Percentages have been rounded.

Based on 1971 data from 132 states, Connor notes that only about 9 percent of states in international relations are genuine nation-states; about 19 percent contain a nation that accounts for 90 percent of the state's total population. Another 19 percent contain a nation that consists of between 75 percent to 89 percent of the population.

In about 23 percent of states, the largest nation accounts for 50 percent to 74 percent of the state's population. Finally, in about 30 percent of states, the largest nation accounts for less than half of the population. Connor, *Ethnonationalism,* p. 96.

23. Brown, "The Causes of Internal Conflict," p. 7.

24. Smith, *The Nation in History,* p. 3.

25. Connor, *Ethnonationalism,* p. xi. Anthony Smith notes the various meanings of the term nationalism. It may be used to describe (1) the process of forming and maintaining states; (2) a consciousness of belonging to the nation, which is the sense in which it is used at present; (3) a language and symbolism of the nation; (4) an ideology or political movement, e.g. Kurdish nationalism; and (5) a movement to achieve the goals of the nation and realize a national will. See Smith, *Myths and Memories of the Nation,* p. 101; and *National Identity,* p. 72.

26. Connor, *Ethnonationalism,* p. 100.

27. Ibid.

28. Ibid., pp. 103–104.

29. John Hutchinson and Anthony D. Smith, eds., *Ethnicity* (New York: Oxford University Press, 1996), pp. 4–5. See also Joshua A. Fishman, Michael H. Gertner, Esther G. Lowy, and William G. Milán, *The Rise and Fall of the Ethnic Revival: Perspectives on Language and Ethnicity* (New York: Mouton Publishers, 1985), pp. 15–38.

30. Hutchinson and Smith, eds., *Ethnicity,* p. 4.

31. Connor notes one difference between the terms. An ethnic group may not be aware of its status as a nation; for example the Slovaks, Croats, and Slovenes under the Habsburg Empire were aware that they were not German or Magyar before they possessed "positive opinions concerning their ethnic or national identity." Thus, an ethnic group may be defined by those outside of the group and may not have a sufficient identity of itself to be considered a nation; however, as mentioned above, a nation is always self-defined and possesses such an identity. Connor, *Ethnonationalism,* p. 103. Paul Brass's definition of ethnicity is similar to my use of the term; he argues that a nation is a politicized ethnic group. Paul R. Brass, *Ethnicity and Nationalism: Theory and Comparison* (London: Sage, 1991), pp. 19–20.

32. Max Weber, *Economy and Society: An Outline of Interpretive Sociology,* 2 Vols., ed. Guenther Roth and Claus Wittich (Berkeley: University of California Press, 1978), Vol. 1, p. 389.

33. Ibid.

34. Ibid., Vol. 1, p. 395.

35. Steven Grosby, "The Verdict of History: The Inexpungeable Tie of Primordiality—A Response to Eller and Coughlan," *Ethnic and Racial Studies,* Vol. 17, No. 1 (January 1994), pp. 164–171, 168.

36. Echoing the remarks of Bush and Clinton is *New York Times* journalist Serge Schmemann, who wrote in May 1992 of Bosnia that its "roll call of warring nationalities invokes some forgotten primer on the Dark Ages," and that its soldiers are fighting "for causes lost in the fog of history." Bush and Schmemann are quoted in Jack Snyder, "Nationalism and the Crisis of the Post-Soviet State," in Michael E. Brown,

ed., *Ethnic Conflict and International Security* (Princeton, N.J.: Princeton University Press, 1993), pp. 79–101, 79 and 98. Clinton's remark is quoted in Brown, "The Causes of Internal Conflict," p. 3. The idea of the prevalence of "ancient hatreds" in the Balkans as the cause of the wars that destroyed Yugoslavia was popularized by Kaplan in his *Balkan Ghosts.*

37. For a discussion of Lee's conception of his identity and the importance of Christianity and Virginia to it, the work of Steven E. Woodworth is particularly important. See his *While God Is Marching On: The Religious World of Civil War Soldiers* (Lawrence: University Press of Kansas, 2001); and *Davis and Lee at War* (Lawrence: University Press of Kansas, 1995).

38. Edward Shils, "Primordial, Personal, Sacred and Civil Ties," *British Journal of Sociology,* Vol. 7 (June 1957), pp. 113–145.

39. Geertz advances this argument in "The Integrative Revolution: Primordial Sentiments and Civil Politics in the New States," in Clifford Geertz, ed., *Old Societies and New States: The Quest for Modernity in Asia and Africa* (New York: Free Press, 1963), pp. 105–157.

40. Ibid., p. 109.

41. Ibid.

42. Grosby, "The Verdict of History," p. 169.

43. Ibid.

44. Ibid.

45. Horowitz also offers other explanations of ethnic conflict. Analyzing many different sources of ethnic conflict drawn from both the primordialist and modernist paradigms, he provides insightful comments and criticisms on each. See Donald L. Horowitz, *The Deadly Ethnic Riot* (Berkeley: University of California Press, 2001), pp. 43–70; and Horowitz, *Ethnic Groups in Conflict.*

46. Horowitz, *The Deadly Ethnic Riot,* p. 48.

47. Ibid.

48. Ibid., p. 47.

49. Connor, *Ethnonationalism,* pp. 202–203. Also see Grosby, "The Verdict of History," p. 166; and Paul C. Stern, "Why Do People Sacrifice for Their Nations?" in John L. Comaroff and Paul C. Stern, eds., *Perspectives on Nationalism and War* (n.p.: Gordon and Breach Publishers, 1995), pp. 99–121.

50. Renan, "What Is a Nation?" p. 19.

51. Conner, *Ethnonationalism,* p. 203.

52. Helen Dukas and Banesh Hoffmann, *Albert Einstein, the Human Side: New Glimpses from His Archives* (Princeton, N.J.: Princeton University Press, 1979), p. 38.

53. John Dunn, *Western Political Theory in the Face of the Future* (Cambridge: Cambridge University Press, 1978), p. 55.

54. Quoted in Connor, *Ethnonationalism,* p. 206.

55. Renan, "What Is a Nation?" p. 18.

56. Connor, *Ethnonationalism,* p. 206.

57. Noting the importance of kin terms in nationalist or patriotic speech are

Gary R. Johnson, Susan H. Ratwik, and Timothy J. Sawyer, "The Evocative Significance of Kin Terms in Patriotic Speech," in Reynolds et al., eds., *The Sociobiology of Ethnocentrism,* pp. 157–174. Important examinations of how nonverbal cues influence political relationships and the media are Denis G. Sullivan and Roger D. Masters, "Biopolitics, the Media, and Leadership: Nonverbal Clues, Emotions, and Trait Attributions in the Evaluation of Leaders," in Albert Somit and Steven A. Peterson, eds., *Research in Biopolitics,* Vol. 2 (Greenwich, Conn.: JAI Press, 1994), pp. 237–273; and Roger D. Masters and Baldwin Way, "Experimental Methods and Attitudes toward Leaders: Nonverbal Displays, Emotion, and Cognition," in Albert Somit and Steven A. Peterson, eds., *Research in Biopolitics,* Vol. 4 (Greenwich, Conn.: JAI Press, 1996), pp. 61–98. Exploring the issue of facial dominance, the degree to which a person is judged from his facial appearance to be dominant or submissive, in leaders is Allan Mazur and Ulrich Mueller, "Facial Dominance," in Somit and Peterson, eds., *Research in Biopolitics,* Vol. 4, pp. 99–111.

58. Quoted in Conner, *Ethnonationalism,* p. 204.

59. Freud quoted in Conner, *Ethnonationalism,* p. 203. Emphasis omitted.

60. Tamir, *Liberal Nationalism,* p. 66.

61. Anthony D. Smith, "The Ethnic Sources of Nationalism," in Brown, ed., *Ethnic Conflict and International Security,* pp. 26–41, 40.

62. Ibid., p. 40.

63. David A. Lake and Donald Rothchild, "Spreading Fear: The Genesis of Transnational Ethnic Conflict," in David A. Lake and Donald Rothchild, eds., *The International Spread of Ethnic Conflict: Fear, Diffusion, and Escalation* (Princeton, N.J.: Princeton University Press, 1998), pp. 3–32, 5. See also Anderson, *Imagined Communities.*

64. Hutchinson and Smith, *Ethnicity,* p. 8.

65. For example, see Lake and Rothchild, "Spreading Fear," in Lake and Rothchild, eds., *The International Spread of Ethnic Conflict,* p. 5.

66. Ibid.

67. Ernest Gellner, *Nations and Nationalism* (New York: Oxford University Press, 1983); and Eric Hobsbawm, *Nations and Nationalism since 1780* (New York: Cambridge University Press, 1990). Smith terms this paradigm the "instrumentalists" in his *Myths and Memories of the Nation,* p. 42; and *Nations and Nationalism in a Global Era* (Cambridge: Polity Press, 1995), p. 30. Rational choice analysis may also be considered an element of this paradigm. Russell Hardin elegantly provides the rational choice explanation for nationalism and ethnic conflict. See Russell Hardin, *One for All: The Logic of Group Conflict* (Princeton, N.J.: Princeton University Press, 1995).

68. Smith, *Myths and Memories of the Nation,* p. 99. Also see Smith, *Nations and Nationalism in a Global Era,* p. 29.

69. Barry Posen describes the impact that the rise of mass armies had on military power and stability in Europe in Barry R. Posen, "Nationalism, the Mass Army, and Military Power," *International Security,* Vol. 18, No. 2 (Fall 1993), pp. 80–124.

70. Smith, *Myths and Memories of the Nation,* p. 100. See also Anderson, *Imagined Communities.*

71. Brown, "The Causes of Internal Conflict," pp. 4–5; and Stephen Van Evera, "Hypotheses on Nationalism and War," *International Security,* Vol. 18, No. 4 (Spring 1994), pp. 5–39, 7.

72. Brown, "The Causes of Internal Conflict," p. 5.

73. Barry R. Posen, "The Security Dilemma and Ethnic Conflict," in Brown, ed., *Ethnic Conflict and International Security,* pp. 103–124. Posen uses two cases, Croats and Serbs in Yugoslavia and the Russian in Ukraine, to test his argument.

74. Brown, "The Causes of Internal Conflict," p. 6.

75. Ibid., p. 7.

76. Van Evera, "Hypotheses on Nationalism and War," p. 18.

77. Smith, *The Ethnic Origins of Nations,* pp. 21–46; and Snyder, "Nationalism and the Crisis of the Post-Soviet State," in Brown, ed., *Ethnic Conflict and International Security,* pp. 79–102. On this distinction, see also Liah Greenfeld, *Nationalism: Five Roads to Modernity* (Cambridge, Mass.: Harvard University Press, 1992).

78. Brown, "The Causes of Internal Conflict," p. 9.

79. Ibid.

80. Ibid., pp. 9–10. Also see V.P. Gagnon Jr., "Ethnic Nationalism and International Conflict: The Case of Serbia," *International Security,* Vol. 19, No. 3 (Winter 1994/95), pp. 130–166.

81. Brown, "The Causes of Internal Conflict," p. 10.

82. Ibid.

83. Ibid., p. 11; and Amita Shastri, "Government Policy and the Ethnic Crisis in Sri Lanka," in Michael E. Brown and Šumit Ganguly, eds., *Government Policies and Ethnic Relations in Asia and the Pacific* (Cambridge, Mass.: MIT Press, 1997), pp. 129–163.

84. Brown, "The Causes of Internal Conflict," p. 11. See also Ted Robert Gurr, *Why Men Rebel* (Princeton, N.J.: Princeton University Press, 1970); Hobsbawm, *Nations and Nationalism since 1780*; Samuel P. Huntington, *Political Order in Changing Societies* (New Haven: Yale University Press, 1968); and Chalmers Johnson, *Revolutionary Change* (Boston: Little, Brown, 1966).

85. Huntington, *Political Order in Changing Societies,* p. 5.

86. Ibid.

87. Ibid.

88. Brown, "The Causes of Internal Conflict," pp. 11–12; and Huntington, *Political Order in Changing Societies,* p. 5.

89. Brown, "The Causes of Internal Conflict," p. 12.

90. Ibid.; and Van Evera, "Hypotheses on Nationalism and War," p. 26.

91. Van Evera, "Hypotheses on Nationalism and War," p. 27.

92. Brown, "The Causes of Internal Conflict," p. 13.

93. Stuart J. Kaufman, *Modern Hatreds: The Symbolic Politics of Ethnic War* (Ithaca, N.Y.: Cornell University Press, 2001), pp. 24–30.

94. Brown, "The Causes of Internal Conflict," p. 13.

95. Van Evera, "Hypotheses on Nationalism and War," p. 29.

96. Ibid.

97. Ibid., p. 23.

98. Ibid., pp. 24–25.

99. Ibid., pp. 30–32.

100. Ibid., p. 33.

101. Van Evera's analysis is useful as well. See Ibid., pp. 10–15.

102. Brown, "The Causes of Internal Conflict," pp. 15–16.

103. Ibid., p. 15; Kanti Bajpai, "Diversity, Democracy, and Devolution in India," in Brown and Ganguly, eds., *Government Policies and Ethnic Relations in Asia and the Pacific,* pp. 33–81; and Šumit Ganguly, "Internal Conflict in South and Southwest Asia," in Michael E. Brown, ed., *The International Dimensions of Internal Conflict* (Cambridge, Mass.: MIT Press, 1996), pp. 141–172.

104. Brown, "The Causes of Internal Conflict," p. 16.

105. Ibid.

106. Michael E. Brown and Šumit Ganguly, "Introduction," in Brown and Ganguly, eds., *Government Policies and Ethnic Relations in Asia and the Pacific,* pp. 1–29, 10.

107. Brown, "The Causes of Internal Conflict," pp. 16–19.

108. Hutchinson and Smith, *Ethnicity,* p. 9.

109. Stern, "Why Do People Sacrifice for Their Nations?" p. 107.

110. A classic anthropological account of the in-group/out-group phenomenon based on his study of twelve hundred Tikopia living on an isolated Pacific island is Raymond Firth, *We, the Tikopia* (London: Allen and Unwin, 1957).

111. On Tajfel's experiments, see Henri Tajfel, *Human Groups and Social Categories* (Cambridge: Cambridge University Press, 1981); and for its relation to ethnic conflict, see Horowitz, *Ethnic Groups in Conflict,* pp. 144–147.

112. Horowitz, *Ethnic Groups in Conflict,* p. 145. Also see Henri Tajfel and John Turner, "An Integrative Theory of Intergroup Conflict," in William G. Austin and Stephen Worchel, eds., *The Social Psychology of Intergroup Relations* (Monterey, Calif.: Brooks/Cole, 1979), pp. 33–47.

113. Robert L. Trivers, "The Evolution of Reciprocal Altruism," *The Quarterly Review of Biology,* Vol. 46, No. 1 (March 1971), pp. 35–57.

114. Vincent S.E. Falger, "Evolutionary World Politics Enriched," in Thompson, ed., *Evolutionary Interpretations of World Politics,* pp. 30–51, 41. A particularly insightful overview is provided by Kristiaan Thienpont and Robert Cliquet, "Introduction: In-Group/Out-Group Studies and Evolutionary Biology," in Kristiaan Thienpont and Robert Cliquet, eds., *In-Group/Out-Group Behaviour in Modern Societies: An Evolutionary Perspective* (Brussels: Vlaamse Gemeenschap, 1999), pp. 3–20.

115. Robert C. Bolles, "Species-Specific Defense Reactions and Avoidance Learning," *Psychological Review,* Vol. 77, No. 1 (January 1970), pp. 32–48, 33. Emphasis in original.

116. Ibid., p. 33.

117. Ibid.

118. Ibid.

119. Ibid.

120. Pierre L. van den Berghe, "Does Race Matter?" *Nations and Nationalism,* Vol. 1, No. 3 (1995), pp. 357–368, 362. For a related argument, see Tatu Vanhanen, *Politics of Ethnic Nepotism: India as an Example* (New Delhi: Sterling Publishers, 1991).

121. Also see Vanhanen, *Ethnic Conflicts Explained by Ethnic Nepotism,* pp. 11–16.

122. Dower notes that crude and vicious racial stereotypes were present in Japanese thought and propaganda as well as among the Western allies. See John W. Dower, *War without Mercy: Race and Power in the Pacific War* (New York: Pantheon Books, 1986); and Dower, *Embracing Defeat: Japan in the Wake of World War II* (New York: New Press, 1999).

123. Dower, *War without Mercy,* p. 34.

124. See Paul Fussell, *Wartime: Understanding and Behavior in the Second World War* (New York: Oxford University Press, 1989).

125. Ibid., p. 120.

126. Ibid., p. 116.

127. Ibid., p. 138.

128. Arthur Miller, "The Face in the Mirror: Anti-Semitism Then and Now," *New York Times Book Review,* October 14, 1984, p. 3. Quoted in Fussell, *Wartime,* p. 138.

129. Ibid.

130. See "Kosovo and Sierra Leone," *Washington Post,* June 15, 1999, p. A32. The editorial also noted the great difference in aid pledged to the two groups. In the summer of 1999, the United States pledged $15 million in aid to Sierra Leone. As of June 1999, it pledged a "down payment" of $13 billion in aid for Kosovo. While the amount of aid has changed since that editorial was written in 1999, the lopsided ratio in aid between Sierra Leone and Kosovo has not.

131. Van den Berghe, "Does Race Matter?" p. 364.

132. Weber, *Economy and Society,* Vol. 1, p. 385.

133. Ibid.

134. Ibid.

135. Ibid., Vol. 1, p. 388.

136. Ibid., Vol. 1, p. 392. In more detail, he argues, "apart from the community of language, which may or may not coincide with objective, or subjectively believed, consanguinity, and apart from common religious belief, which is also independent of consanguinity, the ethnic differences that remain" are on the one hand the product of "esthetically conspicuous differences of the physical appearance . . . and, on the other hand and of equal weight, the perceptible differences in the conduct of everyday life." Ibid., Vol. 1, p. 390. Emphasis omitted.

137. Jack Snyder and Karen Ballentine discuss the important role of the media in exacerbating ethnic conflict by creating nationalist mythmaking in emerging democracies, the former Yugoslavia, and Rwanda, in their "Nationalism and the Marketplace of Ideas," *International Security,* Vol. 21, No. 2 (Fall 1996), pp. 5–40.

138. Implicitly reflecting upon the power of television and film images for creating in-group identity, *New York Times* editorialist Brent Staples notes that "movies in the contemporary United States tend to be rigorously multicultural, with black, Asian, and Latino faces streaming across the screen. But those faces become noticeably whiter when the time frame shifts backward." He continues, "movie makers commonly treat the past as a place where black Americans did not exist." As a result, "the movies have left the mistaken impression that heroism in the 'greatest generation' came exclusively with a white face." Brent Staples, "Celebrating World War II—and the Whiteness of American History," *New York Times,* December 9, 2001, <http://www.nytimes.com/2001/12/09/opinion/09SUN3.html>.

139. United Kingdom Home Office, *Community Cohesion: A Report of the Independent Review Team,* <http://www.homeoffice.gov.uk/reu/community_cohesion.pdf>, pp. 16–17, 33–37.

140. See Anthony Downs, *An Economic Theory of Democracy* (New York: Harper and Row, 1957); and Mancur Olson, *The Logic of Collective Action: Public Goods and the Theory of Groups* (Cambridge, Mass.: Harvard University Press, 1965).

141. Robert D. Putnam, *Bowling Alone: The Collapse and Revival of American Community* (New York: Simon and Schuster, 2000).

142. Van den Berghe, "Does Race Matter?" p. 361.

143. Ibid.

144. A good introduction to xenophobia in animals is Charles H. Southwick et al., "Xenophobia among Free-Ranging Rhesus Groups in India," in Ralph L. Holloway, ed., *Primate Aggression, Territoriality, and Xenophobia: A Comparative Perspective* (New York: Academic Press, 1974), pp. 185–209. Also see Holloway, "Introduction," in Holloway, ed., *Primate Aggression, Territoriality, and Xenophobia,* pp. 1–9, 7; and Fred H. Willhoite Jr., "Evolution and Collective Intolerance," *The Journal of Politics,* Vol. 39, No. 3 (August 1977), pp. 667–684, 674–675. For explicit application to humans, see Ralph B. Taylor, *Human Territorial Functioning: An Empirical, Evolutionary Perspective on Individual and Small Group Territorial Cognitions, Behaviors, and Consequences* (Cambridge: Cambridge University Press, 1988).

145. For an excellent survey of this literature, see van der Dennen, "Ethnocentrism and In-Group/Out-Group Differentiation," in Reynolds et al., eds., *The Sociobiology of Ethnocentrism,* pp. 20–22. Also see J.A.R.A.M. van Hooff, "Intergroup Competition and Conflict in Animals and Man," in J. van der Dennen and V. Falger, eds., *Sociobiology and Conflict: Evolutionary Perspectives on Competition, Cooperation, Violence and Warfare* (London: Chapman and Hall, 1990), pp. 23–54.

146. Peter A. Corning, "Synergy Goes to War," paper prepared for the annual meeting of the Association for Politics and the Life Sciences, Charleston, S.C., October 2001, p. 9. Corning further developed this argument in his *Nature's Magic: Synergy in Evolution and the Fate of Humankind* (New York: Cambridge University Press, 2003). Also see Ralph Adolphs, Daniel Tranel, and Antonio R. Damasio, "The Human Amygdala in Social Judgment," *Nature,* Vol. 393 (June 4, 1998), 470–474; Antonio Damasio, *Looking for Spinoza: Joy, Sorrow, and the Feeling Brain* (New York: Harcourt,

2003), pp. 60–62; and Joseph LeDoux, *The Emotional Brain: The Mysterious Underpinnings of Emotional Life* (New York: Simon and Schuster, 1996).

147. David P. Barash, *Sociobiology and Behavior* (New York: Elsevier, 1977), p. 219.

148. John L. Fuller, "Genes, Brains, and Behavior," in Michael S. Gregory, Anita Silvers, and Diane Sutch, eds., *Sociobiology and Human Nature* (San Francisco: Jossey-Bass, 1978), pp. 98–115, 111.

149. Jared Diamond, *The Third Chimpanzee: The Evolution and Future of the Human Animal* (New York: HarperCollins, 1992), pp. 220–221.

150. Johan Matheus Gerradus van der Dennen, *The Origin of War: The Evolution of a Male-Coalitional Reproductive Strategy,* 2 Vols. (Groningen, Netherlands: Origin Press, 1995), pp. 457–458.

151. See Frans B.M. de Waal, "Introduction," in Frans B.M. de Waal, ed., *Tree of Origin: What Primate Behavior Can Tell Us about Human Social Evolution* (Cambridge, Mass.: Harvard University Press, 2001), pp. 1–8, 2; and Alison Jolly, *Lucy's Legacy: Sex and Intelligence in Human Evolution* (Cambridge, Mass.: Harvard University Press, 1999), p. 159.

152. Noting the importance of limited resources is Joel R. Peck, "The Evolution of Outsider Exclusion," *Journal of Theoretical Biology,* Vol. 142, No. 4 (February 22, 1990), pp. 565–571.

153. Diamond, *The Third Chimpanzee,* p. 220.

154. Ibid.

155. The powerful effect of disease on human society is documented in William H. McNeill, *Plagues and Peoples* (Garden City, N.Y.: Anchor Press, 1976).

156. Donald L. McEachron and Darius Baer, "A Review of Selected Sociobiological Principles: Application to Hominid Evolution II. The Effects of Intergroup Conflict," *Journal of Social and Biological Structures,* Vol. 5, No. 2 (April 1982), pp. 121–139, 134.

157. See chapter 2 and also James L. Boone, "Competition, Conflict, and the Development of Social Hierarchies," in Eric Alden Smith and Bruce Winterhalder, eds., *Evolutionary Ecology and Human Behavior* (New York: Aldine de Gruyter, 1992), pp. 301–337; and K.E. Moyer, "The Biological Basis of Dominance and Aggression," in Diane McGuinness, ed., *Dominance, Aggression and War* (New York: Paragon House, 1987), pp. 1–34, for a summary of the ethological argument on this point.

158. McEachron and Baer, "A Review of Selected Sociobiological Principles: Application to Hominid Evolution II," pp. 134–135.

159. Alexander, *Darwinism and Human Affairs,* pp. 126–127; and Vincent S.E. Falger, "From Xenophobia to Xenobiosis?" in Reynolds et al., eds., *The Sociobiology of Ethnocentrism,* pp. 235–250.

160. Alexander, *Darwinism and Human Affairs,* p. 127.

161. Ibid.

162. Prunier, *The Rwandan Crisis,* pp. 5–9, 249; and Taylor, *Sacrifice as Terror,* pp. 40–41. Also see van den Berghe, *The Ethnic Phenomenon,* pp. 73–74. Despite the

physical differences, there were some cases of mistaken identity: "During the genocide some persons who were legally Hutu were killed as Tutsi because they looked Tutsi." Des Forges, *Leave None to Tell the Story,* p. 33.

163. Diamond, *The Third Chimpanzee,* p. 220.

164. As Irenäus Eibl-Eibesfeldt argues, "xenophobia is a universal phenomenon," but it should be emphasized that "fear of strangers is not to be confused with hatred of strangers." He accurately notes that proximate causation is needed in addition to xenophobia to explain ethnic conflict. Irenäus Eibl-Eibesfeldt, "Us and the Others," in Eibl-Eibesfeldt and Frank Kemp Salter, eds., *Indoctrinability, Ideology, and Warfare: Evolutionary Perspectives* (New York: Berghahn Books, 1998), pp. 21–53, 33.

165. Excellent accounts are found in van der Dennen, "Ethnocentrism and In-Group/Out-Group Differentiation," in Reynolds et al., eds., *The Sociobiology of Ethnocentrism,* pp. 8–17, 37–47; Robin I.M. Dunbar, "Sociobiological Explanations and the Evolution of Ethnocentrism," in Reynolds et al., eds., *The Sociobiology of Ethnocentrism,* pp. 48–59; and van den Berghe, *The Ethnic Phenomenon,* pp. 18–36. Also see Corning, "Synergy Goes to War," p. 60.

166. Ian Vine, "Inclusive Fitness and the Self-System," in Reynolds et al., eds., *The Sociobiology of Ethnocentrism,* pp. 60–80. More formally, ethnocentrism comprises four discrete aspects of group behavior: In-group integration, the hyper-evaluation of the in-group, hostile relations between in-group and out-group, and derogatory stereotyping of out-group individuals and characteristics. See Brewer and Campbell, *Ethnocentrism and Intergroup Attitudes,* p. 74.

167. Also making this point is Dunbar, "Sociobiological Explanations and the Evolution of Ethnocentrism," in Reynolds et al., eds., *The Sociobiology of Ethnocentrism,* p. 56.

168. Eibl-Eibesfeldt, "Us and the Others," p. 41.

169. Ibid., pp. 42–46. Also see Eibl-Eibesfeldt, "Warfare, Man's Indoctrinability and Group Selection," *Zeitschrift für Tierpsychologie,* Vol. 60, No. 3 (1982), pp. 177–198.

170. Edward O. Wilson, *On Human Nature* (Cambridge, Mass.: Harvard University Press, 1978), p. 111.

171. Ibid.

172. Walker Conner, "Eco- or Ethno-Nationalism?" *Ethnic and Racial Studies,* Vol. 7, No. 3 (1984), pp. 242–259.

173. See Posen, "The Security Dilemma and Ethnic Conflict," in Brown, ed., *Ethnic Conflict and International Security,* pp. 103–124; Sandra Halperin, "The Spread of Ethnic Conflict in Europe: Some Comparative-Historical Reflections," in Lake and Rothchild, eds., *The International Spread of Ethnic Conflict,* pp. 151–184; and Milton J. Esman, *Ethnic Politics* (Ithaca, N.Y.: Cornell University Press, 1994), pp. 244–245.

174. Michael E. Brown, "The Impact of Government Policies on Ethnic Relations," in Brown and Ganguly, eds., *Government Policies and Ethnic Relations in Asia and the Pacific,* pp. 511–575, 515. Brown is careful to note the impact of modernization and colonialism, and regional and international influences, as well as demographic and economic factors as causes of ethnic conflict.

Conclusion

1. Edward O. Wilson, *Biophilia* (Cambridge, Mass.: Harvard University Press, 1984), p. 22.

2. Mark Twain (Samuel Langhorne Clemens), *Following the Equator: A Journey around the World,* 2 Vols. (New York: Harper and Brothers, 1899), Vol. 1, p. 264.

3. Irving L. Janis, *Victims of Groupthink: A Psychological Study of Foreign-Policy Decisions and Fiascoes* (Boston: Houghton Mifflin, 1972), pp. 8–9. He revised and enlarged this work in *Groupthink: Psychological Studies of Policy Decisions and Fiascoes* (Boston: Houghton Mifflin, 1982).

4. Janis, *Victims of Groupthink,* p. 8.

5. Ian Stewart, *Life's Other Secret: The New Mathematics of the Living World* (New York: John Wiley and Sons, 1998), p. x.

6. John Tooby and Irven DeVore, "The Reconstruction of Hominid Evolution through Strategic Modeling," in Warren G. Kinzey, ed., *The Evolution of Human Behavior: Primate Models* (Albany, N.Y.: SUNY Press, 1987), pp. 183–237, 235.

7. Gísli Pálsson and Paul Rabinow, *Anthropology Today,* Vol. 15, No. 5 (October 1999), pp. 14–18, 18.

8. Sophocles, *Antigone,* Ode I, strophe 1; from Sophocles, *The Oedipus Cycle,* trans. by Dudley Pitts and Robert Fitzgerald (San Diego: Harcourt Brace Jovanovich Publishers, 1967), p. 199.

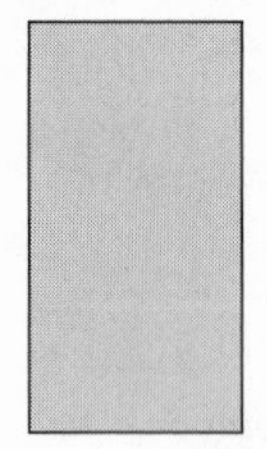

Bibliography

Adams, David B. "Why Are There So Few Women Warriors," *Behavior Science Research,* Vol. 18, No. 3 (1983), pp. 196–212.

Adams, Eldridge S. "Boundary Disputes in the Territorial Ant *Azteca trigona:* Effects of Asymmetries in Colony Size," *Animal Behaviour,* Vol. 39, No. 2 (February 1990), pp. 321–328.

Adolphs, Ralph, Daniel Tranel, Antonio R. Damasio. "The Human Amygdala in Social Judgment," *Nature,* Vol. 393 (June 4, 1998), 470–474.

Ainslie, George, and R.J. Herrnstein. "Preference Reversal and Delayed Reinforcement," *Animal Learning and Behavior,* Vol. 9, No. 4 (1981), pp. 476–482.

Albert, Bruce. "Yanomami 'Violence': Inclusive Fitness or Ethnographer's Representation," *Current Anthropology,* Vol. 30, No. 4 (August–October 1989), pp. 637–640.

Alchian, Armen A. "Uncertainty, Evolution, and Economic Theory," *The Journal of Political Economy,* Vol. 58, No. 3 (June 1950), pp. 211–221.

Alcock, John. *The Triumph of Sociobiology* (New York: Oxford University Press, 2001).

Alexander, Richard D. *Darwinism and Human Affairs* (Seattle: University of Washington Press, 1979).

Alexander. *The Biology of Moral Systems* (New York: Aldine de Gruyter, 1987).

Alexander. "Evolution of the Human Psyche," in Paul Mellars and Chris Stringer, eds., *The Human Revolution: Behavioural and Biological Perspectives on the Origins of Modern Humans* (Princeton, N.J.: Princeton University Press, 1989), pp. 455–513.

Alibek, Ken, with Stephen Handelman. *Biohazard: The Chilling True Story of the Largest Covert Biological Weapons Program in the World—Told from Inside by the Man Who Ran It* (New York: Delta, 1999).

Almagor, Uri. "Raiders and Elders: A Confrontation of Generations among the Dassanetch," in Katsuyoshi Fukui and David Turton, eds., *Warfare among East African Herders* (Osaka: National Museum of Ethnology, 1979), pp. 119–145.

Alves, Abel A. *Brutality and Benevolence: Human Ethology, Culture, and the Birth of Mexico* (Westport, Conn.: Greenwood Press, 1996).

Ambrose, Stephen E. *Citizen Soldiers: The U.S. Army from the Normandy Beaches to the*

Bulge to the Surrender of Germany, June 7, 1944–May 7, 1945 (New York: Simon and Schuster, 1997).

Anderson, Benedict. *Imagined Communities: Reflections on the Origins and Spread of Nationalism* (London: Verso, 1983).

Andrews, Peter, and Louise Humphrey. "African Miocene Environments and the Transition to Early Hominines," in Timothy G. Bromage and Friedemann Schrenk, eds., *African Biogeography, Climate Change, and Human Evolution* (New York: Oxford University Press, 1999), pp. 282–300.

Angier, Natalie. "For Monkeys, a Millipede a Day Keeps Mosquitoes Away," *New York Times,* December 5, 2000, <http://www.nytimes.com/2000/12/05/science/05MONK.html>.

Ardrey, Robert. *The Territorial Imperative: A Personal Inquiry into the Animal Origins of Property and Nations* (New York: Atheneum, 1966).

Aron, Raymond. *Peace and War: A Theory of International Relations,* trans. by Richard Howard and Annette Baker Fox (Garden City, N.Y.: Doubleday, 1966).

Arrow, Kenneth J. *Social Choice and Individual Values* (New Haven: Yale University Press, 1951).

Avise, John C. *The Genetic Gods: Evolution and Belief in Human Affairs* (Cambridge, Mass.: Harvard University Press, 1998).

Axelrod, Robert. *The Evolution of Cooperation* (New York: Basic Books, 1984).

Axelrod, and William D. Hamilton. "The Evolution of Cooperation," *Science,* Vol. 211 (March 27, 1981), pp. 1390–1396.

Baer, Darius, and Donald L. McEachron. "A Review of Selected Sociobiological Principles: Application to Hominid Evolution I. The Development of Group Structure," *Journal of Social and Biological Structures,* Vol. 5, No. 1 (January 1982), pp. 69–90.

Bagehot, Walter. *Physics and Politics: Or Thoughts on the Application of the Principles of "Natural Selection" and "Inheritance" to Political Society* (New York: D. Appleton, 1887).

Baillargeon, Renée, Elizabeth S. Spelke, and Stanley Wasserman. "Object Permanence in Five-Month-Old Infants," *Cognition,* Vol. 20, No. 3 (April 1985), pp. 191–208.

Bajpai, Kanti. "Diversity, Democracy, and Devolution in India," in Michael E. Brown and Šumit Ganguly, eds., *Government Policies and Ethnic Relations in Asia and the Pacific* (Cambridge, Mass.: MIT Press, 1997), pp. 33–81.

Balée, William. "The Ecology of Ancient Tupi Warfare," in R. Brian Ferguson, ed., *Warfare, Culture, and Environment* (Orlando, Fla.: Academic Press, 1984), pp. 241–265.

Balikci, Asen. "The Netsilik Eskimos: Adaptive Processes," in Richard B. Lee and Irven DeVore, eds., *Man the Hunter* (Chicago: Aldine, 1968), pp. 78–82.

Bamforth, Douglas B. "Indigenous People, Indigenous Violence: Precontact Warfare on the North American Great Plains," *Man* (N.S. [New Series]), Vol. 29, No. 1 (March 1994), pp. 95–115.

Bibliography

Barash, David P. *Sociobiology and Behavior* (New York: Elsevier, 1977).

Barkow, Jerome H., Leda Cosmides, and John Tooby, eds. *The Adapted Mind: Evolutionary Psychology and the Generation of Culture* (New York: Oxford University Press, 1994).

Barnett, Michael. *Eyewitness to a Genocide: The United Nations and Rwanda* (Ithaca, N.Y.: Cornell University Press, 2002).

Bateson, Patrick, and Paul Martin. *Design for a Life: How Biology and Psychology Shape Human Behavior* (New York: Simon and Schuster, 2000).

Baum, W., and H. Rachilin. "Choice as Time Allocation," *Journal of the Experimental Analysis of Behavior*, Vol. 12 (1969), pp. 453–467.

Baxter, P.T.W. "Boran Age-Sets and Warfare," in Katsuyoshi Fukui and David Turton, eds., *Warfare among East African Herders* (Osaka: National Museum of Ethnology, 1979), pp. 69–95.

Bazerman, Charles. "Intertextual Self-Fashioning: Gould and Lewontin's Representations of Literature," in Jack Selzer, ed., *Understanding Scientific Prose* (Madison: University of Wisconsin Press, 1993), pp. 20–41.

Beatty, John. "Fitness: Theoretical Contexts," in Evelyn Fox Keller and Elisabeth A. Lloyd, eds., *Keywords in Evolutionary Biology* (Cambridge, Mass.: Harvard University Press, 1992), pp. 115–119.

Becker, Gary. "The Economic Approach to Human Behavior," in Jon Elster, ed., *Rational Choice* (New York: New York University Press, 1986), pp. 108–122.

Belden, Jack. *Still Time to Die* (New York: Harper and Brothers, 1943).

Bell, Graham. *Selection: The Mechanism of Evolution* (London: Chapman and Hall, 1997).

Berenbaum, Michael, ed. *A Mosaic of Victims: Non-Jews Persecuted and Murdered by the Nazis* (New York: New York University Press, 1990).

Berger, Samuel R. "A Foreign Policy for the Global Age," Remarks at the Intercultural Center, Georgetown University, Washington, D.C., October 19, 2000, released by the Office of the Press Secretary, The White House, <http://www.pub.whitehouse.gov/uri-res/I2R?urn:pdi://oma.eop.gov.us/2000/10/20/14.text.1>.

Bergerud, Eric. *Touched with Fire: The Land War in the South Pacific* (New York: Viking Penguin, 1996).

Berkeley, Bill. *The Graves Are Not Yet Full: Race, Tribe and Power in the Heart of Africa* (New York: Basic Books, 2001).

Betzig, Laura L. *Despotism and Differential Reproduction: A Darwinian View of History* (New York: Aldine, 1986).

Bidermann, Gottlob Herbert. *In Deadly Combat: A German Soldier's Memoir of the Eastern Front,* trans. by Derek S. Zumbro (Lawrence: University Press of Kansas, 2000).

Bigelow, Robert. *The Dawn Warriors: Man's Evolution toward Peace* (Boston: Little, Brown, 1969).

Bigelow. "The Role of Competition and Cooperation in Human Evolution," in Martin A. Nettleship, R. Dalegivens, and Anderson Nettleship, eds., *War, Its Causes and Correlates* (The Hague: Mouton Publishers, 1975), pp. 235–258.

Biolsi, Thomas. "Ecological and Cultural Factors in Plains Indian Warfare," in R. Brian Ferguson, ed., *Warfare, Culture, and Environment* (Orlando, Fla.: Academic Press, 1984), pp. 141–168.

Blainey, Geoffrey. *The Causes of War,* 3rd ed. (New York: Free Press, 1988).

Blick, Jeffrey P. "Genocidal Warfare in Tribal Societies as a Result of European-Induced Culture Conflict," *Man* (N.S.), Vol. 23, No. 4 (December 1988), pp. 654–670.

Boas, Franz. *Tsimshian Mythology* (New York: Johnson Reprint Co., 1970).

Bock, Kenneth. *Human Nature and History: A Response to Sociobiology* (New York: Columbia University Press, 1980).

Boehm, Christopher. *Hierarchy in the Forest: The Evolution of Egalitarian Behavior* (Cambridge, Mass.: Harvard University Press, 1999).

Boesch, Christophe, and Hedwige Boesch. "Hunting Behavior of Wild Chimpanzees in the Taï National Park," *American Journal of Physical Anthropology,* Vol. 78, No. 4 (April 1989), pp. 547–573.

Boesch, and Boesch-Achermann. *The Chimpanzees of the Taï Forest: Behavioural Ecology and Evolution* (New York: Oxford University Press, 2000).

Bolles, Robert C. "Species-Specific Defense Reactions and Avoidance Learning," *Psychological Review,* Vol. 77, No. 1 (January 1970), pp. 32–48.

Boone, James L. "Competition, Conflict, and the Development of Social Hierarchies," in Eric Alden Smith and Bruce Winterhalder, eds., *Evolutionary Ecology and Human Behavior* (New York: Aldine de Gruyter, 1992), pp. 301–337.

Bourke, Joanna. *An Intimate History of Killing: Face-to-Face Killing in Twentieth-Century Warfare* (New York: Basic Books, 1999).

Bowden, Mark. *Black Hawk Down: A Story of Modern War* (New York: Atlantic Monthly Press, 1999).

Bowler, Peter J. *Evolution: The History of an Idea,* rev. ed. (Berkeley: University of California Press, 1989).

Boyd, Robert, and Peter J. Richardson. "Sociobiology, Culture and Economic Theory," *Journal of Economic Behavior and Organization,* Vol. 1, No. 2 (June 1980), pp. 97–121.

Boyes, Roger. "Nazis Planned to Use Virus against Britain," *The Times,* March 12, 2001, <http://www.thetimes.co.uk/article/0,,2–97518,00.html>.

Brandon, Robert N. "The Levels of Selection: A Hierarchy of Interactors," in David L. Hull and Michael Ruse, eds., *The Philosophy of Biology* (New York: Oxford University Press, 1998), pp. 176–197.

Brandon, and Richard M. Burian, eds. *Genes, Organisms, Populations: Controversies over the Units of Selection* (Cambridge, Mass.: MIT Press, 1984).

Brass, Paul R. *Ethnicity and Nationalism: Theory and Comparison* (London: Sage, 1991).

Brennan, Geoffrey. "Comment: What Might Rationality Fail to Do?" in Karen Schweers Cook and Margaret Levi, eds., *The Limits of Rationality* (Chicago: University of Chicago Press, 1990), pp. 51–59.

Brewer, Marilynn B., and Donald T. Campbell. *Ethnocentrism and Intergroup Attitudes* (New York: Wiley, 1976).

Brewer. "The Role of Ethnocentrism in Intergroup Conflict," in William G. Austin and Stephen Worchel, eds., *The Social Psychology of Intergroup Relations* (Monterey, Calif.: Brooks/Cole, 1979), pp. 71–84.

Brodie, Bernard. *War and Politics* (New York: Macmillan, 1973).

Bromage, Timothy G., and Friedemann Schrenk, eds. *African Biogeography, Climate Change, and Human Evolution* (New York: Oxford University Press, 1999).

Brown, David. "On the Trail of the 1918 Influenza Epidemic," *Washington Post,* February 27, 2001, p. A3.

Brown, D.J.J. "The Structure of Polopa Feasting and Warfare," *Man* (N.S.), Vol. 14, No. 4 (December 1979), pp. 712–733.

Brown, Donald E. *Human Universals* (New York: McGraw-Hill, 1991).

Brown, E.S. "Immature Nutfall of Coconuts in the Solomon Islands: II, Changes in Ant Populations, and Their Relation to Vegetation," *Bulletin of Entomological Research,* Vol. 50, No. 3 (1959), pp. 523–558.

Brown, Harry. *A Walk in the Sun* (New York: Knopf, 1944).

Brown, Michael E. "Introduction," in Michael E. Brown, ed., *The International Dimensions of Internal Conflict* (Cambridge, Mass.: MIT Press, 1996), pp. 1–31.

Brown. "The Causes of Internal Conflict: An Overview," in Michael E. Brown, Owen R. Coté Jr., Sean M. Lynn-Jones, and Steven E. Miller, eds., *Nationalism and Ethnic Conflict* (Cambridge, Mass.: MIT Press, 1997), pp. 3–25.

Brown. "The Impact of Government Policies on Ethnic Relations," in Michael E. Brown and Šumit Ganguly, eds., *Government Policies and Ethnic Relations in Asia and the Pacific* (Cambridge, Mass.: MIT Press, 1997), pp. 511–575.

Brown and Šumit Ganguly. "Introduction," in Michael E. Brown and Šumit Ganguly, eds., *Government Policies and Ethnic Relations in Asia and the Pacific* (Cambridge, Mass.: MIT Press, 1997), pp. 1–29.

Broyles, William, Jr. "Why Men Love War," *Esquire* (November 1984), pp. 55–65.

Brubaker, Rogers. "National Minorities, Nationalizing States, and External National Homelands in the New Europe," *Daedalus* (Spring 1995), pp. 107–132.

Buchanan, James M., and Gordon Tullock. *The Calculus of Consent: Logical Foundations of Constitutional Democracy* (Ann Arbor: University of Michigan Press, 1962).

Bunge, Mario, ed. *The Methodological Unity of Science* (Dordrecht, Holland: Reidel, 1973).

Burg, Steven L., and Paul S. Shoup. *The War in Bosnia-Herzegovina: Ethnic Conflict and International Intervention* (Armonk, N.Y.: M.E. Sharpe, 1999).

Bush, George W. "Remarks by the President during Announcement of Proposals for Global Fund to Fight HIV/AIDS, Malaria and Tuberculosis," The White House, May 11, 2001, <http://www.whitehouse.gov/news/releases/2001/05/20010511-1.html>.

Buss, David M. *The Evolution of Desire: Strategies of Human Mating* (New York: Basic Books, 1994).

Buss. *The Dangerous Passion: Why Jealousy Is as Necessary as Love and Sex* (New York: Free Press, 2000).

Buss, and Todd K. Shackelford. "Human Aggression in Evolutionary Psychological Perspective," *Clinical Psychological Review,* Vol. 17, No. 6 (1997), pp. 605–619.

Byrne, Richard W., and Andrew Whiten. "Machiavellian Intelligence," in Andrew Whiten and Richard W. Byrne, eds., *Machiavellian Intelligence II: Extensions and Evaluations* (New York: Cambridge University Press, 1997), pp. 1–23.

Campbell, Donald T. "On the Genetics of Altruism and the Counter-Hedonic Components in Human Culture," *Journal of Social Issues,* Vol. 28, No. 3 (1972), pp. 21–37.

Cannon, Aubrey. "Conflict and Salmon on the Interior Plateau of British Columbia," in Brian Hayden, ed., *A Complex Culture of the British Columbia Plateau: Traditional* Stl'átl'imx *Resource Use* (Vancouver: UBC Press, 1992), pp. 506–524.

Cantor, Norman F. *In the Wake of the Plague: The Black Death and the World It Made* (New York: The Free Press, 2001).

Caplan, Arthur L., ed. *The Sociobiology Debate* (New York: Harper and Row, 1978).

Caputo, Philip. *A Rumor of War* (New York: Holt, Rinehart and Winston, 1977).

Carneiro, Robert L. "A Theory of the Origin of the State," *Science,* August 21, 1970, pp. 734–737.

Carr, E.H. *The Twenty Years' Crisis, 1919–1939: An Introduction to the Study of International Relations,* 2nd ed. (London: Macmillan, 1946).

Cartwright, John. *Evolution and Human Behavior: Darwinian Perspectives on Human Nature* (Cambridge, Mass.: MIT Press, 2000).

Cavalli-Sforza, Luigi Luca. *Genes, Peoples, and Languages* (New York: North Point Press, 2000).

Cavalli-Sforza, Paolo Menozzi, and Alberto Piazza. *The History and Geography of Human Genes* (Princeton, N.J.: Princeton University Press, 1994).

Cavalli-Sforza, and Francesco Cavalli-Sforza. *The Great Human Diasporas: The History of Diversity and Evolution,* trans. by Sarah Thorne (Reading, Mass.: Addison-Wesley, 1995).

Caws, Peter. *Structuralism: The Art of the Intelligible* (London: Humanities Press International, 1988).

Central Intelligence Agency. *World Factbook 2000* <http://www.odci.gov/cia/publications/factbook/index.html>.

Chagnon, Napoleon A. *Yanomamö,* 4th ed. (Fort Worth: Harcourt Brace Jovanovich, 1992).

Chagnon, and Paul Bugos. "Kin Selection and Conflict: An Analysis of a Yanomamö Ax Fight," in Napoleon A. Chagnon and William Irons, eds., *Evolutionary Biology and Human Social Behavior* (North Scituate, Mass.: Duxbury Press, 1979), pp. 213–238.

Chagnon, and William Irons, eds. *Evolutionary Biology and Human Social Behavior: An Anthropological Perspective* (North Scituate, Mass.: Duxbury Press, 1979).

Chang, Kenneth. "Scientists Find Signs of Meteor Crash That Led to Extinctions in

Era before Dinosaurs," *New York Times,* February 23, 2001, <http://www.nytimes.com/2001/02/23/science/23EXTI.html>.

Chapais, Bernard. "Alliances as a Means of Competition in Primates: Evolutionary, Developmental, and Cognitive Aspects," *Yearbook of Physical Anthropology,* Vol. 38 (New York: Wiley-Liss, 1995), pp. 115–136.

Chatters, James C. *Ancient Encounters: Kennewick Man and the First Americans* (New York: Simon and Schuster, 2001).

Cheney, Dorothy L. "Interactions and Relationships between Groups," in Barbara B. Smuts, Dorothy L. Cheney, Robert M. Seyfarth, Richard W. Wrangham, and Thomas T. Struhsaker, eds., *Primate Societies* (Chicago: University of Chicago Press, 1987), pp. 267–281.

Cheney, and Robert M. Seyfarth. *How Monkeys See the World: Inside the Mind of Another Species* (Chicago: University of Chicago Press, 1990).

Chorover, Stephen L. *From Genesis to Genocide: The Meaning of Human Nature and Power of Behavior Control* (Cambridge, Mass.: MIT Press, 1979).

Choucri, Nazli, and Robert C. North. *Nations in Conflict: National Growth and International Violence* (San Francisco: W.H. Freeman, 1975).

Christopher, George W., Theodore J. Cieslak, Julie A. Pavlin, and Edward M. Eitzen Jr. "Biological Warfare: A Historical Perspective," in Joshua Lederberg, ed., *Biological Weapons: Limiting the Threat* (Cambridge, Mass.: MIT Press, 1999), pp. 17–35.

Chung, Shin-Ho, and Richard Herrnstein. "Choice and Delay of Reinforcement," *Journal of the Experimental Analysis of Behavior,* Vol. 10, No. 1 (January 1967), pp. 67–74.

Churchill, Winston S. *The Story of the Malakand Field Force: An Episode of Frontier War* (New York: Norton, 1990 [1898]).

Churchland, Patricia S. *Neurophilosophy* (Cambridge, Mass.: Bradford/MIT Press, 1986).

Ciochon, Russell L., and John G. Fleagle. *Primate Evolution and Human Origins* (New York: Aldine de Gruyter, 1987).

Cioffi-Revilla, Claudio. "Origins and Evolution of War and Politics," *International Studies Quarterly,* Vol. 40, No. 1 (March 1996), pp. 1–22.

Cioffi-Revilla. "Ancient Warfare: Origins and Systems," in Manus I. Midlarsky, ed., *Handbook of War Studies II* (Ann Arbor: University of Michigan Press, 2000), pp. 59–89.

Clark, William R., and Michael Grunstein. *Are We Hardwired? The Role of Genes in Human Behavior* (New York: Oxford University Press, 2000).

Clinton, William Jefferson. "Clinton: U.S. Is Fighting for the World's Future," *New York Times,* April 17, 1999, p. A6.

Coase, R.H. "Adam Smith's View of Man," *The Journal of Law and Economics,* Vol. 19, No. 3 (October 1976), pp. 529–546.

Coen, Enrico. *The Art of Genes: How Organisms Make Themselves* (New York: Oxford University Press, 1999).

Cohen, Rich. *The Avengers: A Jewish War Story* (New York: Knopf, 2000).
Conklin, Harold C. "An Ethnoecological Approach to Shifting Agriculture," *Transactions of the New York Academy of Sciences,* 2nd series, Vol. 17 (New York: New York Academy of Sciences, 1954), pp. 133–142.
Conner, Richard C., Rachel A. Smolker, and Andrew F. Richards. "Dolphin Alliances and Coalitions," in Alexander H. Harcourt and Frans B.M. de Waal, eds., *Coalitions and Alliances in Humans and Other Animals* (New York: Oxford University Press, 1992), pp. 415–443.
Connor, Walker. "Eco- or Ethno-Nationalism?" *Ethnic and Racial Studies,* Vol. 7, No. 3 (1984), pp. 242–259.
Connor. *Ethnonationalism: The Quest for Understanding* (Princeton, N.J.: Princeton University Press, 1994).
Conze, Susanne, and Beate Fieseler. "Soviet Women as Comrades-in-Arms: A Blind Spot in the History of the War," in Robert W. Thurston and Bernd Bonwetsch, eds., *The People's War: Responses to World War II in the Soviet Union* (Urbana: University of Illinois Press, 2000), pp. 211–234.
Cook, Karen Schweers, and Margaret Levi, eds. *The Limits of Rationality* (Chicago: University of Chicago Press, 1990).
Cook, Noble David. *Born to Die: Disease and New World Conquest, 1492–1650* (New York: Cambridge University Press, 1998).
Cook-Deegan, Robert. *The Gene Wars: Science, Politics, and the Human Genome* (New York: Norton, 1995).
Cooper, Belton Y. *Death Traps: The Survival of an American Armored Division in World War II* (Novato, Calif.: Presidio, 1998).
Corning, Peter A. "The Biological Bases of Behavior and Some Implications for Political Science," *World Politics,* Vol. 23, No. 3 (April 1971), pp. 321–370.
Corning. "An Evolutionary Paradigm for the Study of Human Aggression," in Martin A. Nettleship, R. Dalegivens, and Anderson Nettleship, eds., *War, Its Causes and Correlates* (The Hague: Mouton Publishers, 1975), pp. 359–387.
Corning. "Biopolitical Economy: A Trail-Guide for an Inevitable Discipline," in Albert Somit and Steven A. Peterson, eds., *Research in Biopolitics,* Vol. 5 (Greenwich, Conn.: JAI Press, 1997), pp. 247–277.
Corning. "Synergy Goes to War," paper prepared for the annual meeting of the Association for Politics and the Life Sciences, Charleston, S.C., October 2001.
Corning. *Nature's Magic: Synergy in Evolution and the Fate of Humankind* (New York: Cambridge University Press, 2003).
Coser, Lewis A. *The Function of Social Conflict* (Glencoe, Ill.: Free Press, 1956).
Cosmides, Leda. "The Logic of Social Exchange: Has Natural Selection Shaped How Humans Reason?" *Cognition,* Vol. 31, No. 3 (April 1989), pp. 187–276.
Cosmides, and John Tooby. "From Evolution to Behavior: Evolutionary Psychology as the Missing Link," in John Dupré, ed., *The Latest on the Best: Essays on Evolution and Optimality* (Cambridge, Mass.: MIT Press, 1987), pp. 277–306.
Cosmides and Tooby. "Cognitive Adaptations for Social Exchange," in Jerome H.

Barkow, Leda Cosmides, and John Tooby, eds., *The Adapted Mind: Evolutionary Psychology and the Generation of Culture* (New York: Oxford University Press, 1992), pp. 163–228.

Cosmides and Tooby. "Better Than Rational: Evolutionary Psychology and the Invisible Hand," *The American Economic Review,* Vol. 84, No. 2 (May 1994), pp. 327–332.

Cronk, Lee. *That Complex Whole: Culture and the Evolution of Human Behavior* (Boulder, Colo.: Westview Press, 1999).

Cronk, Napoleon Chagnon, and William Irons, eds. *Adaptation and Human Behavior: An Anthropological Perspective* (New York: Aldine de Gruyter, 2000).

Crook, Paul. *Darwinism, War and History: The Debate over the Biology of War from the "Origin of Species" to the First World War* (New York: Cambridge University Press, 1994).

Crosby, Alfred W. *The Columbian Exchange: Biological and Cultural Consequences of 1492* (Westport, Conn.: Greenwood Press, 1972).

Crosby. *Ecological Imperialism: The Biological Expansion of Europe, 900–1900* (New York: Cambridge University Press, 1986).

Crosby. *America's Forgotten Pandemic: The Influenza of 1918* (New York: Cambridge University Press, 1989).

Crosby. *Germs, Seeds, and Animals: Studies in Ecological History* (Armonk, N.Y.: M.E. Sharpe, 1994).

Cummins, Denise Dellarosa. "Social Norms and Other Minds," in Dellarosa Cummins and Colin Allen, eds., *The Evolution of Mind* (New York: Oxford University Press, 1998), pp. 30–50.

Curtin, Philip D. *Death by Migration: Europe's Encounter with the Tropical World in the Nineteenth Century* (New York: Cambridge University Press, 1989).

Curtin. *Disease and Empire: The Health of European Troops in the Conquest of Africa* (New York: Cambridge University Press, 1998).

Curtis, Edward S. *The North American Indian, Volume IX: The Salish* (Cambridge, Mass.: The University Press, 1913).

Daalder, Ivo H., and Michael E. O'Hanlon. *Winning Ugly: NATO's War to Save Kosovo* (Washington, D.C.: Brookings Institution Press, 2000).

Daly, Martin, and Margo Wilson. *Homicide* (New York: Aldine de Gruyter, 1988).

Damasio, Antonio R. *Descartes' Error: Emotion, Reason, and the Human Brain* (New York: Grosset/Putnam, 1994).

Damasio. *Looking for Spinoza: Joy, Sorrow, and the Feeling Brain* (New York: Harcourt, 2003).

Damuth, John, and I. Lorraine Heisler. "Alternative Formulations of Multilevel Selection," *Biology and Philosophy,* Vol. 3 (1988), pp. 407–430.

Danzig, Richard, and Pamela B. Berkowsky. "Why Should We Be Concerned about Biological Warfare," in Joshua Lederberg, ed., *Biological Weapons: Limiting the Threat* (Cambridge, Mass.: MIT Press, 1999), pp. 9–14.

Darwin, Charles. *The Autobiography of Charles Darwin, 1809–1882,* ed. by Nora Barlow (London: Collins, 1958).

Darwin. *The Descent of Man, and Selection in Relation to Sex,* 2 Vols. (Princeton, N.J.: Princeton University Press, 1981 [1871]).

Darwin. *On the Origin of Species by Means of Natural Selection, or the Preservation of Favoured Races in the Struggle for Life* (Cambridge, Mass.: Harvard University Press, 1964 [1859]).

Dawkins, Richard. "Replicators and Vehicles," King's College Sociobiology Study Group, ed., *Current Problems in Sociobiology* (Cambridge: Cambridge University Press, 1982), pp. 45–64.

Dawkins. Review of *Not in Our Genes, New Scientist* (January 24, 1985), pp. 59–60.

Dawkins. *The Blind Watchmaker: Why the Evidence of Evolution Reveals a Universe without Design* (New York: Norton, 1986).

Dawkins. *The Selfish Gene,* new ed. (Oxford: Oxford University Press, 1989).

Dawkins. "Darwin Triumphant: Darwinism as a Universal Truth," in Michael H. Robinson and Lionel Tiger, eds., *Man and Beast Revisited* (Washington, D.C.: Smithsonian Institution Press, 1991), pp. 23–39.

Dawkins. *The Extended Phenotype: The Long Reach of the Gene,* rev. ed. (Oxford: Oxford University Press, 1999).

Dawkins, and M. Ridley, eds. *Oxford Surveys in Evolutionary Biology,* Vol. 2 (New York: Oxford University Press, 1985).

Degler, Carl N. *In Search of Human Nature: The Decline and Revival of Darwinism in American Social Thought* (New York: Oxford University Press, 1991).

Denevan, William. "Native American Populations in 1492: Recent Research and a Revised Hemispheric Estimate," in William M. Denevan, ed., *The Native Population of the Americas in 1492,* 2nd ed. (Madison: University of Wisconsin Press, 1992), pp. xvii–xxix.

Denig, Edwin T. *Indian Tribes of the Upper Missouri* (Washington, D.C.: Bureau of American Ethnology, 1930).

Dennett, Daniel C. *Darwin's Dangerous Idea: Evolution and the Meanings of Life* (New York: Simon and Schuster, 1995).

Dentan, Robert Knox. "Band-Level Eden: A Mystifying Chimera," *Cultural Anthropology,* Vol. 3, No. 3 (August 1988), pp. 276–284.

Denton, George H. "Cenozoic Climate Change," in Timothy G. Bromage and Friedemann Schrenk, eds., *African Biogeography, Climate Change, and Human Evolution* (New York: Oxford University Press, 1999), pp. 94–114.

Desch, Michael C. "The Keys That Lock up the World: Identifying American Interests in the Periphery," *International Security,* Vol. 14, No. 1 (Summer 1989), pp. 86–121.

Des Forges, Alison. *Leave None to Tell the Story: Genocide in Rwanda* (New York: Human Rights Watch, 1999).

Destexhe, Alain. *Rwanda and Genocide in the Twentieth Century* (New York: New York University Press, 1995).

Deudney, Daniel H., and Richard A. Matthew, eds. *Contested Grounds: Security and Conflict in the New Environmental Politics* (Albany: State University of New York Press, 1999).

De Waal, Frans B.M. *Peacemaking among Primates* (Cambridge, Mass.: Harvard University Press, 1989).

De Waal. *Good Natured: The Origins of Right and Wrong in Humans and Other Animals* (Cambridge, Mass.: Harvard University Press, 1996).

De Waal. *Bonobo: The Forgotten Ape* (Berkeley: University of California Press, 1997).

De Waal. *Chimpanzee Politics: Power and Sex among Apes,* rev. ed. (Baltimore, Md.: Johns Hopkins University Press, 1998).

De Waal, "Introduction," in Frans B.M. de Waal, ed., *Tree of Origin: What Primate Behavior Can Tell Us about Human Social Evolution* (Cambridge, Mass.: Harvard University Press, 2001), pp. 1–8.

De Waal, and Lesleigh M. Luttrell. "Mechanisms of Social Reciprocity in Three Primate Species: Symmetrical Relationship Characteristics or Cognition?" *Ethology and Sociobiology,* Vol. 9, No. 2 (1988), pp. 101–118.

De Zayas, Alfred-Maurice. *A Terrible Revenge: The Ethnic Cleansing of the East European Germans, 1944–1950* (New York: St. Martin's Press, 1994).

Diamond, Jared. *The Third Chimpanzee: The Evolution and Future of the Human Animal* (New York: HarperCollins Publishers, 1992).

Diamond. *Guns, Germs, and Steel: The Fates of Human Societies* (New York: Norton, 1997).

Divale, William Tulio. *Warfare in Primitive Societies: A Bibliography* (Santa Barbara, Calif.: American Bibliographic Center—Clio Press, 1973).

Divale, and Marvin Harris. "Population, Warfare, and the Male Supremacist Complex," *American Anthropologist,* Vol. 78, No. 3 (September 1976), pp. 521–538.

Dobyns, Henry F., with the assistance of William R. Swagerty. *Their Number Become Thinned* (Knoxville: University of Tennessee Press, 1983).

Dobzhansky, Theodosius. *Genetics and the Origin of Species,* 3rd ed., rev. (New York: Columbia University Press, 1961).

Dobzhansky. *Mankind Evolving: The Evolution of the Human Species* (New Haven: Yale University Press, 1962).

Dobzhansky. "On Some Fundamental Concepts of Darwinian Biology," in Theodosius Dobzhansky, Max K. Hecht, William C. Steere, eds., *Evolutionary Biology,* Vol. 2 (New York: Appleton-Century-Crofts, 1968).

Dobzhansky. *Genetics of the Evolutionary Process* (New York: Columbia University Press, 1970).

Dower, John W. *War without Mercy: Race and Power in the Pacific War* (New York: Pantheon Books, 1986).

Dower. *Embracing Defeat: Japan in the Wake of World War II* (New York: New Press, 1999).

Downs, Anthony. *An Economic Theory of Democracy* (New York: Harper and Row, 1957).

Drinnon, Richard. *Facing West: The Metaphysics of Indian-Hating and Empire-Building* (Minneapolis: University of Minnesota Press, 1980).

Dugatkin, Lee Alan. *The Imitation Factor: Evolution beyond the Gene* (New York: Free Press, 2000).

Dukas, Helen, and Banesh Hoffmann. *Albert Einstein, the Human Side: New Glimpses from His Archives* (Princeton, N.J.: Princeton University Press, 1979).

Dunbar, Robin. "The Sociobiology of War," *Medicine and War,* Vol. 1 (1985), pp. 201–208.

Dunbar. "Sociobiological Explanations and the Evolution of Ethnocentrism," in Vernon Reynolds, Vincent Falger, and Ian Vine, eds., *The Sociobiology of Ethnocentrism* (London: Croom Helm, 1987), pp. 48–59.

Dunbar. "Culture, Honesty and the Freerider Problem," in Robin Dunbar, Chris Knight, and Camilla Power, eds., *The Evolution of Culture: An Interdisciplinary View* (Edinburgh: Edinburgh University Press, 1999), pp. 194–213.

Dunn, John. *Western Political Theory in the Face of the Future* (Cambridge: Cambridge University Press, 1978).

Dupré, John, ed. *The Latest on the Best: Essays on Evolution and Optimality* (Cambridge, Mass.: MIT Press, 1987).

Durham, William H. "Resource Competition and Human Aggression. Part I: A Review of Primitive War," *The Quarterly Review of Biology,* Vol. 51, No. 3 (September 1976), pp. 385–415.

Durham. *Scarcity and Survival in Central America: Ecological Origins of the Soccer War* (Stanford, Calif.: Stanford University Press, 1979).

Durham. "Toward a Coevolutionary Theory of Human Biology and Culture," in Napoleon A. Chagnon and William Irons, eds., *Evolutionary Biology and Human Social Behavior: An Anthropological Perspective* (North Scituate, Mass.: Duxbury Press, 1979), pp. 39–59.

Durham. *Coevolution: Genes, Culture, and Human Diversity* (Stanford, Calif.: Stanford University Press, 1991).

Durkheim, Émile. *The Rules of Sociological Method,* trans. by Sarah A. Solovay and John H. Mueller (New York: Free Press, 1938).

Durkheim. *Suicide: A Study in Sociology,* trans. by John A. Spaulding and George Simpson (New York: Free Press, 1951).

Eberstadt, Nicholas. "The Future of AIDS," *Foreign Affairs,* Vol. 81, No. 6 (November/December 2002), pp. 22–45.

Edgerton, Robert B. *Warrior Women: The Amazons of Dahomey and the Nature of War* (Boulder, Colo.: Westview Press, 2000).

Ehrenreich, Barbara. *Blood Rites: Origins and History of the Passions of War* (New York: Metropolitan, 1997).

Ehrlich, Paul R. *Human Natures: Genes, Cultures, and the Human Prospect* (Washington, D.C.: Island Press, 2000).

Eibl-Eibesfeldt, Irenäus. "The Myth of the Aggression-Free Hunter and Gatherer Society," in Ralph L. Holloway, ed., *Primate Aggression, Territoriality, and Xenophobia: A Comparative Perspective* (New York: Academic Press, 1974), pp. 435–457.

Eibl-Eibesfeldt. "Human Ethology: Concepts and Implications for the Sciences of Man," *The Behavioral and Brain Sciences,* Vol. 2 (1979), pp. 1–26.

Eibl-Eibesfeldt. "Warfare, Man's Indoctrinability and Group Selection," *Zeitschrift für Tierpsychologie,* Vol. 60, No. 3 (1982), pp. 177–198.

Eibl-Eibesfeldt. *Human Ethology* (New York: Aldine de Gruyter, 1989).

Eibl-Eibesfeldt. "Us and the Others," in Irenäus Eibl-Eibesfeldt and Frank Kemp Salter, eds., *Indoctrinability, Ideology, and Warfare: Evolutionary Perspectives* (New York: Berghahn Books, 1998), pp. 21–53.

Ekman, Paul. "All Emotions Are Basic," in Paul Ekman and Richard J. Davidson, eds., *The Nature of Emotion: Fundamental Questions* (New York: Oxford University Press, 1994), pp. 15–19.

Eldredge, Niles. *Reinventing Darwin: The Great Debate at the High Table of Evolutionary Theory* (New York: John Wiley, 1995).

Eldredge. *The Triumph of Evolution and the Failure of Creationism* (New York: W.H. Freeman, 2000).

Eldredge, and Stephen Jay Gould. "Punctuated Equilibria: An Alternative to Phyletic Gradualism," in Thomas J.M. Schopf, ed., *Models in Paleobiology* (San Francisco: Freeman Cooper, 1972), pp. 82–115.

Elias, Norbert. "On Transformations of Aggressiveness," *Theory and Society,* Vol. 5, No. 2 (March 1978), pp. 229–242.

Elster, Jon. "Introduction," in Jon Elster, ed., *Rational Choice* (New York: New York University Press, 1986), pp. 1–33.

Elster. *The Cement of Society: A Study of Social Order* (New York: Cambridge University Press, 1989).

Elster. *Nuts and Bolts for the Social Sciences* (New York: Cambridge University Press, 1989).

Elster. "When Rationality Fails," in Karen Schweers Cook and Margaret Levi, eds., *The Limits of Rationality* (Chicago: University of Chicago Press, 1990), pp. 19–51.

Ember, Carol R. "Myths about Hunter-Gatherers," *Ethnology,* Vol. 17, No. 4 (October 1978), pp. 439–448.

Ember, and Melvin Ember. *Cultural Anthropology,* 6th ed. (Englewood Cliffs, N.J.: Prentice Hall, 1990).

Ember and Ember. "Resource Unpredictability, Mistrust, and War," *The Journal of Conflict Resolution,* Vol. 36, No. 2 (June 1992), pp. 242–262.

Ember and Ember. "Warfare, Aggression, and Resource Problems: Cross-Cultural Codes," *Behavior Science Research,* Vol. 26, Nos. 1–4 (1992), pp. 169–226.

Ember, Ember, and Bruce Russett. "Peace between Participatory Polities: A Cross-Cultural Test of the 'Democracies Rarely Fight Each Other' Hypothesis," *World Politics,* Vol. 44, No. 4 (July 1992), pp. 573–599.

Ember, Melvin. "Statistical Evidence for an Ecological Explanation of Warfare," *American Anthropologist,* Vol. 84, No. 3 (June 1982), pp. 645–649.

Enquist, Magnus, and Olof Leimar. "The Evolution of Fatal Fighting," *Animal Behaviour,* Vol. 39, No. 1 (January 1990), pp. 1–9.

Erikson, Erik H. *Gandhi's Truth: On the Origins of Militant Nonviolence* (New York: Norton, 1969).

Esman, Milton J. *Ethnic Politics* (Ithaca, N.Y.: Cornell University Press, 1994).

Fagan, Brian. *The Little Ice Age: How Climate Made History, 1300–1850* (New York: Basic Books, 2000).

Falger, Vincent S.E. "From Xenophobia to Xenobiosis?" in Vernon Reynolds, Vincent Falger, and Ian Vine, eds., *The Sociobiology of Ethnocentrism* (London: Croom Helm, 1987), pp. 235–250.

Falger. "The Missing Link in International Relations Theory?" *Journal of Social and Biological Structures,* Vol. 14, No. 1 (1991), pp. 73–77.

Falger. "Biopolitics and the Study of International Relations," in Albert Somit and Steven A. Peterson, eds., *Research in Biopolitics,* Vol. 2 (Greenwich, Conn.: JAI Press, 1994), pp. 115–134.

Falger. "Human Nature in Modern International Relations: Part I. Theoretical Backgrounds," in Albert Somit and Steven A. Peterson, eds., *Research in Biopolitics,* Vol. 5*: Recent Explorations in Biology and Politics* (Greenwich, Conn.: JAI Press, 1997), pp. 155–175.

Falger. "Evolutionary World Politics Enriched," in William R. Thompson, ed., *Evolutionary Interpretations of World Politics* (New York: Routledge, 2001), pp. 30–51.

Falger. "Evolution in International Relations?" in Steven A. Peterson and Albert Somit, eds., *Research in Biopolitics,* Vol. 8 (Amsterdam: Elsevier, 2001), pp. 91–118.

Falkenrath, Richard A., Robert D. Newman, and Bradley A. Thayer. *America's Achilles' Heel: Nuclear, Biological, and Chemical Terrorism and Covert Attack* (Cambridge, Mass.: MIT Press, 1998).

Fearon, James D. "Domestic Political Audiences and the Escalation of International Disputes," *American Political Science Review,* Vol. 88, No. 3 (September 1994), pp. 577–592.

Fearon. "Commitment Problems and the Spread of Ethnic Conflict," in David A. Lake and Donald Rothchild, eds., *The International Spread of Ethnic Conflict* (Princeton, N.J.: Princeton University Press, 1998), pp. 107–126.

Fenn, Elizabeth Anne. *Pox Americana: The Great Smallpox Epidemic of 1775–82* (New York: Hill and Wang, 2001).

Fenner, Frank. "Human Monkeypox, A Newly Discovered Human Virus Disease," in Stephen S. Morse, ed., *Emerging Viruses* (New York: Oxford University Press, 1993), pp. 176–183.

Ferguson, Niall. *The Pity of War: Explaining World War I* (New York: Basic Books, 1999).

Ferguson, R. Brian. "Introduction: Studying War," in R. Brian Ferguson, ed., *Warfare, Culture, and Environment* (Orlando, Fla.: Academic Press, 1984), pp. 1–81.

Ferguson. "A Reexamination of the Causes of Northwest Coast Warfare," in R. Brian Ferguson, ed., *Warfare, Culture, and Environment* (Orlando, Fla.: Academic Press, 1984), pp. 267–328.

Ferguson, ed. *Warfare, Culture, and Environment* (Orlando, Fla.: Academic Press, 1984).

Ferguson. *Yanomami Warfare: A Political History* (Santa Fe, N.Mex.: School of American Research Press, 1995).

Ferguson, and Neil L. Whitehead. "The Violent Edge of Empire," in R. Brian Ferguson and Neil L. Whitehead, eds., *War in the Tribal Zone* (Santa Fe, N.Mex.: School of American Research Press, 1992), pp. 1–30.

Ferguson and Whitehead, eds. *War in the Tribal Zone: Expanding States and Indigenous Warfare* (Santa Fe, N.Mex.: School of American Research Press, 1992).

Firth, Raymond. *We, the Tikopia* (London: Allen and Unwin, 1957).

Fisher, Ronald A. *The Genetical Theory of Natural Selection* (Oxford: Clarendon Press, 1930).

Fisher. *The Genetical Theory of Natural Selection,* 2nd ed., rev. (New York: Dover Publications, 1958).

Fishman, Joshua A., Michael H. Gertner, Esther G. Lowy, and William G. Milán. *The Rise and Fall of the Ethnic Revival: Perspectives on Language and Ethnicity* (New York: Mouton Publishers, 1985).

Foley, R.A. "Evolutionary Ecology of Fossil Hominids," in Eric Alden Smith and Bruce Winterhalder, eds., *Evolutionary Ecology and Human Behavior* (New York: Aldine de Gruyter, 1992), pp. 131–164.

Foley, Robert. *Humans before Humanity: An Evolutionary Perspective* (Oxford: Blackwell, 1995).

Fowler, Brenda. "After 5,300 Years, Mystery of Iceman's Death Is Solved," *New York Times,* August 7, 2001, at <http://www.nytimes.com/2001/08/07/science/social/07ICEM.html>.

Fox, Robin, ed. *Biosocial Anthropology* (New York: Wiley and Sons, 1975).

Frank, Robert H. *Passions within Reason: The Strategic Role of the Emotions* (New York: Norton, 1988).

Fraser, George MacDonald. *Quartered Safe Out Here: A Recollection of the War in Burma* (London: Harvill, 1992).

Fratkin, Elliot. "A Comparison of the Role of Prophets in Samburu and Maasai Warfare," in Katsuyoshi Fukui and David Turton, eds., *Warfare among East African Herders* (Osaka: National Museum of Ethnology, 1979), pp. 53–67.

Freud, Sigmund. *The Ego and the Id,* trans. by Joan Riviere (New York: Norton, 1962).

Freud. "Why War?" in Melvin Small and J. David Singer, eds., *International War: An Anthology* (Homewood, Ill.: Dorsey Press, 1985), pp. 158–164.

Fridlund, Alan J. *Human Facial Expression: An Evolutionary View* (New York: Academic Press, 1994).

Fried, Milton, Marvin Harris, and Robert Murphy, eds. *War* (Garden City, N.Y.: Natural History Press, 1968).

Friedman, Milton. *Essays in Positive Economics* (Chicago: University of Chicago Press, 1953).

Fromm, Erich. *The Anatomy of Human Destructiveness* (London: Jonathan Cape, 1974).

Fukui, Katsuyoshi, and David Turton, eds. *Warfare among East African Herders* (Osaka: National Museum of Ethnology, 1979).

Fukuyama, Francis. *The Great Disruption: Human Nature and the Reconstitution of Social Order* (New York: Free Press, 1999).

Fukuyama. "Natural Rights and Human History," *The National Interest*, No. 64 (Summer 2001), pp. 19–30.

Fuller, John L. "Genes, Brains, and Behavior," in Michael S. Gregory, Anita Silvers, and Diane Sutch, eds., *Sociobiology and Human Nature* (San Francisco: Jossey-Bass, 1978), pp. 98–115.

Fussell, Paul. *Wartime: Understanding and Behavior in the Second World War* (New York: Oxford University Press, 1989).

Gadagkar, Raghavendra. *Survival Strategies: Cooperation and Conflict in Animal Societies* (Cambridge, Mass.: Harvard University Press, 1997).

Gagnon, V.P., Jr. "Ethnic Nationalism and International Conflict: The Case of Serbia," *International Security*, Vol. 19, No. 3 (Winter 1994/95), pp. 130–166.

Gallagher, Nancy L. *Breeding Better Vermonters: The Eugenics Project in the Green Mountain State* (Hanover, N.H.: University Press of New England, 1999).

Ganguly, Šumit. "Internal Conflict in South and Southwest Asia," in Michael E. Brown, ed., *The International Dimensions of Internal Conflict* (Cambridge, Mass.: MIT Press, 1996), pp. 141–172.

Gat, Azar. "The Pattern of Fighting in Simple, Small-Scale, Prestate Societies," *Journal of Anthropological Research*, Vol. 55, No. 4 (Winter 1999), pp. 563–583.

Gat. "The Human Motivational Complex: Evolutionary Theory and the Causes of Hunter-Gatherer Fighting, Part I. Primary Somatic and Reproductive Causes," *Anthropological Quarterly*, Vol. 73, No. 1 (January 2000), pp. 20–34.

Gat. "The Human Motivational Complex: Evolutionary Theory and the Causes of Hunter-Gatherer Fighting, Part II. Proximate, Subordinate, and Derivative Causes," *Anthropological Quarterly*, Vol. 73, No. 2 (April 2000), pp. 74–88.

Gazzaniga, Michael S. *Nature's Mind: The Biological Roots of Thinking, Emotions, Sexuality, Language, and Intelligence* (New York: Basic Books, 1992).

Geertz, Clifford. "The Integrative Revolution: Primordial Sentiments and Civil Politics in the New States," in Clifford Geertz, ed., *Old Societies and New States: The Quest for Modernity in Asia and Africa* (New York: Free Press, 1963), pp. 105–157.

Geertz. *The Interpretation of Cultures* (New York: Basic Books, 1973).

Geertz. *Local Knowledge: Further Essays in Interpretative Anthropology* (New York: Basic Books, 1983).

Geissler, Erhard, and John Ellis van Courtland Moon, eds. *Biological and Toxin Weapons: Research, Development and Use from the Middle Ages to 1945* (Oxford: Oxford University Press, 1999).

Gellner, Ernest. *Nations and Nationalism* (New York: Oxford University Press, 1983).

George, Alexander L., and Timothy J. McKeown. "Case Studies and Theories of Organizational Decision Making," in Robert F. Coulam and Richard A. Smith, eds., *Advances in Information Processing in Organizations: A Research Annual* (Greenwich, Conn.: JAI Press, 1985), Vol. 2, pp. 34–41.

Gerth, H.H., and C. Wright Mills, eds. *From Max Weber: Essays in Sociology* (New York: Oxford University Press, 1946).

Ghiglieri, Michael P. *The Dark Side of Man: Tracing the Origins of Male Violence* (Reading, Mass.: Perseus Books, 1999).

Gilpin, Robert. *U.S. Power and the Multinational Corporation: The Political Economy of Foreign Direct Investment* (New York: Basic Books, 1975).

Gilpin. *War and Change in World Politics* (Cambridge: Cambridge University Press, 1981).

Gilpin. "The Richness of the Tradition of Political Realism," in Robert O. Keohane, ed., *Neorealism and Its Critics* (New York: Columbia University Press, 1986), pp. 301–321.

The Global Infectious Disease Threat and Its Implications for the United States, National Intelligence Estimate NIE 99–17D, January 2000, available at <http://www.odci.gov/cia/publications/nie/report/nie99–17d.html>.

Goldstein, Joshua S. "The Emperor's New Genes," *International Studies Quarterly,* Vol. 31, No. 1 (March 1987), pp. 33–43.

Goldstein. *War and Gender: How Gender Shapes the War System and Vice Versa* (Cambridge: Cambridge University Press, 2001).

Goodall, Jane. *The Chimpanzees of Gombe: Patterns of Behavior* (Cambridge, Mass.: Harvard University Press, 1986).

Goodall. *Through a Window: My Thirty Years with the Chimpanzees of Gombe* (Boston: Houghton Mifflin, 1990).

Gould, Stephen Jay. "Sociobiology: The Art of Storytelling," *New Scientist,* Vol. 80, No. 1129 (November 16, 1978), pp. 530–533.

Gould. "Sociobiology and the Theory of Natural Selection," in George W. Barlow and James Silverberg, eds., *Sociobiology: Beyond Nature/Nurture?* (Boulder, Colo.: Westview Press, 1980), pp. 257–269.

Gould. *The Panda's Thumb* (New York: Norton, 1980).

Gould. *Time's Arrow, Time's Cycle: Myth and Metaphor in the Discovery of Geological Time* (Cambridge, Mass.: Harvard University Press, 1987).

Gould. *The Mismeasure of Man,* rev. and exp. (New York: Norton, 1996).

Gould. "More Things in Heaven and Earth," in Hilary Rose and Steven Rose, eds., *Alas, Poor Darwin: Arguments against Evolutionary Psychology* (London: Jonathan Cape, 2000), pp. 85–105.

Gould. *The Structure of Evolutionary Theory* (Cambridge, Mass.: Harvard University Press, 2002).

Gould, and Richard C. Lewontin. "The Spandrels of San Marco and the Panglossian Paradigm: A Critique of the Adaptationist Programme," in Elliott Sober, ed., *Conceptual Issues in Evolutionary Biology,* 2nd ed. (Cambridge, Mass.: MIT Press, 1994), pp. 73–90.

Grafen, Alan. "Natural Selection, Kin Selection and Group Selection," in J. Krebs and

N.B. Davies, eds., *Behavioural Ecology,* 2nd ed. (Oxford: Blackwell Scientific Publications, 1984), pp. 62–84.

Graves, Robert. *Goodbye to All That* (London: Cassell and Co., 1929).

Gray, J. Glenn. *The Warriors: Reflections on Men in Battle* (New York: Harcourt, Brace and Company, 1959).

Green, Donald P., and Ian Shapiro. *The Pathology of Rational Choice: A Critique of Applications in Political Science* (New Haven: Yale University Press, 1994).

Greenfeld, Liah. *Nationalism: Five Roads to Modernity* (Cambridge, Mass.: Harvard University Press, 1992).

Gregor, Thomas, and Clayton A. Robarchek. "Two Paths to Peace: Semai and Mehinaku Nonviolence," in Thomas Gregor, ed., *A Natural History of Peace* (Nashville, Tenn.: Vanderbilt University Press, 1996), pp. 159–188.

Griffin, Donald R. *Animal Minds* (Chicago: University of Chicago Press, 1992).

Grinnell, George Bird. *The Cheyenne Indians: Their History and Ways of Life,* 2 Vols. (Lincoln: University of Nebraska Press, 1972).

Groom, W.H.A. *Poor Bloody Infantry: A Memoir of the First World War* (London: William Kimber, 1976).

Grosby, Steven. "The Verdict of History: The Inexpungeable Tie of Primordiality—A Response to Eller and Coughlan," *Ethnic and Racial Studies,* Vol. 17, No. 1 (January 1994), pp. 164–171.

Grossman, Dave. *On Killing: The Psychological Cost of Learning to Kill in War and Society* (Boston: Little, Brown, 1995).

Gugliotta, Guy. "Comet Tied to Mass Extinction," *Washington Post,* February 23, 2001, p. A3.

Guillemin, Jeanne. *Anthrax: The Investigation of a Deadly Outbreak* (Berkeley: University of California Press, 1999).

Gurr, Ted Robert. *Why Men Rebel* (Princeton, N.J.: Princeton University Press, 1970).

Gutmann, Stephanie. *The Kinder, Gentler Military: Can America's Gender-Neutral Fighting Force Still Win Wars?* (New York: Scribner, 2000).

Hackworth, David H., and Julie Sherman. *About Face* (New York: Simon and Schuster, 1989).

Haldane, J.B.S. *The Causes of Evolution* (London: Longmans Green, 1932).

Hallam, A., and P.B. Wignall. *Mass Extinctions and Their Aftermath* (New York: Oxford University Press, 1997).

Halperin, Sandra. "The Spread of Ethnic Conflict in Europe: Some Comparative-Historical Reflections," in David A. Lake and Donald Rothchild, eds., *The International Spread of Ethnic Conflict: Fear, Diffusion, and Escalation* (Princeton, N.J.: Princeton University Press, 1998), pp. 151–184.

Hamilton, W.D. "The Genetical Evolution of Social Behavior. I," *Journal of Theoretical Biology,* Vol. 7, No. 1 (July 1964), pp. 1–16.

Hamilton. "The Genetical Evolution of Social Behavior. II," *Journal of Theoretical Biology,* Vol. 7, No. 1 (July 1964), pp. 17–52.

Hamilton. "Innate Social Aptitudes of Man: An Approach from Evolutionary Genetics," in Robin Fox, ed., *Biosocial Anthropology* (New York: Wiley and Sons, 1975), pp. 133–155.

Hansen, Randall, and Desmond King. "Eugenic Ideas, Political Interests, and Policy Variance: Immigration and Sterilization Policy in Britain and the U.S.," *World Politics,* Vol. 53, No. 2 (January 2001), pp. 237–263.

Hanson, Victor Davis. *The Western Way of Warfare: Infantry Battle in Classical Greece* (New York: Oxford University Press, 1989).

Hanson. *Carnage and Culture: Landmark Battles in the Rise of Western Power* (New York: Doubleday, 2001).

Harcourt, A.H. "Strategies of Emigration and Transfer by Primates, with Particular Reference to Gorillas," *Zeitschrift für Tierpsychologie,* Vol. 48 (1978), pp. 401–420.

Harcourt, A.H., and Frans B.M. de Waal. "Cooperation in Conflict: From Ants to Anthropoids," in Alexander H. Harcourt and Frans B.M. de Waal, eds., *Coalitions and Alliances in Humans and Other Animals* (New York: Oxford University Press, 1992), pp. 493–510.

Harcourt, Alexander H., and Frans B.M. de Waal, eds. *Coalitions and Alliances in Humans and Other Animals* (Oxford: Oxford University Press, 1992).

Hardin, Russell. *Collective Action* (Baltimore, Md.: Johns Hopkins University Press, 1982).

Hardin. *One for All: The Logic of Group Conflict* (Princeton, N.J.: Princeton University Press, 1995).

Harlan, Richard. *Superstructuralism: The Philosophy of Structuralism and Post-Structuralism* (London: Routledge, 1987).

Harries, Sheldon H. *Factories of Death: Japanese Biological Warfare, 1932–45, and the American Cover-Up* (New York: Routledge, 1994).

Hauser, Marc, and Susan Carey. "Building a Cognitive Creature from a Set of Primitives: Evolutionary and Developmental Insights," in Denise Dellarosa Cummins and Colin Allen, eds., *The Evolution of Mind* (New York: Oxford University Press, 1998), pp. 51–106.

Hausfater, Glenn, and Sarah Blaffer Hrdy, eds. *Infanticide: Comparative and Evolutionary Perspectives* (New York: Aldine, 1984).

Hausmann, Ricardo. "Prisoners of Geography," *Foreign Policy,* No. 122 (January/February 2001), pp. 44–53.

Hedges, Chris. *War Is a Force That Gives Us Meaning* (New York: PublicAffairs, 2002).

Heider, Karl G. *The Dugum Dani: A Papuan Culture in the Highlands of West New Guinea,* Viking Fund Publications in Anthropology Number Forty-Nine (New York: Wenner-Gren Foundation for Anthropological Research, 1970).

Heider. *Grand Valley Dani: Peaceful Warriors* (New York: Holt, Rinehart and Winston, 1979).

Hempel, Carl G. *Philosophy of Natural Science* (Englewood Cliffs, N.J.: Prentice-Hall, 1966).

Hempel, and Paul Oppenheim. "Studies in the Logic of Explanation," *Philosophy of Science,* Vol. 15 (1948), pp. 135–148. Reprinted in Carl G. Hempel, *Aspects of Scientific Explanation* (New York: Free Press, 1965), pp. 245–290.

Herrnstein, Richard. "On the Law of Effect," *Journal of the Experimental Analysis of Behavior,* Vol. 13, No. 2 (March 1970), pp. 242–266.

Hill, Adrian V.S., and Arno G. Motulsky. "Genetic Variation and Human Disease: The Role of Natural Selection," in Stephen C. Stearns, ed., *Evolution in Health and Disease* (New York: Oxford University Press, 1999), pp. 50–61.

Himmelfarb, Gertrude. *Darwin and the Darwinian Revolution* (Garden City, N.Y.: Doubleday, 1962).

Hinde, Robert A. *Animal Behavior* (New York: McGraw-Hill, 1970).

Hinde. *Ethology: Its Nature and Relations with Other Sciences* (New York: Oxford University Press, 1982).

Hinde. "Aggression and War: Individuals, Groups, and States," in Philip E. Tetlock, Jo. L. Husbands, Robert Jervis, Paul C. Stern, and Charles Tilly, eds., *Behavior, Society, and International Conflict,* Vol. 3 (New York: Oxford University Press, 1993), pp. 8–70.

Hirshleifer, J. "Natural Economy versus Political Economy," *Journal of Social and Biological Structures,* Vol. 1 (1978), pp. 319–337.

Hirshleifer. "On the Emotions as Guarantors of Threats and Promises," in John Dupré, ed., *The Latest on the Best: Essays on Evolution and Optimality* (Cambridge, Mass.: MIT Press, 1987), pp. 307–326.

Hobbes, Thomas. *Leviathan,* ed. C.B. Macpherson (Harmondsworth: Penguin, 1985 [1651]).

Hobsbawm, Eric. *Nations and Nationalism since 1780* (New York: Cambridge University Press, 1990).

Hobson, J.A. *Imperialism: A Study* (London: George Allen and Unwin, 1938).

Hoffman, Bruce. *Inside Terrorism* (New York: Columbia University Press, 1998).

Hofstadter, Richard. *Social Darwinism in American Thought* (Boston: Beacon Press, 1955).

Hokanson, Jack E. "The Effect of Frustration and Anxiety on Overt Aggression," *Journal of Abnormal and Social Psychology,* Vol. 62, No. 2 (March 1961), pp. 346–351.

Hokanson, and Sanford Shetler. "The Effect of Overt Aggression on Physiological Arousal," *Journal of Abnormal and Social Psychology,* Vol. 63, No. 2 (September 1961), pp. 446–448.

Hölldobler, Bert. "Foraging and Spatiotemporal Territories in the Honey Ant *Myrmecocystus mimicus* Wheeler (Hymenoptera: Formicidae)," *Behavioral Ecology and Sociobiology,* Vol. 9 (1981), pp. 301–314.

Hölldobler, and Wilson, Edward O. *Journey to the Ants: A Story of Scientific Exploration* (Cambridge, Mass.: Harvard University Press, 1994).

Holloway, Ralph L. "Introduction," in Ralph L. Holloway, ed., *Primate Aggression,*

Territoriality, and Xenophobia: A Comparative Perspective (New York: Academic Press, 1974), pp. 1–9.

Holmes, Richard. *Acts of War: The Behavior of Men in Battle* (New York: Free Press, 1986).

Homer-Dixon, Thomas F. "On the Threshold: Environmental Changes as Causes of Acute Conflict," *International Security,* Vol. 16, No. 2 (Fall 1991), pp. 76–116.

Homer-Dixon. "Environmental Scarcities and Violent Conflict: Evidence from Cases," *International Security,* Vol. 19, No. 1 (Summer 1994), pp. 5–40.

Homer-Dixon. *Environment, Scarcity, and Violence* (Princeton, N.J.: Princeton University Press, 1999).

Homer-Dixon. *The Ingenuity Gap* (New York: Knopf, 2000).

Hopkins, Donald R. *Princes and Peasants: Smallpox in History* (Chicago: University of Chicago Press, 1983).

Horowitz, Donald L. *Ethnic Groups in Conflict* (Berkeley: University of California Press, 1985).

Horowitz. *The Deadly Ethnic Riot* (Berkeley: University of California Press, 2001).

Hrdy, Sarah Blaffer. *Mother Nature: A History of Mothers, Infants, and Natural Selection* (New York: Pantheon, 1999).

Hrdy, and Glenn Hausfater. "Comparative and Evolutionary Perspectives on Infanticide," in Glenn Hausfater and Sarah Blaffer Hrdy, eds., *Infanticide: Comparative and Evolutionary Perspectives* (New York: Aldine, 1984).

Hughes, Austin L. *Adaptive Evolution of Genes and Genomes* (New York: Oxford University Press, 1999).

Hughes, Daniel J., ed. *Moltke on the Art of War: Selected Writings,* trans. by Daniel J. Hughes and Harry Bell (Novato, Calif.: Presidio, 1993).

Hull, David L. "Individuality and Selection," *Annual Review of Ecology and Systematics,* Vol. 11 (Palo Alto, Calif.: Annual Reviews, Inc., 1980), pp. 311–332.

Hull. "Interactors versus Vehicles," in H.C. Plotkin, ed., *The Role of Behavior in Evolution* (Cambridge, Mass.: MIT Press, 1988), pp. 19–50.

Hull, and Michael Ruse, eds. *The Philosophy of Biology* (New York: Oxford University Press, 1998).

Hume, David. *A Treatise of Human Nature,* ed. L.A. Selby-Bigge (Oxford: Clarendon Press, 1964 [1739]).

Huntington, Samuel P. *Political Order in Changing Societies* (New Haven: Yale University Press, 1968).

Huntington. *The Clash of Civilizations and the Remaking of World Order* (New York: Simon and Schuster, 1996).

Hutchinson, John, and Anthony D. Smith, eds. *Ethnicity* (New York: Oxford University Press, 1996).

Huxley, Julian. *Evolution: The Modern Synthesis* (New York: Harper and Brothers, 1942).

Irons, William, and Lee Cronk. "Two Decades of a New Paradigm," in Lee Cronk, Napoleon Chagnon, and William Irons, eds., *Adaptation and Human Behavior: An Anthropological Perspective* (New York: Aldine de Gruyter, 2000), pp. 3–26.

Isaacs, Harold R. *Idols of the Tribe: Group Identity and Political Change* (New York: Harper and Row, 1975).

Itani, Junichiro. "Intraspecific Killing among Non-Human Primates," *Journal of Social and Biological Structures,* Vol. 5, No. 4 (1982), pp. 361–368.

Jacob, François. "Evolution and Tinkering," *Science,* Vol. 196 (June 10, 1977), pp. 1161–1166.

Jacob. *Of Flies, Mice, and Men,* trans. by Giselle Weiss (Cambridge, Mass.: Harvard University Press, 1998).

Jacobs, Alan H. "Maasai Inter-Tribal Relations: Belligerent Herdsmen or Peaceable Pastoralists," in Katsuyoshi Fukui and David Turton, eds., *Warfare among East African Herders* (Osaka: National Museum of Ethnology, 1979), pp. 33–52.

James, William. *Memories and Studies* (New York: Longmans Green, 1911).

Jensen, Marvin. *Strike Swiftly!: The 70th Tank Battalion—From North Africa to Germany* (Novato, Calif.: Presidio Press, 1997).

Jervis, Robert. *The Logic of Images in International Relations* (Princeton, N.J.: Princeton University Press, 1970).

Jervis. *Perception and Misperception in International Politics* (Princeton, N.J.: Princeton University Press, 1976).

Jervis. "Cooperation under the Security Dilemma," *World Politics,* Vol. 30, No. 2 (January 1978), pp. 167–214.

Jervis. "Domino Beliefs and Strategic Behavior," in Robert Jervis and Jack Snyder, eds., *Dominoes and Bandwagons: Strategic Beliefs and Great Power Competition in the Eurasian Rimland* (New York: Oxford University Press, 1991), pp. 20–50.

Johnson, Chalmers. *Revolutionary Change* (Boston: Little, Brown, 1966).

Johnson, Gary R. "The Evolutionary Origins of Government and Politics," in Albert Somit and Steven A. Peterson, eds., *Research in Biopolitics,* Vol. 3 (Greenwich, Conn.: JAI Press, 1995), pp. 243–305.

Johnson, Susan H. Ratwik, and Timothy J. Sawyer. "The Evocative Significance of Kin Terms in Patriotic Speech," in Vernon Reynolds, Vincent Falger, and Ian Vine, eds., *The Sociobiology of Ethnocentrism: Evolutionary Dimensions of Xenophobia, Discrimination, Racism and Nationalism* (London: Croom Helm, 1987), pp. 157–174.

Jolly, Allison. *Lucy's Legacy: Sex and Intelligence in Human Evolution* (Cambridge, Mass.: Harvard University Press, 1999).

Jorgensen, Joseph G. *Western Indians: Comparative Environments, Languages, and Cultures of 172 Western American Indian Tribes* (San Francisco: W.H. Freeman, 1980).

Kaplan, David E. "Aum Shinrikyo (1995)," in Jonathan B. Tucker, ed., *Toxic Terror: Assessing Terrorist Use of Chemical and Biological Weapons* (Cambridge, Mass.: MIT Press, 2000), pp. 207–226.

Kaplan, Robert. *Balkan Ghosts: A Journey through History* (New York: St. Martin's Press, 1993).

Karlen, Arno. *Man and Microbes: Disease and Plagues in History and Modern Times* (New York: Simon and Schuster, 1995).

Katz, Steven T. *The Holocaust in Historical Context,* 3 Vols. (New York: Oxford University Press, 1994).

Kauffman, Stuart. *At Home in the Universe: The Search for Laws of Self-Organization and Complexity* (New York: Oxford University Press, 1995).

Kaufman, Stuart J. *Modern Hatreds: The Symbolic Politics of Ethnic War* (Ithaca: Cornell University Press, 2001).

Keegan, John. *The Face of Battle: A Study of Agincourt, Waterloo, and the Somme* (New York: Viking Press, 1976).

Keeley, Lawrence H. *War before Civilization: The Myth of the Peaceful Savage* (New York: Oxford University Press, 1996).

Keith, Sir Arthur. *The Place of Prejudice in Modern Civilization* (London: Williams and Norgate, Ltd., 1931).

Keller, Laurent, ed. *Levels of Selection in Evolution* (Princeton, N.J.: Princeton University Press, 1999).

Keller, and H. Kern Reeve. "Dynamics of Conflicts within Insect Societies," in Laurent Keller, ed., *Levels of Selection in Evolution* (Princeton, N.J.: Princeton University Press, 1999), pp. 153–175.

Kellett, Anthony. *Combat Motivation: The Behavior of Soldiers in Battle* (Boston: Kluwer, 1982).

Kellman, Henry Gleitman, and Elizabeth S. Spelke. "Object and Observer Motion in the Perception of Objects by Infants," *Journal of Experimental Psychology,* Vol. 13, No. 4 (November 1987), pp. 586–593.

Kellman, Philip J., and Elizabeth S. Spelke. "Perception of Partly Occluded Objects in Infancy," *Cognitive Psychology,* Vol. 15, No. 4 (October 1983), pp. 483–524.

Kennan, George F. *Realities of American Foreign Policy* (Princeton, N.J.: Princeton University Press, 1954).

Kennedy, David M. *Over Here: The First World War and American Society* (New York: Oxford University Press, 1980).

Khong, Yuen Foong. *Analogies at War: Korea, Munich, Dien Bien Phu, and the Vietnam Decisions of 1965* (Princeton: Princeton University Press, 1992).

Kimura, Motoo. *The Neutral Theory of Molecular Evolution* (New York: Cambridge University Press, 1983).

King, Desmond. *Making Americans: Immigration, Race, and the Origins of the Diverse Democracy* (Cambridge, Mass.: Harvard University Press, 2000).

King, Gary, Robert O. Keohane, and Sidney Verba. *Designing Social Inquiry: Scientific Inference in Qualitative Research* (Princeton, N.J.: Princeton University Press, 1994).

Kingdon, Jonathan. *Self-Made Man: Human Evolution from Eden to Extinction?* (New York: John Wiley and Sons, 1993).

Kinzey, Warren G., ed. *The Evolution of Human Behavior: Primate Models* (Albany: State University of New York Press, 1987).

Kipling, Rudyard. *Just So Stories* (Garden City, N.Y.: Doubleday, 1952).

Kitcher, Philip. *Vaulting Ambition: Sociobiology and the Quest for Human Nature* (Cambridge, Mass.: MIT Press, 1985).

Kitcher. *Lives to Come: The Genetic Revolution and Human Possibilities* (New York: Simon and Schuster, 1996).

Klare, Michael T. *Resource Wars: The New Landscape of Global Conflict* (New York: Metropolitan Books, 2001).

Klare. "The New Geography of Conflict," *Foreign Affairs,* Vol. 80, No. 3 (May/June 2001), pp. 49–61.

Knauft, Bruce M. "Violence and Sociality in Human Evolution," *Current Anthropology,* Vol. 32, No. 4 (August–October 1991), pp. 391–409.

Kolata, Gina. "Kill All the Bacteria!" *New York Times,* January 7, 2001, <http://www.nytimes.com/2001/01/07/weekinreview/07KOLA.html>.

Koshiro, Yukiko. *Trans-Pacific Racisms and the U.S. Occupation of Japan* (New York: Columbia University Press, 1999).

"Kosovo and Sierra Leone," *Washington Post,* June 15, 1999, p. A32.

Krech, Shepard, III. "Genocide in Tribal Society," *Nature,* Vol. 371, No. 6492 (September 1, 1994), pp. 14–15.

Kroeber, Clifton B., and Bernard L. Fontana. *Massacre on the Gila: An Account of the Last Major Battle between American Indians, with Reflections on the Origin of War* (Tucson: University of Arizona Press, 1986).

Kruuk, H. [Hans]. *The Spotted Hyena: A Study of Predation and Social Behavior* (Chicago: University of Chicago Press, 1972).

Kuperman, Alan J. "Rwanda in Retrospect," *Foreign Affairs,* Vol. 79, No. 1 (January/February 2000), pp. 94–118.

Labs, Eric J. "Beyond Victory: Offensive Realism and the Expansion of War Aims," *Security Studies,* Vol. 6, No. 4 (Summer 1997), pp. 1–49.

Lake, David A., and Donald Rothchild. "Spreading Fear: The Genesis of Transnational Ethnic Conflict," in David A. Lake and Donald Rothchild, eds., *The International Spread of Ethnic Conflict: Fear, Diffusion, and Escalation* (Princeton, N.J.: Princeton University Press, 1998), pp. 3–32.

Lancaster, Jane B., and Beatrix A. Hamburg, eds. *School-Age Pregnancy and Parenthood: Biosocial Dimensions* (New York: Aldine, 1986).

Landes, David S. *The Wealth and Poverty of Nations: Why Some Are So Rich and Some So Poor* (New York: Norton, 1999).

Larson, Deborah Welch. *The Origins of Containment: A Psychological Explanation* (Princeton, N.J.: Princeton University Press, 1985).

Lathrap, Donald W. "The 'Hunting' Economics of the Tropical Forest Zone of South America: An Attempt at Historical Perspective," in Richard B. Lee and Irven DeVore, eds., *Man the Hunter* (Chicago: Aldine, 1968), pp. 23–29.

Lathrap. *The Upper Amazon* (New York: Praeger Publishers, 1970).

Layne, Christopher. "Miscalculations and Blunders Lead to War," in Ted Galen Carpenter, ed., *NATO's Empty Victory: A Postmortem on the Balkan War* (Washington, D.C.: Cato Institute, 2000), pp. 11–20.

Layne. *The Peace of Illusions: International Relations Theory and American Grand Strat-*

egy in the Post–Cold War Era [tentative title] (Ithaca, N.Y.: Cornell University Press, forthcoming).

Lederberg, Joshua. "Viruses and Humankind: Intracellular Symbiosis and Evolutionary Competition," in Stephen S. Morse, ed., *Emerging Viruses* (New York: Oxford University Press, 1993), pp. 3–9.

Lederberg, ed. *Biological Weapons: Limiting the Threat* (Cambridge, Mass.: MIT Press, 1999).

LeDoux, Joseph. *The Emotional Brain: The Mysterious Underpinnings of Emotional Life* (New York: Simon and Schuster, 1996).

Lee, Richard B., and Irven DeVore, eds. *Man the Hunter* (Chicago: Aldine, 1968).

Lerner, Richard M. *Final Solutions: Biology, Prejudice, and Genocide* (University Park, Pa.: Pennsylvania State University Press, 1992).

Levine, Robert A., and Donald T. Campbell. *Ethnocentrism: Theories of Conflict, Ethnic Attitudes and Group Behavior* (New York: Wiley, 1972).

Levi-Strauss, Claude. *The Elementary Structures of Kinship*, rev. ed., trans. by James Harle Bell and John Richard von Sturmer (London: Eyre and Spottiswoode, 1969).

Levy, Jack S. "The Causes of War: A Review of Theories and Evidence," in Philip E. Tetlock, Jo L. Husbands, Robert Jervis, Paul C. Stern, and Charles Tilly, eds., *Behavior, Society and Nuclear War: Volume One* (New York: Oxford University Press, 1989), pp. 209–333.

Levy. "Domestic Politics and War," in Robert I. Rotberg and Theodore K. Rabb, eds., *The Origin and Prevention of Major Wars* (New York: Cambridge University Press, 1989), pp. 79–99.

Lewis, Jonathan, and Ben Steele. *Hell in the Pacific: From Pearl Harbor to Hiroshima and Beyond* (London: Channel Four Books, 2001).

Lewontin, R.C., Steven Rose, and Leon J. Kamin. *Not in Our Genes: Biology, Ideology, and Human Nature* (New York: Pantheon Books, 1984).

Lewontin, Richard. *It Ain't Necessarily So: The Dream of the Human Genome and Other Illusions* (New York: New York Review of Books, 2000).

Lieber, Keir A. "Grasping the Technological Peace: The Offense-Defense Balance and International Security," *International Security*, Vol. 25, No. 1 (Summer 2000), pp. 71–104.

Linderman, Gerald F. *The World within War: America's Combat Experience in World War II* (New York: Free Press, 1997).

Lizot, J. "Population, Resources and Warfare among the Yanomami," *Man* (N.S.), Vol. 12, Nos. 3/4 (December 1977), pp. 497–517.

Lopreato, Joseph. *Human Nature and Biocultural Evolution* (Boston: Allen and Unwin, 1984).

Lopreato, and F.P.A. Green. "The Evolutionary Foundations of Revolution," in J. van der Dennen and V. Falger, eds., *Sociobiology and Conflict: Evolutionary Perspectives on Competition, Cooperation, Violence and Warfare* (London: Chapman and Hall, 1990), pp. 107–122.

Lorenz, Konrad. *On Aggression,* trans. by Marjorie Kerr Wilson (New York: Harcourt, Brace and World, 1966).

Lovejoy, Thomas E. "Global Change and Epidemiology: Nasty Synergies," in Stephen S. Morse, ed., *Emerging Viruses* (New York: Oxford University Press, 1993), pp. 261–268.

Low, Bobbi S. "An Evolutionary Perspective on War," in William Zimmerman and Harold K. Jacobson, eds., *Behavior, Culture, and Conflict in World Politics* (Ann Arbor: University of Michigan Press, 1993), pp. 13–55.

Low. *Why Sex Matters: A Darwinian Look at Human Behavior* (Princeton, N.J.: Princeton University Press, 2000).

Ludwig, Arnold M. *King of the Mountain: The Nature of Political Leadership* (Lexington: University Press of Kentucky, 2002).

Lumsden, Charles J., and Edward O. Wilson. *Genes, Minds, and Culture: The Coevolutionary Process* (Cambridge, Mass.: Harvard University Press, 1981).

Lumsden and Wilson. *Promethean Fire: Reflections on the Origin of Mind* (Cambridge, Mass.: Harvard University Press, 1983).

Lynn-Jones, Sean M. "Offense-Defense Theory and Its Critics," *Security Studies,* Vol. 4, No. 4 (Summer 1995), pp. 660–691.

Lynn-Jones. "Realism and America's Rise: A Review Essay," *International Security,* Vol. 23, No. 2 (Fall 1998), pp. 157–182.

Mackinder, Sir Halford J. *Democratic Ideals and Reality* (New York: Henry Holt and Co., 1919).

MacLeod, William Christie. "Economic Aspects of Indigenous American Slavery," *American Anthropologist,* Vol. 30 (1928), pp. 632–650.

Malinowski, Bronislaw. "War and Weapons among the Natives of the Trobriand Islands," *Man,* Vol. 20, No. 5 (January 1920), pp. 10–12.

Mamdani, Mahmood. *When Victims Become Killers: Colonialism, Nativism, and the Genocide in Rwanda* (Princeton, N.J.: Princeton University Press, 2001).

Manchester, William. *Goodbye Darkness: A Memoir of the Pacific War* (New York: Little, Brown, 1980).

Mandelbaum, Michael. "A Perfect Failure," *Foreign Affairs,* Vol. 78, No. 5 (September/October 1999), pp. 2–8.

Mangold, Tom, and Jeff Goldberg. *Plague Wars: A True Story of Biological Warfare* (New York: St. Martin's Press, 1999).

Mansbridge, Jane J. "The Rise and Fall of Self-Interest in the Explanation of Political Life," in Jane J. Mansbridge, ed., *Beyond Self-Interest* (Chicago: University of Chicago Press, 1990).

Manson, Joseph H., and Richard W. Wrangham. "Intergroup Aggression in Chimpanzees and Humans," *Current Anthropology,* Vol. 32, No. 4 (August–October 1991), pp. 369–390.

Marder, Arthur J. *The Anatomy of British Seapower: A History of British Naval Policy in the Pre-Dreadnought Era, 1880–1905* (New York: Knopf, 1940).

Mark, Robert. "Architecture and Evolution," *The American Scientist,* Vol. 84, No. 4 (July–August 1996), pp. 383–389.

Marshall, S.L.A. *Men against Fire: The Problem of Battle Command in Future War* (New York: William Morrow and Co., 1947).

Martin, Paul S., and Richard G. Klein. *Quaternary Extinctions: A Prehistoric Revolution* (Tucson: University of Arizona Press, 1984).

Masters, Roger D. "The Biological Nature of the State," *World Politics,* Vol. 35, No. 2 (January 1983), pp. 161–193.

Masters. *The Nature of Politics* (New Haven: Yale University Press, 1989).

Masters. "Political Science," in Mary Maxwell, ed., *The Sociobiological Imagination* (Albany: State University of New York, 1991), pp. 141–156.

Masters, and Baldwin Way. "Experimental Methods and Attitudes toward Leaders: Nonverbal Displays, Emotion, and Cognition," in Albert Somit and Steven A. Peterson, eds., *Research in Biopolitics,* Vol. 4 (Greenwich, Conn.: JAI Press, 1996), pp. 61–98.

Maynard Smith, John. "Group Selection and Kin Selection," *Nature,* Vol. 201, No. 4924 (March 14, 1964), pp. 1145–1147.

Maynard Smith. *Evolution and the Theory of Games* (New York: Cambridge University Press, 1982).

Maynard Smith. *Did Darwin Get It Right? Essays on Games, Sex, and Evolution* (New York: Chapman and Hall, 1989).

Maynard Smith. *The Theory of Evolution,* Canto ed. (New York: Cambridge University Press, 1993).

Maynard Smith. "Commentary," in Patricia Adair Gowaty, ed., *Feminism and Evolutionary Biology: Boundaries, Intersections, and Frontiers* (New York: Chapman and Hall, 1997), pp. 522–526.

Maynard Smith. "The Origin of Altruism," review of *Unto Others, Nature,* Vol. 393 (1998), pp. 639–640.

Maynard Smith, and G.R. Price. "The Logic of Animal Conflict," *Nature,* Vol. 246 (November 2, 1973), pp. 15–18.

Mayr, Ernst. *The Growth of Biological Thought: Diversity, Evolution, and Inheritance* (Cambridge, Mass.: Belknap Press of Harvard University Press, 1982).

Mayr. *Toward a New Philosophy of Biology: Observations of an Evolutionist* (Cambridge, Mass.: Harvard University Press, 1988).

Mayr. *One Long Argument: Charles Darwin and the Genesis of Modern Evolutionary Thought* (Cambridge, Mass.: Harvard University Press, 1991).

Mayr. *This Is Biology: The Science of the Living World* (Cambridge, Mass.: Harvard University Press, 1997).

Mayr. *What Evolution Is* (New York: Basic Books, 2001).

Mazur, Allan, and Ulrich Mueller. "Facial Dominance," in Albert Somit and Steven A. Peterson, eds., *Research in Biopolitics,* Vol. 4 (Greenwich, Conn.: JAI Press, 1996), pp. 99–111.

McCook, Henry C. *The Honey Ants of the Garden of the Gods, and Occident Ants of the American Plains* (Philadelphia: J.B. Lippincott, 1882).

McEachron, Donald L., and Darius Baer. "A Review of Selected Sociobiological Principles: Application to Hominid Evolution II. The Effects of Intergroup Conflict," *Journal of Social and Biological Structures,* Vol. 5, No. 2 (April 1982), pp. 121–139.

McKee, Jeffrey K. *The Riddled Chain: Chance, Coincidence, and Chaos in Human Evolution* (New Brunswick, N.J.: Rutgers University Press, 2000).

McNeill, William H. *Plagues and Peoples* (Garden City, N.Y.: Anchor Press/Doubleday, 1976).

McNeill. *The Global Condition: Conquerors, Catastrophes, and Community* (Princeton, N.J.: Princeton University Press, 1992).

McNeill. "Patterns of Disease Emergence in History," in Stephen S. Morse, ed., *Emerging Viruses* (New York: Oxford University Press, 1993), pp. 29–36.

Mead, Margaret. "Alternatives to War," in Milton Fried, Marvin Harris, and Robert Murphy, eds., *War* (Garden City, N.Y.: Natural History Press, 1968), pp. 215–228.

Mealey, Linda, Christopher Daood, and Michael Krage. "Enhanced Memory for Faces of Cheaters," *Ethology and Sociobiology,* Vol. 17, No. 2 (March 1996), pp. 119–128.

Mearsheimer, John J. "The False Promise of International Institutions," *International Security,* Vol. 19, No. 3 (Winter 1994/95), pp. 5–49.

Mearsheimer. *The Tragedy of Great Power Politics* (New York: Norton, 2001).

Mech, L. David, et al. *The Wolves of Denali* (Minneapolis: University of Minnesota Press, 1998).

Meggitt, M.J. *The Lineage System of the Mae-Enga of New Guinea* (New York: Barnes and Noble, 1965).

Meggitt. *Blood Is Their Argument: Warfare among the Mae Enga Tribesmen of the New Guinea Highlands* (Palo Alto, Calif.: Mayfield Publishing, 1977).

Meinecke, Friedrich. *Machiavellism: The Doctrine of Raison d'Etat and Its Place in Modern History* (Boulder, Colo.: Westview, 1984).

Mellaart, James. *Earliest Civilizations of the Near East* (New York: McGraw-Hill, 1965).

Mellars, Paul. "Technological Change across the Middle-Upper Palaeolithic Transition: Economic, Social and Cognitive Perspectives," in Paul Mellars and Chris Stringer, eds., *The Human Revolution: Behavioural and Biological Perspectives on the Origins of Modern Humans* (Princeton, N.J.: Princeton University Press, 1989), pp. 338–365.

Meyer, Peter. "Warfare in Social Evolution: A Look at Its Evolutionary Underpinnings," paper presented at the Association for Politics and the Life Sciences, annual meeting, Charleston, S.C., October 2001.

Midgley, Mary. *Beast and Man: The Roots of Human Nature* (Ithaca, N.Y.: Cornell University Press, 1978).

Milgram, Stanley. *Obedience to Authority: An Experimental View* (New York: Harper and Row, 1974).

Miller, Arthur. "The Face in the Mirror: Anti-Semitism Then and Now," *New York Times Book Review,* October 14, 1984, pp. 3–5.

Miller, Arthur G. *The Obedience Experiments: A Case Study of Controversy in Social Science* (New York: Praeger, 1986).

Miller, David. *On Nationality* (New York: Oxford University Press, 1995).

Miller, Geoffrey F. "Sexual Selection for Cultural Displays," in Robin Dunbar, Chris Knight, and Camilla Power, eds., *The Evolution of Culture: An Interdisciplinary View* (Edinburgh: Edinburgh University Press, 1999), pp. 71–91.

Miller. *The Mating Mind: How Sexual Choice Shaped the Evolution of Human Nature* (New York: Doubleday, 2000).

Miller, Judith, Stephen Engelberg, and William Broad. *Germs: Biological Weapons and America's Secret War* (New York: Simon and Schuster, 2001).

Miller, William Ian. *The Mystery of Courage* (Cambridge, Mass.: Harvard University Press, 2000).

Millett, Allan R., and Peter Maslowski. *For the Common Defense: A Military History of the United States of America,* rev. and exp. (New York: Free Press, 1994).

Mills, Susan K., and John Beatty. "The Propensity Interpretation of Fitness," in Elliott Sober, ed., *Conceptual Issues in Evolutionary Biology* (Cambridge, Mass.: MIT Press, 1994), pp. 3–23.

Milner, George R., Eve Anderson, and Virginia G. Smith. "Warfare in Late Prehistoric West-Central Illinois," *American Antiquity,* Vol. 56, No. 4 (1991), pp. 581–603.

Montagu, Ashley, ed. *Sociobiology Examined* (New York: Oxford University Press, 1980).

Moon, Katherine H.S. *Sex among Allies: Military Prostitution in U.S.-Korea Relations* (New York: Columbia University Press, 1997).

Moore, George Edward. *Principia Ethica* (Cambridge: Cambridge University Press, 1903).

Moore, Lt. Gen. Harold G., USA (Ret.), and Joseph L. Galloway. *We Were Soldiers Once . . . and Young: Ia Drang: The Battle That Changed the War in Vietnam* (New York: Random House, 1992).

Morange, Michel. *The Misunderstood Gene,* trans. by Matthew Cobb (Cambridge, Mass.: Harvard University Press, 2001).

Morgenthau, Hans J. *Scientific Man vs. Power Politics* (Chicago: University of Chicago Press, 1946).

Morgenthau. *Politics among Nations: The Struggle for Power and Peace,* 5th ed., rev. (New York: Knopf, 1978).

Morren, George E.B., Jr. "Warfare on the Highland Fringe of New Guinea: The Case of the Mountain Ok," in R. Brian Ferguson, ed., *Warfare, Culture, and Environment* (Orlando, Fla.: Academic Press, 1984), pp. 169–207.

Morrow, James D. *Game Theory for Political Scientists* (Princeton, N.J.: Princeton University Press, 1994).

Morrow. "Modeling International Institutions," *International Organization,* Vol. 48, No. 3 (Summer 1994), pp. 387–423.

Morse, Stephen S., ed. *Emerging Viruses* (New York: Oxford University Press, 1993).

Moyer, K.E. "The Biological Basis of Dominance and Aggression," in Diane McGuinness, ed., *Dominance, Aggression and War* (New York: Paragon House, 1987), pp. 1–34.

Murphy, Robert F. "Intergroup Hostility and Social Cohesion," *American Anthropologist,* Vol. 59, No. 6 (December 1957), pp. 1018–1035.

National Intelligence Council, *Global Trends 2015: A Dialogue about the Future with Nongovernment Experts,* <http://www.odci.gov/cia/reports/globaltrends2015/index.html>.

Nelson, Richard R., and Sidney G. Winter. *An Evolutionary Theory of Economic Change* (Cambridge, Mass.: Belknap Press of Harvard University Press, 1982).

Nettleship, Martin A., R. Dalegivens, and Anderson Nettleship, eds. *War, Its Causes and Correlates* (The Hague: Mouton Publishers, 1975).

Niebuhr, Reinhold. *Faith and History: A Comparison of Christian and Modern Views of History* (London: Nisbet, 1938).

Niebuhr. *Christianity and Power Politics* (New York: Charles Scribner's Sons, 1940).

Niebuhr. *The Nature and Destiny of Man: A Christian Interpretation,* 2 Vols. (New York: Charles Scribner's Sons, 1941, 1943).

Niebuhr. *The Children of Light and the Children of Darkness: A Vindication of Democracy and a Critique of Its Traditional Defence* (New York: Charles Scribner's Sons, 1944).

Nishida, Toshisada. "The Social Structure of Chimpanzees of the Mahale Mountains," in David A. Hamburg and Elizabeth R. McCown, eds., *The Great Apes* (Menlo Park, Calif.: Cummings, 1979), pp. 73–121.

Nishida, Mariko Hiraiwa-Hasegawa, Toshikazu Hasegawa, and Yukio Takahata. "Group Extinction and Female Transfer in Wild Chimpanzees in the Mahale National Park, Tanzania," *Zeitschrift für Tierpsychologie,* Vol. 67 (1985), pp. 284–301.

Nishida, Hiroyuki Takasaki, and Yukio Takahata. "Demography and Reproductive Profiles," in Toshisada Nishida, ed., *The Chimpanzees of the Mahale Mountains* (Tokyo: University of Tokyo Press, 1990), pp. 63–97.

Numbers, Ronald L. *Darwinism Comes to America* (Cambridge, Mass.: Harvard University Press, 1998).

Nunney, Len. "Female-Biased Sex Ratios: Individual or Group Selection," *Evolution,* Vol. 39, No. 2 (March 1985), pp. 349–361.

Nunney. "Group Selection, Altruism, and Structured-Deme Models," *The American Naturalist,* Vol. 126, No. 2 (August 1985), pp. 212–230.

Oberg, Kalervo. "Types of Social Structure among the Lowland Tribes of South and Central America," *American Anthropologist,* Vol. 57 (1955), pp. 472–487.

O'Donnell, Patrick K. *Into the Rising Sun: In Their Own Words, World War II's Pacific Veterans Reveal the Heart of Combat* (New York: Free Press, 2002).

Olson, Mancur. *The Logic of Collective Action: Public Goods and the Theory of Groups* (Cambridge, Mass.: Harvard University Press, 1965).

Oppenheimer, Franz. *The State,* trans. by John Gitterman (New York: Free Life Editions, 1975).

Otterbein, Keith F. *The Evolution of War: A Cross-Cultural Study* (New Haven: HRAF Press, 1970).

Otterbein. *The Evolution of War: A Cross-Cultural Study,* 2nd ed. (New Haven: HRAF Press, 1985).

Otterbein. "The Origins of War," *Critical Review,* Vol. 11, No. 2 (Spring 1997), pp. 251–277.

Owen-Smith, Norman. "Ecological Links between African Savanna Environments, Climate Change, and Early Hominid Evolution," in Timothy G. Bromage and Friedemann Schrenk, eds., *African Biogeography, Climate Change, and Human Evolution* (New York: Oxford University Press, 1999), pp. 138–149.

The Oxford Dictionary of Quotations, 5th ed., ed. by Elizabeth Knowles (New York: Oxford University Press, 1999).

Oye, Kenneth A., ed. *Cooperation under Anarchy* (Princeton, N.J.: Princeton University Press, 1986).

Packer, Charles. "The Ecology of Sociality in Felids," in Daniel I. Rubenstein and Richard W. Wrangham, eds., *Ecological Aspects of Social Evolution: Birds and Mammals* (Princeton, N.J.: Princeton University Press, 1986), pp. 429–451.

Palumbi, Stephen R. *The Evolution Explosion: How Humans Cause Rapid Evolutionary Change* (New York: Norton, 2001).

Parish, Colin R. "Canine Parvovirus 2: A Probable Example of Interspecies Transfer," in Stephen S. Morse, ed., *Emerging Viruses* (New York: Oxford University Press, 1993), pp. 194–202.

Parker, Geoffrey. *The Military Revolution: Military Innovation and the Rise of the West, 1500–1800* (New York: Cambridge University Press, 1988).

Pearson, Raymond. *The Longman Companion to European Nationalism, 1789–1920* (London: Longman, 1994).

Peck, Joel R. "The Evolution of Outsider Exclusion," *Journal of Theoretical Biology,* Vol. 142, No. 4 (February 22, 1990), pp. 565–571.

Peluso, Nancy Lee, and Michael Watts. "Violent Environments," in Nancy Lee Peluso and Michael Watts, eds., *Violent Environments* (Ithaca, N.Y.: Cornell University Press, 2001), pp. 3–38.

Pennington, Reina. *Wings, Women, and War: Soviet Airwomen in World War II Combat* (Lawrence: University Press of Kansas, 2002).

Peterson, Steven A., and Albert Somit, eds. *Research in Biopolitics,* Vol. 8 (Amsterdam: Elsevier, 2001).

Phelps, Edmund S., ed. *Altruism, Morality, and Economic Theory* (New York: Russell Sage, 1975).

Pinker, Steven. *The Language Instinct: The New Science of Language and Mind* (New York: William Morrow, 1994).

Pinker. *How the Mind Works* (New York: Norton, 1997).

Pinker. *The Blank Slate: The Modern Denial of Human Nature* (New York: Viking, 2002).

Pinker, and Paul Bloom. "Natural Language and Natural Selection," in Jerome H. Barkow, Leda Cosmides, and John Tooby, eds., *The Adapted Mind: Evolutionary Psychology and the Generation of Culture* (New York: Oxford University Press, 1994), pp. 451–493.

Pitt, Roger. "Warfare and Hominid Brain Evolution," *Journal of Theoretical Biology*, Vol. 72, No. 3 (June 1978), pp. 551–575.

Plotkin, H.C., ed. *The Role of Behavior in Evolution* (Cambridge, Mass.: MIT Press, 1988).

Popper, Karl R. *The Logic of Scientific Discovery* (London: Hutchinson, 1959).

Popper. "Intellectual Autobiography," in Paul Arthur Schilpp, ed., *The Philosophy of Karl Popper*, 2 Vols. (La Salle, Ill.: Open Court, 1974).

Popper. "Natural Selection and the Emergence of Mind," *Dialectica*, Vol. 32, Nos. 3–4 (1978), pp. 339–355.

Popper. *Objective Knowledge*, rev. ed. (Oxford: Clarendon Press, 1979).

Popper. "Evolution," letter, *New Scientist*, Vol. 87, No. 1215 (August 21, 1980), p. 611.

Porter, Bruce D. *War and the Rise of the State: The Military Foundations of Modern Politics* (New York: Free Press, 1994).

Posen, Barry R. "The Security Dilemma and Ethnic Conflict," in Michael E. Brown, ed., *Ethnic Conflict and International Security* (Princeton, N.J.: Princeton University Press, 1993), pp. 103–124.

Posen. "Nationalism, the Mass Army, and Military Power," *International Security*, Vol. 18, No. 2 (Fall 1993), pp. 80–124.

Posey, D.A. "Do Amazonian Indians Conserve Their Environment or Not? And Who Are We to Know If They Do or Do Not?" *Abstracts, 92nd Annual Meeting of the American Anthropological Association* (Washington, D.C.: American Anthropological Association, 1993).

Pospisil, Leopold. *The Kapauku Papuans and Their Law* (New Haven: Human Relations Area Files Press, 1971).

Powell, Robert. "Absolute and Relative Gains in International Relations," *American Political Science Review*, Vol. 85, No. 4 (December 1991), pp. 1303–1320.

Price, Barbara J. "Competition, Production Intensification, and Ranked Society: Speculations from Evolutionary Theory," in R. Brian Ferguson, ed., *Warfare, Culture, and Environment* (Orlando, Fla.: Academic Press, 1984), pp. 209–240.

Prunier, Gérard. *The Rwandan Crisis: History of a Genocide* (New York: Columbia University Press, 1997).

Putnam, Robert D. *Bowling Alone: The Collapse and Revival of American Community* (New York: Simon and Schuster, 2000).

Queller, David C. "The Spaniels of St. Marx and the Panglossian Paradox: A Critique of a Rhetorical Programme," *The Quarterly Review of Biology,* Vol. 70, No. 4 (December 1995), pp. 485–489.

Ratcliffe, Susan, ed. *The Oxford Dictionary of Thematic Quotations* (New York: Oxford University Press, 2000).

Reeve, H. Kern, and Laurent Keller. "Levels of Selection: Burying the Units-of-Selection Debate and Unearthing the Crucial New Issues," in Laurent Keller, ed., *Levels of Selection in Evolution* (Princeton, N.J.: Princeton University Press, 1999), pp. 3–14.

Regis, Ed. *The Biology of Doom: The History of America's Secret Germ Warfare Project* (New York: Henry Holt and Co., 1999).

Renan, Ernest. "What Is a Nation?" in Homi Bhabha, ed., *Nation and Narration* (New York: Routledge, 1990), pp. 8–22.

Rex, John, and David Mason, eds. *Theories of Race and Ethnic Relations* (New York: Cambridge University Press, 1986).

Richards, Robert J. *Darwin and the Emergence of Evolutionary Theories of Mind and Behavior* (Chicago: University of Chicago Press, 1987).

Richburg, Keith B. "Case Closed on the 'Iceman' Mystery," *Washington Post,* July 26, 2001, p. A1.

Ridley, Matt. *The Origins of Virtue: Human Instincts and the Evolution of Cooperation* (New York: Viking, 1996).

Ridley. *Genome: The Autobiography of a Species in 23 Chapters* (New York: HarperCollins, 1999).

Ristau, Carolyn A. "Cognitive Ethology: The Minds of Children and Animals," in Denise Dellarosa Cummins and Colin Allen, eds., *The Evolution of Mind* (New York: Oxford University Press, 1998), pp. 127–161.

Robbins, Sterling. *Auyana: Those Who Held onto Home* (Seattle: University of Washington Press, 1982).

Roberts, Michael. "The Military Revolution, 1560–1660," in Clifford J. Rogers, ed., *The Military Revolution Debate: Readings on the Military Transformation of Early Modern Europe* (Boulder, Colo.: Westview Press, 1995), pp. 13–35.

Rodman, Margaret, and Matthew Cooper, eds. *The Pacification of Melanesia* (Ann Arbor: University of Michigan Press, 1979).

Roper, Marilyn Keyes. "Evidence of Warfare in the Near East from 10,000–4,300 B.C.," in Martin A. Nettleship, R. Dalegivens, and Anderson Nettleship, eds., *War, Its Causes and Correlates* (The Hague: Mouton Publishers, 1975), pp. 299–340.

Rose, Michael R. "The Mental Arms Race Amplifier," *Human Ecology,* Vol. 8, No. 3 (1980), pp. 285–293.

Rose, Steven, ed. *Against Biological Determinism* (London: Allison and Busby, 1982).

Rose, ed. *Towards a Liberatory Biology* (London: Allison and Busby, 1982).

Rose. "Escaping Evolutionary Psychology," in Hilary Rose and Steven Rose, eds., *Alas, Poor Darwin: Arguments against Evolutionary Psychology* (London: Jonathan Cape, 2000), pp. 247–265.

Rosecrance, Richard N. *Action and Reaction in World Politics* (Boston: Little, Brown, 1963).

Rosenberg, Alexander. *The Structure of Biological Science* (New York: Cambridge University Press, 1985).

Rosenberg. *Darwinism in Philosophy, Social Science and Policy* (New York: Cambridge University Press, 2000).

Ross, Marc Howard. "Political Decision Making and Conflict: Additional Cross-Cultural Codes and Scales," *Ethnology,* Vol. 22, No. 2 (April 1983), pp. 169–192.

Rubenstein, Daniel I., and Richard W. Wrangham, eds. *Ecological Aspects of Social Evolution: Birds and Mammals* (Princeton, N.J.: Princeton University Press, 1986).

Rummel, R.J. *Death by Government* (New Brunswick, N.J.: Transaction Publishers, 1997).

Ruse, M. "Confirmation and Falsification of Theories in Evolution," *Scientia,* Vol. 104, Nos. 7–8 (1969), pp. 329–357.

Ruse, Michael. *Monad to Man: The Concept of Progress in Evolutionary Biology* (Cambridge, Mass.: Harvard University Press, 1996).

Ruse. *Taking Darwin Seriously: A Naturalistic Approach to Philosophy,* 2nd ed. (New York: Prometheus Books, 1998).

Ryan, Frank. *Virus X: Tracking the New Killer Plagues* (Boston: Little, Brown, 1997).

Saffirio, John, and Raymond Hames. "The Forest and the Highway," in *The Impact of Contact: Two Yanomamö Case Studies* (Bennington, Vt.: Working Papers on South American Indians, No. 6, 1983), pp. 1–52.

Sahlins, Marshall D. *Tribesmen* (Englewood Cliffs, N.J.: Prentice-Hall, 1968).

Sahlins. *The Use and Abuse of Biology* (Ann Arbor: University of Michigan Press, 1976).

Sahlins. *How "Natives" Think: About Captain Cook, for Example* (Chicago: University of Chicago Press, 1995).

Sajer, Guy. *The Forgotten Soldier* (New York: Harper and Row, 1971).

Salmon, Wesley C. *Four Decades of Scientific Explanation* (Minneapolis: University of Minnesota Press, 1989).

Salter, Frank Kemp. "Indoctrination as Institutionalized Persuasion," in Irenäus Eibl-Eibesfeldt and Frank Kemp Salter, eds., *Indoctrinability, Ideology, and Warfare: Evolutionary Perspectives* (New York: Berghahn Books, 1998), pp. 421–452.

Sanderson, Stephen K. *The Evolution of Human Sociality: A Darwinian Conflict Perspective* (Lanham, Md.: Rowman and Littlefield Publishers, 2001).

Santoli, Al, ed. *Everything We Had: An Oral History of the Vietnam War by Thirty-Three American Soldiers Who Fought It* (New York: Random House, 1981).

Schelling, Thomas C. *Strategy of Conflict* (Cambridge, Mass.: Harvard University Press, 1960).

Schelling. *Arms and Influence* (New Haven: Yale University Press, 1966).

Schelling. *Micromotives and Macrobehavior* (New York: Norton, 1978).

Schmookler, Andrew Bard. *The Parable of the Tribes: The Problem of Power in Social Evolution* (Berkeley: University of California Press, 1984).

Schwartz, Barry. *The Battle for Human Nature: Science, Morality and Modern Life* (New York: Norton, 1986).

Scott, J. Paul. "The Biological Basis of Warfare," in J. Martin Ramirez, Robert A. Hinde, and Jo Groebel, eds., *Essays on Violence* (Sevilla: Publicaciones de la Universidad de Sevilla, 1987), pp. 21–43.

Segerstråle, Ullica. *Defenders of the Truth: The Battle for Science in the Sociology Debate and Beyond* (New York: Oxford University Press, 2000).

Selzer, Jack. "Introduction," in Jack Selzer, ed., *Understanding Scientific Prose* (Madison: University of Wisconsin Press, 1993), pp. 3–19.

Sen, Amartya K. "Rational Fools," in Jane Mansbridge, ed., *Beyond Self-Interest* (Chicago: University of Chicago Press, 1990), pp. 25–43.

Seton-Watson, Hugh. *Nations and States* (London: Methuen, 1977).

Shankman, Paul. "Culture Contact, Cultural Ecology, and Dani Warfare," *Man* (N.S.), Vol. 26, No. 2 (June 1991), pp. 299–321.

Shastri, Amita. "Government Policy and the Ethnic Crisis in Sri Lanka," in Michael E. Brown and Šumit Ganguly, eds., *Government Policies and Ethnic Relations in Asia and the Pacific* (Cambridge, Mass.: MIT Press, 1997), pp. 129–163.

Shaw, R. Paul, and Yuwa Wong. "Ethnic Mobilization and the Seeds of Warfare: An Evolutionary Perspective," *International Studies Quarterly*, Vol. 31, No. 1 (March 1987), pp. 5–31.

Shaw and Wong. *Genetic Seeds of Warfare: Evolution, Nationalism, and Patriotism* (Boston: Unwin Hyman, 1989).

Sherman, Paul W. "The Levels of Analysis," *Animal Behaviour*, Vol. 36, No. 2 (April 1988), pp. 616–619.

Shils, Edward. "Primordial, Personal, Sacred and Civil Ties," *British Journal of Sociology*, Vol. 7 (June 1957), pp. 113–145.

Sillitoe, Paul. "Big Men and War in New Guinea," *Man* (N.S.), Vol. 13, No. 2 (June 1978), pp. 252–271.

Simon, Herbert A. *Administrative Behavior: A Study of Decision-Making Processes in Administrative Organization*, 3rd ed. (New York: Free Press, 1976).

Simpson, George G. *Tempo and Mode in Evolution* (New York: Columbia University Press, 1944).

Simpson. *The Meaning of Evolution* (New Haven: Yale University Press, 1949).

Singer, J. David. "The Level-of-Analysis Problem in International Relations," *World Politics*, Vol. 14, No. 1 (October 1961), pp. 77–92.

Singer, Peter. *Animal Liberation* (New York: HarperCollins, 2002).

Sledge, E.B. *With the Old Breed at Peleliu and Okinawa* (New York: Oxford University Press, 1990).

Slotnik, Jay V., Catherine H. Kannenberg, David A. Eckerman, and Marcus B. Waller. "An Experimental Analysis of Impulsivity and Impulse Control in Humans," *Learning and Motivation*, Vol. 11, No. 1 (February 1980), pp. 61–77.

Small, Melvin, and J. David Singer, eds. *International War: An Anthology* (Homewood, Ill.: Dorsey Press, 1985).

Smith, Adam. *The Theory of Moral Sentiments* (Indianapolis, Ind.: Liberty Fund, 1982 [1759]).
Smith. *An Inquiry into the Nature and Causes of the Wealth of Nations* (Chicago: University of Chicago Press, 1976 [1776]).
Smith, Anthony D. *Theories of Nationalism,* 2nd ed. (New York: Holmes and Meier, 1983).
Smith. *The Ethnic Origins of Nations* (Oxford: Blackwell, 1986).
Smith. *National Identity* (Reno: University of Nevada Press, 1991).
Smith. "The Ethnic Sources of Nationalism," in Michael E. Brown, ed., *Ethnic Conflict and International Security* (Princeton, N.J.: Princeton University Press, 1993), pp. 26–41.
Smith. *Nations and Nationalism in a Global Era* (Cambridge: Polity Press, 1995).
Smith. *Myths and Memories of the Nation* (New York: Oxford University Press, 1999).
Smith. *The Nation in History: Historiographical Debates about Ethnicity and Nationalism* (Hanover, N.H.: University Press of New England, 2000).
Smith, Edward, and Walter Sapp, eds. *Plain Talk about the Human Genome Project: A Tuskegee University Conference on Its Promise and Perils . . . and Matters of Race* (Tuskegee, Ala.: Tuskegee University, 1997).
Smith, Eric Alden, and Bruce Winterhalder. "Natural Selection and Decision-Making," in Erin Alden Smith and Bruce Winterhalder, eds., *Evolutionary Ecology and Human Behavior* (New York: Aldine de Gruyter, 1992), pp. 25–60.
Smith, and Bruce Winterhalder, eds. *Evolutionary Ecology and Human Behavior* (New York: Aldine de Gruyter, 1992).
Smith, Rogers M. *Civic Ideas: Conflicting Visions of Citizenship in U.S. History* (New Haven: Yale University Press, 1997).
Snyder, Jack. *The Ideology of the Offensive: Military Decision Making and the Disasters of 1914* (Ithaca, N.Y.: Cornell University Press, 1984).
Snyder. *Myths of Empire: Domestic Politics and International Ambition* (Ithaca, N.Y.: Cornell University Press, 1991).
Snyder. "Nationalism and the Crisis of the Post-Soviet State," in Michael E. Brown, ed., *Ethnic Conflict and International Security* (Princeton, N.J.: Princeton University Press, 1993), pp. 79–101.
Snyder, and Karen Ballentine. "Nationalism and the Marketplace of Ideas," *International Security,* Vol. 21, No. 2 (Fall 1996), pp. 5–40.
Sober, Elliott. *Philosophy of Biology* (Boulder, Colo.: Westview, 1993).
Sober, ed. *Conceptual Issues in Evolutionary Biology,* 2nd ed. (Cambridge, Mass.: MIT Press, 1994).
Sober. "Six Sayings about Adaptationism," in David L. Hull and Michael Ruse, eds., *The Philosophy of Biology* (New York: Oxford University Press, 1998), pp. 72–86.
Sober. "What Is Evolutionary Altruism?" in David L. Hull and Michael Ruse, eds., *The Philosophy of Biology* (New York: Oxford University Press, 1998), pp. 459–478.
Sober, and David Sloan Wilson. *Unto Others: The Evolution and Psychology of Unselfish Behavior* (Cambridge, Mass.: Harvard University Press, 1998).

Soltis, Joseph, Robert Boyd, and Peter J. Richerson. "Can Group-Functional Behaviors Evolve by Cultural Group Selection?" *Current Anthropology,* Vol. 36, No. 3 (June 1995), pp. 473–483.

Somit, Albert. "Biopolitics," *British Journal of Political Science,* Vol. 2, Part 2 (April 1972), pp. 209–238.

Somit. "Human Nature as the Central Issue in Political Philosophy," in Elliott White, ed., *Sociobiology and Human Politics* (Lexington, Mass.: D.C. Heath, 1981), pp. 167–180.

Somit. "Humans, Chimps, and Bonobos: The Biological Bases of Aggression, War, and Peacemaking," *The Journal of Conflict Resolution,* Vol. 34, No. 3 (September 1990), pp. 553–582.

Somit, and Steven A. Peterson. *Biopolitics and Mainstream Political Science: A Master Bibliography* (DeKalb, Ill.: Program for Biosocial Research, 1990).

Somit and Peterson, eds. *The Dynamics of Evolution: The Punctuated Equilibrium Debate in the Natural and Social Sciences* (Ithaca: Cornell University Press, 1992).

Somit and Peterson, eds. *Research in Biopolitics,* Vol. 2 (Greenwich, Conn.: JAI Press, 1994).

Somit and Peterson, eds. *Research in Biopolitics,* Vol. 3 (Greenwich, Conn.: JAI Press, 1995).

Somit and Peterson, eds. *Research in Biopolitics,* Vol. 4 (Greenwich, Conn.: JAI Press, 1996).

Somit and Peterson. *Darwinism, Dominance, and Democracy: The Biological Bases of Authoritarianism* (Greenwich, Conn.: Praeger, 1997).

Somit and Peterson, eds. *Research in Biopolitics,* Vol. 5 (Greenwich, Conn.: JAI Press, 1997).

Somit and Peterson, eds. *Research in Biopolitics,* Vol. 6: *Sociobiology and Politics* (Stamford, Conn.: JAI Press, 1998).

Sophocles. *The Oedipus Cycle,* trans. by Dudley Pitts and Robert Fitzgerald (San Diego: Harcourt Brace Jovanovich Publishers, 1967).

Sorokin, Pitirim A. *Social and Cultural Dynamics,* 4 Vols. (New York: Bedminster Press, 1962).

Southwick, Charles H., et al. "Xenophobia among Free-Ranging Rhesus Groups in India," in Ralph L. Holloway, ed., *Primate Aggression, Territoriality, and Xenophobia: A Comparative Perspective* (New York: Academic Press, 1974), pp. 185–209.

Sparrow, Bartholomew H. *From the Outside In: World War II and the American State* (Princeton, N.J.: Princeton University Press, 1996).

Sponsel, Leslie E. "The Natural History of Peace: The Positive View of Human Nature and Its Potential," in Thomas Gregor, ed., *A Natural History of Peace* (Nashville, Tenn.: Vanderbilt University Press, 1996), pp. 95–125.

Stanford, Craig B. *Chimpanzee and Red Colobus: The Ecology of Predator and Prey* (Cambridge, Mass.: Harvard University Press, 1998).

Stannard, David E. *American Holocaust: The Conquest of the New World* (New York: Oxford University Press, 1992).

Staples, Brent. "Celebrating World War II—and the Whiteness of American History," *New York Times,* December 9, 2001, <http://www.nytimes.com/2001/12/09/opinion/09SUN3.html>.

Stearn, E. Wagner, and Allen E. Stearn. *The Effect of Smallpox on the Destiny of the Amerindian* (Boston: Bruce Humphries, Inc., 1945).

Sterling-Folker, Jennifer. "Realism and the Constructivist Challenge: Rejecting, Reconstructing, or Rereading," *International Studies Review,* Vol. 4, No. 1 (Spring 2002), pp. 73–97.

Stern, Jessica Eve. "Larry Wayne Harris (1998)," in Jonathan B. Tucker, ed., *Toxic Terror: Assessing Terrorist Use of Chemical and Biological Weapons* (Cambridge, Mass.: MIT Press, 2000), pp. 227–246.

Stern, Paul C. "Why Do People Sacrifice for Their Nations?" in John L. Comaroff and Paul C. Stern, eds., *Perspectives on Nationalism and War* (N.p.: Gordon and Breach Publishers, 1995), pp. 99–121.

Stewart, Ian. *Life's Other Secret: The New Mathematics of the Living World* (New York: John Wiley and Sons, 1998).

Stigler, George J. "Smith's Travels on the Ship of State," in Andrew S. Skinner and Thomas Wilson, eds., *Essays on Adam Smith* (Oxford: Clarendon Press, 1975).

Stigler, and Gary S. Becker. "De Gustibus Non Est Disputandum," in Karen Schweers Cook and Margaret Levi, eds., *The Limits of Rationality* (Chicago: University of Chicago Press, 1990), pp. 191–221.

Stoler, Mark A. *Allies and Adversaries: The Joint Chiefs of Staff, the Grand Alliance, and U.S. Strategy in World War II* (Chapel Hill: University of North Carolina Press, 2000).

Strahler, Alan, and Arthur Strahler. *Physical Geography: Science and Systems of the Human Environment* (New York: John Wiley and Sons, 1997).

Strate, John. "The Role of War in the Evolution of Political Systems and the Functional Priority of Defense," *Humboldt Journal of Social Relations,* Vol. 12, No. 2 (Spring/Summer 1985), pp. 87–114.

Streit, Christian. "The Fate of the Soviet Prisoners of War," in Michael Berenbaum, ed., *A Mosaic of Victims: Non-Jews Persecuted and Murdered by the Nazis* (New York: New York University Press, 1990), pp. 142–149.

Streri, Arlette, and Elizabeth S. Spelke. "Haptic Perception of Objects in Infancy," *Cognitive Psychology,* Vol. 20, No. 1 (January 1988), pp. 1–23.

Stringer, C.B. "The Origin of Early Modern Humans: A Comparison of the European and Non-European Evidence," in Paul Mellars and Chris Stringer, eds., *The Human Revolution: Behavioural and Biological Perspectives on the Origins of Modern Humans* (Princeton, N.J.: Princeton University Press, 1989), pp. 232–244.

Stringer, Christopher, and Clive Gamble. *In Search of the Neanderthals: Solving the Puzzle of Human Origins* (New York: Thames and Hudson, 1993).

Sullivan, Denis G., and Roger D. Masters. "Biopolitics, the Media, and Leadership: Nonverbal Clues, Emotions, and Trait Attributions in the Evaluation of Leaders," in Albert Somit and Steven A. Peterson, eds., *Research in Biopolitics,* Vol. 2 (Greenwich, Conn.: JAI Press, 1994), pp. 237–273.

Sulloway, Frank J. *Born to Rebel: Birth Order, Family Dynamics, and Creative Lives* (New York: Pantheon, 1996).

Suplee, Curt. "Dolphins May Communicate Individually," *Washington Post,* August 25, 2000, p. A3.

Symons, Donald. "An Evolutionary Approach: Can Darwin's View of Life Shed Light on Human Sexuality?" in James H. Geer and William T. O'Donohue, eds., *Theories of Human Sexuality* (New York: Plenum Press, 1987), pp. 91–125.

Symons. "On the Use and Misuse of Darwinism in the Study of Human Behavior," in Jerome H. Barkow, Leda Cosmides, and John Tooby, eds., *The Adapted Mind: Evolutionary Psychology and the Generation of Culture* (New York: Oxford University Press, 1994), pp. 137–159.

Szathmáry, Eörs. "The First Replicators," in Laurent Keller, ed., *Levels of Selection in Evolution* (Princeton, N.J.: Princeton University Press, 1999), pp. 31–52.

Szayna, Thomas S., ed. *Identifying Potential Ethnic Conflict: Application of a Process Model* (Santa Monica, Calif.: Rand Corp., 2000).

Tajfel, Henri. *Human Groups and Social Categories* (Cambridge: Cambridge University Press, 1981).

Tajfel, and John Turner. "An Integrative Theory of Intergroup Conflict," in William G. Austin and Stephen Worchel, eds., *The Social Psychology of Intergroup Relations* (Monterey, Calif.: Brooks/Cole, 1979), pp. 33–47.

Tamir, Yael. *Liberal Nationalism* (Princeton, N.J.: Princeton University Press, 1993).

Tanaka, Yuki. *Hidden Horrors: Japanese War Crimes in World War II* (Boulder, Colo.: Westview Press, 1996).

Tattersall, Ian. *The Fossil Trail: How We Know What We Think We Know about Human Evolution* (New York: Oxford University Press, 1995).

Taylor, Christopher C. *Sacrifice as Terror: The Rwandan Genocide of 1994* (New York: Berg, 1999).

Taylor, Ralph B. *Human Territorial Functioning: An Empirical, Evolutionary Perspective on Individual and Small Group Territorial Cognitions, Behaviors, and Consequences* (Cambridge: Cambridge University Press, 1988).

Tellis, Ashley Joachim. "The Drive to Domination: Towards a Pure Realist Theory of Politics" (Chicago: Ph.D. diss., University of Chicago, 1994).

Tellis, Thomas S. Szayna, and James A. Winnefeld. *Anticipating Ethnic Conflict* (Santa Monica, Calif.: Rand Corp., 1997).

Thayer, Bradley A. "Creating Stability in New World Orders" (Chicago: Ph.D. diss., University of Chicago, 1996).

Thayer. "Bringing in Darwin: Evolutionary Theory, Realism, and International Politics," *International Security,* Vol. 25, No. 2 (Fall 2000), pp. 124–151.

Thayer. "The Political Effects of Information Warfare: Why New Military Capabilities Cause Old Political Dangers," *Security Studies,* Vol. 10, No. 1 (Autumn 2000), pp. 46–91.

Thienpont, Kristiaan, and Robert Cliquet. "Introduction: In-Group/Out-Group Studies and Evolutionary Biology," in Kristiaan Thienpont and Robert Cliquet, eds., *In-Group/Out-Group Behaviour in Modern Societies: An Evolutionary Perspective* (Brussels: Vlaamse Gemeenschap, 1999), pp. 3–20.

Thompson, Marilyn W. *The Killer Strain: Anthrax and a Government Exposed* (New York: HarperCollins, 2003).

Thompson, William R. "Evolving Toward an Evolutionary Perspective," in William R. Thompson, ed., *Evolutionary Interpretations of World Politics* (New York: Routledge, 2001), pp. 1–14.

Thornhill, Randy, and Craig T. Palmer. *A Natural History of Rape: Biological Bases of Sexual Coercion* (Cambridge, Mass.: MIT Press, 2000).

Thornton, Russell. *American Indian Holocaust and Survival: A Population History Since 1492* (Norman: University of Oklahoma Press, 1987).

Tiger, Lionel. *The Decline of Males* (New York: Golden Books, 1999).

Tiger, and Robin Fox. *The Imperial Animal* (New York: Holt, Reinhart, and Winston, 1971).

Tilly, Charles. "Reflections on the History of European State-Making," in Charles Tilly, ed., *The Formation of National States in Western Europe* (Princeton, N.J.: Princeton University Press, 1975), pp. 3–83.

Tinbergen, Niko. "On the Aims and Methods of Ethology," *Zeitschrift für Tierpsychologie,* Vol. 20 (1963), pp. 410–433.

Tinbergen. "On War and Peace in Animals and Man," *Science,* Vol. 160 (June 28, 1968), pp. 1411–1418.

Tomikawa, Morimichi. "The Migrations and Inter-Tribal Relations of the Pastoral Datoga," in Katsuyoshi Fukui and David Turton, eds., *Warfare among East African Herders* (Osaka: National Museum of Ethnology, 1979), pp. 15–31.

Tooby, John, and Irven DeVore. "The Reconstruction of Hominid Evolution through Strategic Modeling," in Warren G. Kinzey, ed., *The Evolution of Human Behavior: Primate Models* (Albany, N.Y.: SUNY Press, 1987), pp. 183–237.

Tooby, and Leda Cosmides. "The Evolution of War and Its Cognitive Foundations," *Proceedings of the Institute of Evolutionary Studies,* Vol. 88 (1988), pp. 1–15.

Tooby and Cosmides. "Evolutionary Psychology and the Generation of Culture, Part I: Theoretical Considerations," *Ethology and Sociobiology,* Vol. 10, No. 1 (January 1989), pp. 29–49.

Tooby and Cosmides. "Evolutionary Psychology and the Generation of Culture, Part II: A Computational Theory of Social Exchange," *Ethology and Sociobiology,* Vol. 10, No. 1 (January 1989), pp. 51–97.

Tooby and Cosmides. "The Innate Versus the Manifest: How Universal Does Universal Have to Be?" *Behavior and Brain Sciences,* Vol. 12, No. 1 (March 1989), pp. 36–37.

Tooby and Cosmides. "The Psychological Foundations of Culture," in Jerome H. Barkow, Leda Cosmides, and John Tooby, eds., *The Adapted Mind: Evolutionary Psychology and the Generation of Culture* (New York: Oxford University Press, 1994), pp. 19–136.

Tornay, Serge. "Armed Conflicts in the Lower Omo Valley, 1970–1976: An Analysis from within Nyangatom Society," in Katsuyoshi Fukui and David Turton, eds., *Warfare among East African Herders* (Osaka: National Museum of Ethnology, 1979), pp. 97–117.

Trevathan, Wenda R. *Human Birth: An Evolutionary Perspective* (New York: Aldine, 1987).

Trivers, Robert L. "The Evolution of Reciprocal Altruism," *The Quarterly Review of Biology,* Vol. 46, No. 1 (March 1971), pp. 35–57.

Trivers. *Social Evolution* (Menlo Park, Calif.: Benjamin/Cummings Publishing, 1985).

Trivers. "As They Would Do to You," Review of Elliott Sober and David Sloan Wilson, *Unto Others, The Skeptic,* Vol. 6, No. 4 (1998), pp. 81–83.

Trivers. "Think for Yourself," Trivers replies to Sober and Wilson, *The Skeptic,* Vol. 6, No. 4 (1998), pp. 86–87.

Tucker, Jonathan B. "Introduction," in Jonathan B. Tucker, ed., *Toxic Terror: Assessing Terrorist Use of Chemical and Biological Weapons* (Cambridge, Mass.: MIT Press, 2000), pp. 1–14.

Tucker. *Scourge: The Once and Future Threat of Smallpox* (New York: Atlantic Monthly Press, 2001).

Turney-High, Harry Holbert. *Primitive War: Its Practice and Concepts* (Columbia: University of South Carolina Press, 1949).

Tuzin, Donald. "The Spectre of Peace in Unlikely Places: Concept and Paradox in the Anthropology of Peace," in Thomas Gregor, ed., *A Natural History of Peace* (Nashville, Tenn.: Vanderbilt University Press, 1996), pp. 3–33.

Tversky, Amos, and Daniel Kahneman. "Judgment under Uncertainty: Heuristics and Biases," in Daniel Kahneman, Paul Slovic, and Amos Tversky, eds., *Judgment under Uncertainty: Heuristics and Biases* (New York: Cambridge University Press, 1982), pp. 3–20.

Twain, Mark (Samuel Langhorne Clemens). *Following the Equator: A Journey around the World,* 2 Vols. (New York: Harper and Brothers, 1899).

United Kingdom Home Office. *Community Cohesion: A Report of the Independent Review Team,* <http://www.homeoffice.gov.uk/reu/community_cohesion.pdf>.

United Nations. "East Timor—UNTAET Background," <http://www.un.org./peace/etimor/UntaetB.htm>.

Van Creveld, Martin. *The Rise and Decline of the State* (New York: Cambridge University Press, 1999).

Van Creveld. *Men, Women and War: Do Women Belong in the Front Line?* (London: Cassell and Co., 2001).

Van den Berghe, Pierre. *The Ethnic Phenomenon* (New York: Elsevier, 1981).

Van den Berghe. "Ethnicity and the Sociobiology Debate," in John Rex and David Mason, eds., *Theories of Race and Ethnic Relations* (New York: Cambridge University Press, 1986), pp. 246–263.

Van den Berghe. "Does Race Matter?" *Nations and Nationalism,* Vol. 1, No. 3 (1995), pp. 357–368.

Van der Dennen, Johan M.G. "Ethnocentrism and In-Group/Out-Group Differentiation," in Vernon Reynolds, Vincent S.E. Falger, and Ian Vine, eds., *The Sociobiology of Ethnocentrism: Evolutionary Dimensions of Xenophobia, Discrimination, Racism and Nationalism* (London: Croom Helm, 1987), pp. 1–47.

Van der Dennen. "Primitive War and the Ethnological Inventory Project," in J. van der Dennen and V. Falger, eds., *Sociobiology and Conflict: Evolutionary Perspectives on Competition, Cooperation, Violence and Warfare* (London: Chapman and Hall, 1990), pp. 247–269.

Van der Dennen. *The Origin of War: The Evolution of a Male-Coalitional Reproductive Strategy,* 2 Vols. (Groningen, Netherlands: Origin Press, 1995).

Van der Dennen. "The Politics of Peace in Preindustrial Societies: The Adaptive Rationale behind Corroboree and Calumet," in Albert Somit and Steven A. Peterson, eds., *Research in Biopolitics,* Vol. 6 (Stamford, Conn.: JAI Press, 1998), pp. 159–192.

Van der Dennen. "Human Evolution and the Origin of War: A Darwinian Heritage," in Johan M.G. van der Dennen, David Smillie, and Daniel R. Wilson, eds., *The Darwinian Heritage and Sociobiology* (Westport, Conn.: Praeger, 1999), pp. 163–185.

Van der Dennen. "Animal Intergroup and Intercoalitional Agonistic Behavior," correspondence with the author, September 2002.

Van der Dennen. "(Evolutionary) Theories of Warfare in Preindustrial (Foraging) Societies," *Neuroendocrinology Letters,* Vol. 23 (Supplement 4) (December 2002), pp. 55–65.

Van der Dennen, J., and V. Falger, eds. *Sociobiology and Conflict: Evolutionary Perspectives on Competition, Cooperation, Violence and Warfare* (London: Chapman and Hall, 1990).

Van Evera, Stephen. "Hypotheses on Nationalism and War," *International Security,* Vol. 18, No. 4 (Spring 1994), pp. 5–39.

Van Evera. *Guide to Methods for Students of Political Science* (Ithaca, N.Y.: Cornell University Press, 1997).

Van Evera. *The Causes of War: Power and the Roots of Conflict* (Ithaca, N.Y.: Cornell University Press, 1999).

Vanhanen, Tatu. *Politics of Ethnic Nepotism: India as an Example* (New Delhi: Sterling Publishers, 1991).

Vanhanen. *Ethnic Conflicts Explained by Ethnic Nepotism,* published as *Research in Biopolitics,* Vol. 7 (Stamford, Conn.: JAI Press, 1999).

Van Hooff, J.A.R.A.M. "Intergroup Competition and Conflict in Animals and Man," in J. van der Dennen and V. Falger, eds., *Sociobiology and Conflict: Evolutionary Perspectives on Competition, Cooperation, Violence and Warfare* (London: Chapman and Hall, 1990), pp. 23–54.

Vayda, A.P. *Maori Warfare* (Wellington: The Polynesian Society, 1960).

Vayda. "Expansion and Warfare among Swidden Agriculturalists," *American Anthropologist,* Vol. 63, No. 2 (April 1961), pp. 346–358.

Vayda. "Phases of the Process of War and Peace among the Marings of New Guinea," *Oceania,* Vol. 42, No. 1 (September 1971), pp. 1–24.

Vayda. *War in Ecological Perspective: Persistence, Change, and Adaptive Processes in Three Oceanian Societies* (New York: Plenum Press, 1976).

Vayda. "Explaining Why Marings Fought," *Journal of Anthropological Research,* Vol. 45, No. 2 (Summer 1989), pp. 159–177.

Vick, Karl. "Plea Bargain Rejected in Bubonic Plague Case," *Washington Post,* April 3, 1996, p. A8.

Vickers, William T. "From Opportunism to Nascent Conservation: The Case of the Siona-Secoya," *Human Nature,* Vol. 5, No. 4 (1994), pp. 307–337.

Vine, Ian. "Inclusive Fitness and the Self-System," in Vernon Reynolds, Vincent S.E. Falger, and Ian Vine, eds., *The Sociobiology of Ethnocentrism: Evolutionary Dimensions of Xenophobia, Discrimination, Racism and Nationalism* (London: Croom Helm, 1987), pp. 60–80.

Vogel, Kathleen M. "Pathogen Proliferation: Threats from the Former Soviet Bioweapons Complex," *Politics and the Life Sciences,* Vol. 19, No. 1 (March 2000), pp. 3–16.

Von Bernhardi, Gen. Friedrich. *Germany and the Next War,* trans. by Allen H. Powles (London: Longman, Green, and Co., 1912).

Von Bernhardi. *Britain as Germany's Vassal,* trans. by J. Ellis Barker (London: Longman, Green and Co., 1914).

Von Clausewitz, Carl. *On War,* ed. and trans. by Michael Howard and Peter Paret (Princeton, N.J.: Princeton University Press, 1976).

Vrba, Elisabeth S. "On the Connections between Paleoclimate and Evolution," in Elisabeth S. Vrba, George H. Denton, Timothy C. Partridge, Lloyd H. Burckle, eds., *Paleoclimate and Evolution, with Emphasis on Human Origins* (New Haven: Yale University Press, 1995), pp. 24–45.

Waage, Jonathan K., and Patricia Adair Gowaty. "Myths of Genetic Determinism," in Patricia Adair Gowaty, ed., *Feminism and Evolutionary Biology: Boundaries, Intersections, and Frontiers* (London: Chapman and Hall, 1997), pp. 592–594.

Walt, Stephen M. *Revolution and War* (Ithaca, N.Y.: Cornell University Press, 1996).

Walt. "Rigor or Rigor Mortis? Rational Choice and Security Studies," *International Security,* Vol. 23, No. 4 (Spring 1999), pp. 5–48.

Walt. "Two Cheers for Clinton's Foreign Policy," *Foreign Affairs,* Vol. 79, No. 2 (March/April 2000), pp. 63–79.

Waltz, Kenneth N. *Man, the State, and War: A Theoretical Analysis* (New York: Columbia University Press, 1959).

Waltz. *Theory of International Politics* (Reading, Mass.: Addison-Wesley, 1979).

Waltz. "Realist Thought and Neorealist Theory," *Journal of International Affairs,* Vol. 44, No. 1 (Spring/Summer 1990), pp. 21–37.

Waltz. "Structural Realism after the Cold War," *International Security,* Vol. 25, No. 1 (Summer 2000), pp. 5–41.

Warnecke, A. Michael, Roger D. Masters, and Guido Kempter, "The Roots of Nationalism: Nonverbal Behavior and Xenophobia," *Ethology and Sociobiology,* Vol. 13, No. 4 (1992), pp. 267–282.

Watts, Sheldon. *Epidemics and History: Disease, Power and Imperialism* (New Haven: Yale University Press, 1997).

Weber, Eugen. *Peasants into Frenchmen: The Modernization of Rural France, 1870–1914* (Stanford, Calif.: Stanford University Press, 1976).

Weber, Max. "Politics as a Vocation," in H.H. Gerth and C. Wright Mills, eds., *From Max Weber: Essays in Sociology* (New York: Oxford University Press, 1946), pp. 77–128.

Weber. *Economy and Society: An Outline of Interpretive Sociology,* 2 Vols., ed. by Guenther Roth and Claus Wittich (Berkeley: University of California Press, 1978).

Wedgwood, Camilla H. "Some Aspects of Warfare in Melanesia," *Oceania,* Vol. 1 (1930), pp. 5–33.

Weiss, Rick. "Clarifying the Facts and Risks of Anthrax," *Washington Post,* October 18, 2001, p. A12.

Weiss. "Germ Tests Point Away From Iraq," *Washington Post,* October 30, 2001, p. A9.

Westing, Arthur H., ed. *Global Resources and International Conflict* (Oxford: Oxford University Press, 1986).

White, Douglas R. "Rethinking Polygyny: Co-Wives, Codes, and Cultural Systems, *Current Anthropology,* Vol. 29, No. 4 (August–October 1988), pp. 529–558.

White, and Michael L. Burton. "Causes of Polygyny: Ecology, Economy, Kinship, and Warfare," *American Anthropologist,* Vol. 90, No. 4 (August 1988), pp. 871–887.

Whiten, Andrew, and Richard W. Byrne, eds. *Machiavellian Intelligence II: Extensions and Evaluations* (New York: Cambridge University Press, 1997).

Wiegele, Thomas C., ed. *Biology and the Social Sciences: An Emerging Revolution* (Boulder, Colo.: Westview Press, 1982).

Wight, Martin. *Power Politics* (London: Royal Institute of International Affairs, 1946).

Willey, P. [Patrick]. *Prehistoric Warfare on the Great Plains: Skeletal Analysis of the Crow Creek Massacre Victims* (New York: Garland Publishing, 1990).

Willhoite, Fred H., Jr. "Primates and Political Authority," *American Political Science Review,* Vol. 70, No. 4 (December 1976), pp. 1110–1126.

Willhoite. "Evolution and Collective Intolerance," *The Journal of Politics,* Vol. 39, No. 3 (August 1977), pp. 667–684.

Williams, George C. *Adaptation and Natural Selection: A Critique of Some Current Evolutionary Thought* (Princeton, N.J.: Princeton University Press, 1966).

Williams. *Sex and Evolution* (Princeton, N.J.: Princeton University Press, 1975).

Williams. "A Defense of Reductionism in Evolutionary Biology," in R. Dawkins and M. Ridley, eds., *Oxford Surveys in Evolutionary Biology,* Vol. 2 (New York: Oxford University Press, 1985), pp. 1–27.

Williams. *Natural Selection: Domains, Levels, and Challenges* (New York: Oxford University Press, 1992).

Williams. *The Pony Fish's Glow: And Other Clues to Plan and Purpose in Nature* (New York: Basic Books, 1997).

Williams, Mary B. "Falsifiable Prediction of Evolutionary Theory," *Philosophy of Science,* Vol. 40, No. 4 (December 1973), pp. 518–537.

Williams. "The Logical Status of the Theory of Natural Selection and Other Evolutionary Controversies," in Mario Bunge, ed., *The Methodological Unity of Science* (Dordrecht, Holland: Reidel, 1973), pp. 84–102.

Wills, Christopher. *Yellow Fever, Black Goddess: The Coevolution of People and Plagues* (Reading, Mass.: Addison Wesley, 1996).

Wilmut, Ian, Keith Campbell, and Colin Tudge. *The Second Creation: Dolly and the Age of Biological Control* (New York: Farrar, Straus and Giroux, 2000).

Wilson, David Sloan. *The Natural Selection of Populations and Communities* (Menlo Park, Calif.: Benjamin/Cummings, 1980).

Wilson, Edward O. *The Insect Societies* (Cambridge, Mass.: Belknap Press of Harvard University Press, 1971).

Wilson. "Competitive and Aggressive Behavior," in J.F. Eisenberg and Wilton S. Dillon, eds., *Man and Beast: Comparative Social Behavior* (Washington, D.C.: Smithsonian Institution Press, 1971), pp. 181–217.

Wilson. *Sociobiology: The New Synthesis* (Cambridge, Mass.: Harvard University Press, 1975).

Wilson. *On Human Nature* (Cambridge, Mass.: Harvard University Press, 1978).

Wilson. "Biology and Anthropology: A Mutual Transformation?" in Napoleon A. Chagnon and William Irons, eds., *Evolutionary Biology and Human Social Behavior: An Anthropological Perspective* (North Scituate, Mass.: Duxbury Press, 1979), pp. 519–521.

Wilson. *Consilience: The Unity of Knowledge* (New York: Knopf, 1998).

Wilson. *The Diversity of Life* (New York: Norton, 1999).

Wilson, James Q. *The Moral Sense* (New York: Free Press, 1993).

Winterhalder, Bruce, and Eric Alden Smith. "Evolutionary Ecology and the Social Sciences," in Eric Alden Smith and Bruce Winterhalder, eds., *Evolutionary Ecology and Human Behavior* (New York: Aldine de Gruyter, 1992), pp. 3–23.

Wolfers, Arnold. *Discord and Collaboration: Essays in International Politics* (Baltimore, Md.: Johns Hopkins University Press, 1965).

Wolpert, Lewis. *Malignant Sadness: The Anatomy of Depression* (New York: Free Press, 1999).

Wolpoff, Milford H. *Paleoanthropology,* 2nd ed. (Boston: McGraw-Hill, 1999).

Woodworth, Steven E. *Davis and Lee at War* (Lawrence: University Press of Kansas, 1995).

Woodworth. *While God Is Marching On: The Religious World of Civil War Soldiers* (Lawrence: University Press of Kansas, 2001).

Wrangham, Richard W. "Evolution of Coalitionary Killing," in *Yearbook of Physical Anthropology,* Vol. 42 (New York: Wiley-Liss, 1999), pp. 1–30.

Wrangham, and Dale Peterson. *Demonic Males: Apes and the Origins of Human Violence* (Boston: Houghton Mifflin, 1996).

Wren, Christopher S. "Era Waning, Holbrooke Takes Stock," *New York Times,* January 14, 2001, <http://www.nytimes.com/2001/01/14/world/14HOLB.html>.

Wright, Quincy. *A Study of War* (Chicago: University of Chicago Press, 1942).

Wright. *A Study of War,* 2nd ed. (Chicago: University of Chicago Press, 1965).

Wright, Robert. *The Moral Animal: Evolutionary Psychology and Everyday Life* (New York: Vintage Books, 1994).

Wright, Sewall. "Comments," in Paul S. Moorhead and Martin M. Kaplan, eds., *Mathematical Challenges to the Neo-Darwinian Interpretation of Evolution* (Philadelphia: Wistar Institute Press, 1967).

Wynne-Edwards, V.C. *Animal Dispersion in Relation to Social Behaviour* (Edinburgh: Oliver and Boyd, 1962).

Young, Crawford, ed. *The Rising Tide of Cultural Pluralism: The Nation-State at Bay?* (Madison: University of Wisconsin Press, 1993).

Zakaria, Fareed. *From Wealth to Power: The Unusual Origins of America's World Role* (Princeton, N.J.: Princeton University Press, 1998).

Zegwaard, Rev. Gerard A. "Headhunting Practices of the Asmat of Netherlands New Guinea," *American Anthropologist,* Vol. 61, No. 6 (December 1959), pp. 1020–1041.

Zubrow, Ezra. "The Demographic Modelling of Neanderthal Extinction," in Paul Mellars and Chris Stringer, eds., *The Human Revolution: Behavioural and Biological Perspectives on the Origins of Modern Humans* (Princeton, N.J.: Princeton University Press, 1989), pp. 212–231.

Zuger, Abigail. "Infectious Diseases Rising Again in Russia," *New York Times,* December 5, 2000, <http://www.nytimes.com/2000/12/05/science/05INFE.html>.

Index

Index

Index

Index

Index

Index

Index

Index

Index